Understanding Antennas for Radar, Communications, and Avionics

VAN NOSTRAND REINHOLD ELECTRICAL/COMPUTER SCIENCE AND ENGINEERING SERIES

Series Editor: Sanjit Mitra

Understanding Antennas for Radar, Communications, and Avionics

Benjamin Rulf

Gregory A. Robertshaw

VNR VAN NOSTRAND REINHOLD COMPANY
New York

Copyright © 1987 by **Van Nostrand Reinhold Company Inc.**
Softcover reprint of the hardcover 1st edition 1987
Library of Congress Catalog Card Number: 86-28297
ISBN 978-94-011-6543-3   ISBN 978-94-011-6541-9 (eBook)
DOI 10.1007/978-94-011-6541-9

Van Nostrand Reinhold Company Inc.
115 Fifth Avenue
New York, New York 10003

Van Nostrand Reinhold Company Limited
Molly Millars Lane
Wokingham, Berkshire RG 11 2PY, England

Van Nostrand Reinhold
480 La Trobe Street
Melbourne, Victoria 3000, Australia

Macmillan of Canada
Division of Canada Publishing Corporation
164 Commander Boulevard
Agincourt, Ontario M1S 3C7, Canada

16 15 14 13 12 11 10 9 8 7 6 5 4 3 2 1

Library of Congress Cataloging-in-Publication Data
Rulf, Benjamin, 1934-
   Understanding antennas for radar, communications,
and avionics.
   (Van Nostrand Reinhold electrical/computer science
and engineering series)
   Bibliography: p.
   Includes index.
   1. Antennas (Electronics)   I. Robertshaw, Gregory Alan, 1950-   II. Title.
III. Series.
TK7871.6.R79   1987   621.38'028'3   86-28297
ISBN 978-94-011-6543-3

Contents

Preface

Antennas are part of every radar, every communications system, and every electronic warfare system. Therefore antennas are an important technical subject. The number of antenna books that have been published in the last 5 years alone attest to this fact. But why another book about antennas? What does this book offer that is different from all the others? The answer is: accessibility. Most of the technical literature on antennas is written for those with extensive backgrounds in electromagnetic theory and familiarity with the mathematical language of vector analysis, differential equations, and special functions. This puts much of the antenna literature out of reach for the nonspecialist who needs to understand the subject but cannot afford the time required to become familiar with all the background material.

With the rapid expansion of technical knowledge, the number of engineering students who find themselves attracted to "classical" subjects, such as electromagnetic theory, is on the decline, making it increasingly difficult to train engineers in areas involving antennas. Some authors in other technical fields have recognized this problem. For example, books on solid-state electronic devices that do not require an extensive background in quantum mechanics are available, useful, and informative. This book fulfills a similar need in the area of antennas. Antenna theory can be understood with rather modest prerequisities: a working knowledge of trigonometry and algebra (including complex numbers), basic differential and integral calculus, electricity, magnetism, and optics—as usually covered in the first two years at engineering and science programs in typical U.S. colleges. Many now working in the areas of radar, communications, avionics, or other military electronics have the background described and have a need to understand antenna principles. This book is intended to fulfill this need.

Understanding Antennas is developed to be useful for both classroom presentations and individual study. Problems are presented throughout the text, many of which are derivation problems. For those problems, detailed steps have been omitted in the derivation of certain formulas. By completing the omitted steps, the reader will find those problems instructive because they require a careful study of the text. More advanced problems are also presented. Formulas that are used frequently are presented in frames for easy reference.

The introduction provides a qualitative overview of antennas in radar, communications, and electronic warfare. The first two chapters are a review of wave propagation and interference, and of transmission lines. Chapter 3 summarizes the basic facts of electromagnetic theory necessary to understand antennas. Plane waves and geometrical optics are covered in more detail in Chapter 5, and Chapter 6 explains waveguides. Chapters 4, 7, 8, 9, 10, and 11 cover the antenna field. Chapter 12 is devoted to radomes. Chapter 13 explains antenna reciprocity, the Friis formula, and antenna noise temperature. Chapter 14, of particular interest to those who work with radars, describes the radar range equation and the radar cross section. Three mathematical appendixes explain the complex number notation used in AC circuit theory, elementary vector algebra, and the basics of Fourier integrals. A fourth appendix discusses the design of parabolic reflector antennas. A bibliography for further study is included at the end of this book. *Understanding Antennas for Radar, Communications, and Avionics*, therefore, is also a complete antenna reference.

Having evolved from a set of notes used for a short "continuing engineering education" course, this material has benefited practicing engineers, many without electromagnetic theory background. The notes have also have been used as a text for a regular one-semester college course, taken by electrical engineering students with a little background in electromagnetic theory.

We wish to express our deep appreciation to Herbert Feldman, of the MITRE Corporation, for many useful discussions and for the material in Appendix D.

BENJAMIN RULF
GREGORY A. ROBERTSHAW

Understanding Antennas for Radar, Communications, and Avionics

Introduction

Antennas are devices that launch energy into space as electromagnetic waves or, in the reverse process, extract energy from an existing electromagnetic field. They are used for

Telecommunications: information is transmitted between cooperating parties;

Active Remote Sensing: electromagnetic waves are transmitted to "illuminate" an object, and returning echoes are detected and analyzed (e.g., radar);

Passive Remote Sensing: waves are detected from distant sources (e.g., radio telescopy);

Energy Transmission: electromagnetic waves are transmitted into matter or space, as in medical applications (dielectric heating) or wireless energy transmission experiments (solar power satellites); and

Electronic Warfare (EW): enemy communications and radar are disrupted, and friendly force capabilities are protected.

Antennas used for such a variety of applications differ from one another in shape, size, and principle of operation. The discussion in this introduction centers on those areas of application and on parameters and properties that are of greatest importance for each application.

Wireless communication was the first technology that required antennas. Until about 1920 most work in wireless was in the low frequencies (below 3 MHz) of the radio spectrum. In Chapter 1, the relation between frequency and wavelength is given as

$$\text{frequency} = \frac{\text{speed of light}}{\text{wavelength}}$$

At frequencies below 1 MHz wavelengths are greater than 0.3 km, and antenna equipment consists mostly of wires, systems of wires, or even towers (classified as wire antennas and discussed in Chap. 4). These wire antennas are usually less than a wavelength in size and are, therefore, not very directional.

As development of radio technologies progressed, higher and higher frequencies were used for communications. Currently, with the proliferation of communications satellites, the spectrum has shifted to microwaves (roughly from 1 GHz to 30 GHz, corresponding to wavelengths of 30 cm to 1.0 cm). In this regime, antennas typically are much larger than a wavelength and are capable of producing narrow, highly directional *pencils* of radiation, necessary for efficient point-to-point communications. Communications systems now operate at frequency bands that range from extremely low frequencies (ELF) with wavelengths of many kilometers to millimeter waves (wavelengths shorter than 1 cm). This wide region includes radio and television broadcasting; ground-based, mobile, airborne, and space vehicle communications; satellite communications; and broadcasting systems. Almost any antenna mentioned in this book can be found in one communications system or another. Wire antennas and linear dipole arrays are most commonly used for the lower frequencies (up to and including very high frequency [VHF]), and the parabolic reflector antenna, which is discussed in Chapter 10, is most commonly used for the higher frequencies (microwaves).

A radar antenna usually must perform one function in addition to transmission and reception: determine the target's direction. To do so requires a well-defined narrow beam of radiation that searches for or tracks a target. Narrow radiation beams are created by means of antennas or antenna arrays that are large compared to a wavelength. This is why the great majority of radars operate at high frequencies (i.e., wavelengths below 1 m). The beam may be moved mechanically by rotating a reflector or a fixed array; however, much faster scanning is feasible in phased arrays, in which the beam is moved by electronic means, without moving parts (discussed in Chaps. 8 and 9).

Another characteristic of radar is the high-power levels that are usually required. As is described in Chapter 14, the power in the echo signal diminishes like $1/R^4$, where R is the target's distance. To achieve long-range detection, an instantaneous power of many megawatts is transmitted by some radar systems. This is much more than is needed for communications, where the signal strength falls off like $1/R^2$.

An important antenna design parameter is the frequency bandwidth of operation. The bandwidth, BW, is usually expressed as a dimensionless ratio.

$$BW = 2\frac{f_h - f_e}{f_h + f_e}$$

in which f_h and f_e are the highest and the lowest frequency in the operating band, respectively. Most radar and communication systems are narrowband systems with bandwidths of a few percent (10% of less). Some modern applications call for wider bands (spread spectrum) to avoid enemy detection or jamming. Even wider bands are used in many electronic warfare (EW) systems, which have to deal with an adversary whose electronic emissions are unknown a priori. EW includes a wide variety of activities. Passive listening for the purpose of extracting information includes direction finding, location, and identification of enemy installations, and extraction of intelligence data from intercepted enemy signals. Active transmission may be used for disruption of enemy radar or communications (jamming) or introduction of misleading signals (spoofing). Both in active and passive EW, octave or even multioctave bandwidths are often required. Systems that can operate over wide frequency bands need antennas that differ greatly from their more common narrowband counterparts. Some of these antennas are described in Chapter 11.

Some confusion arises because antennas can be both transmitters and receivers. In some cases the classification is clear: a television station has a transmit-only antenna, and a home television unit has a receive-only antenna, with the attendant difference in power levels. In many uses, notably radar and two-way, point-to-point communications, the same antenna performs both functions. An antenna's capability for transmitting and receiving is based on the reciprocity property, discussed in Chapter 13.

Directional antennas or arrays that are designed to produce a narrow beam inevitably produce a certain amount of stray radiation (*side lobes*). Stray radiation is sometimes very undesirable: in communications it may cause *crosstalk*, or increase system noise. In radar, side lobes illuminate undesired regions, increasing both noise and clutter. Side lobes also contribute to system detectability and vulnerability to jamming, which are necessary to avoid in EW applications. Side lobe levels and their control are discussed in Chapters 4, 8, and 10.

Many other design considerations arise from the environment in which antennas operate. In airborne and space applications, for example, weight and available volume are prime considerations for antennas of all types. In ground installations the cost may be a function of location and environmental factors, such as wind loads, precipitation, and extreme temperatures. At times, it is desirable to cover an antenna by a radome. Radomes and windows (discussed in Chap. 12) are often used to protect both communications, radar, and EW antennas. In some military applications the radome may also serve to hide the antenna from the enemy, because information about a system can be derived by merely looking at the antennas.

The antenna design for a specific application may take into account directivity requirements; power handling; frequency band and bandwidth; environmental conditions; available space; weight; cost; and other factors, such as impedance matching and side lobes.

To summarize, antennas are the interface between an electronic system and the outside world, and so are similar to loudspeakers in an accoustic system. Just as it takes expertise and understanding to select the right speakers for a particular sound system, it takes understanding of antenna principles and properties to select an antenna or a radome for a specific electronic system. Likewise, some working knowledge in the areas of complex number algebra, vector algebra, and Fourier integrations is indispensible in the study of electromagnetic (EM) wave phenomena. Appendixes A, B, and C are intended to assist in assimilating the text. Taken alone, they are not adequate introductions to the respective subjects.

This book should provide sufficient background for the understanding of antenna specifications, the selection of an antenna for certain systems using vendor supplied data, or for the use of antenna design handbooks to design simple antennas.

CHAPTER 1

Sinusoidal Waves

All antennas act as transducers between a transmission line and space, (i.e., launch or receive electromagnetic waves).

Wave phenomena frequently encountered in the physical world, include one-dimensional waves, three-dimensional (spherical) waves, and wave interference. The various types of waves—electromagnetic waves, sound waves (in the air or water), waves in the earth's crust (earthquakes), the deflection of a taut string, and the wave functions that describe atomic and molecular interactions (quantum mechanics)—share a common mathematical foundation. Therefore, the mathematics used to describe basic wave phenomena are reviewed.

Waves can be periodic in time. For example, a long note emitted by a flute can be envisioned as a wave train that has many thousands of peaks and valleys. In contrast, other sounds, such as the report of a firearm, are not periodic in time. Most of the signals encountered in antenna theory are long wave trains that may be considered sinusoidal; that is, they can be characterized by a well-defined frequency, f, measured in hertz (cycles per second). However, for linear systems, more complicated and even nonperiodic disturbances can be represented as a sum (or superposition) of constant frequency sinusoidal waves. The systems considered in subsequent chapters (transmission lines, antennas, waveguides, etc.) are assumed to be linear.

The decomposition of periodic or even nonperiodic functions into sinusoidal wave components is the subject of Fourier analysis. Some antenna characteristics are conveniently described by Fourier integrals, (reviewed in Appendix C).

1.1 BASICS OF SINUSOIDAL WAVE MOTION IN ONE DIMENSION

A one-dimensional sinusoidal wave propagating along the x axis, as shown in Figure 1-1, may be expressed

$$u(x,t) = a_0 \cos[(\omega t \mp kx) - \phi_0]. \qquad (1\text{-}1)$$

Here, $u(x,t)$ is the wave function, which may describe air pressure (for sound waves), voltage (for waves on a transmission line), or the deflection of a string from its equilibrium position. The angular frequency is a_0, $\omega = 2\pi f$, k is the wave number, and ϕ_0 is a reference phase. Throughout this book, the more convenient complex notation for Eq. (1-1) is used,

$$u(x,t) = Ae^{j\psi} = a_0 e^{-j\phi_0} e^{j(\omega t \mp kx)}, \qquad (1\text{-}2)$$

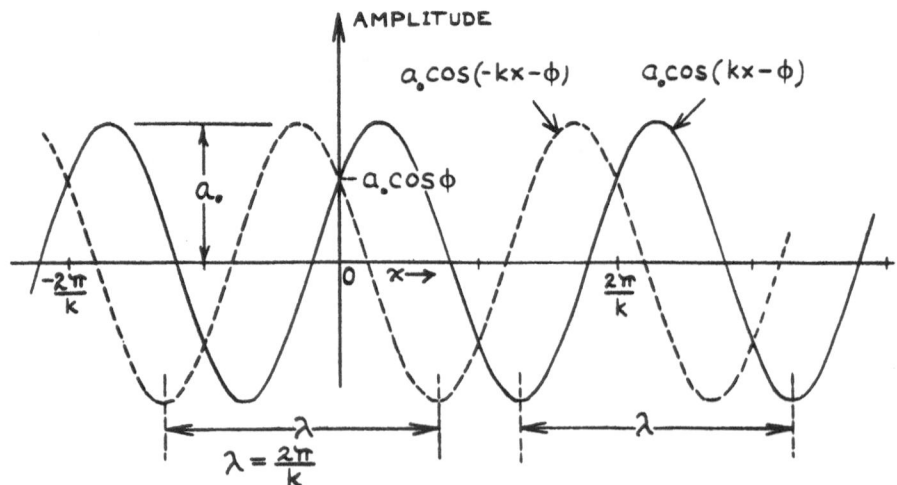

Figure 1-1.   Spatial variation of one-dimensional waves.

in which

$$\psi = \omega t \mp k x$$

and

$$A = a_0 e^{-j\phi_0} \qquad \textbf{(1-3)}$$

are the wave's phase and complex amplitude. The equivalence of Eq. (1-1) and (1-2) is explained in Appendix A, for those unfamiliar with the complex representation.

In considering any point of constant phase (and therefore constant u) along the wave, in order for ψ to remain constant, x and t must change simultaneously such that $\omega \Delta t \mp k \Delta x = 0$ or, in the limit

$$\frac{dx}{dt} = \pm \frac{\omega}{k} = c_0 \qquad \textbf{(1-4)}$$

in which c_0 is the phase velocity of the wave (i.e., the velocity of a point of constant phase). Note that when a minus sign precedes k, the wave's points of constant phase (or simply, the waves points) propagate in the positive x direction, and a plus sign indicates propagation along the negative x direction.

Examination of Figure 1-1 and either Eq. (1-1) or (1-2) reveals that the wave function is periodic with period 2π (i.e., the argument of the cosine or exponential can change by 2π, or any integral multiple of 2π radians, without changing the value of u). For example, if x is held constant, a 2π increase in ψ represents one temporal period, T:

$$\omega(t + T) = \omega t + 2\pi,$$

or

$$T = 2\pi/\omega = 1/f \qquad \textbf{(1-5)}$$

On the other hand, if t is fixed, a 2π increase in ψ corresponds to one spatial period or wavelength, λ:

$$k(x + \lambda) = kx + 2\pi$$

or

$$k = 2\pi/\lambda \qquad \textbf{(1-6)}$$

By use of Eqs. (1-4), (1-5), and (1-6),

$$\frac{\omega}{k} = \lambda f = c_0 \qquad \textbf{(1-7)}$$

is obtained, which is the basic relationship between wavelength, frequency, and phase velocity for sinusoidal waves.

In writing wave functions, the explicit time factor, $e^{j\omega t}$, and the reference phase, ϕ_0 are often omitted. Thus, a wave propagating in the $+x$ direction is expressed

$$u_+ = Ae^{-jkx}, \qquad \textbf{(1-8)}$$

a wave propagating in the $-x$ direction is expressed

$$u_- = Be^{jkx}, \qquad \textbf{(1-9)}$$

and the time factor remains implicit. The complex factor, e^{+jkx}, like the time factor, can be represented graphically as a phasor of unit length. The wave amplitude (A or B) is constant (i.e., does not depend upon x or t).

The example of a one-dimensional wave that propagates along the x axis was a choice of convenience. A wave in three-dimensional space that propagates along an arbitrary but constant direction is defined by its wave vector*

$$\mathbf{k} = k_1\mathbf{x}_0 + k_2\mathbf{y}_0 + k_3\mathbf{z}_0, \qquad \textbf{(1-10)}$$

where $\mathbf{x}_0$, $\mathbf{y}_0$, and $\mathbf{z}_0$ are unit vectors in the x, y, and z directions, respectively. The length, or magnitude, of $\mathbf{k}$ is denoted by k and is given by

$$k = \sqrt{k_1^2 + k_2^2 + k_3^2}. \qquad \textbf{(1-11)}$$

---

*Bold face denotes a vector. Thus, $\mathbf{k}$ in Eq. (1-10) is a vector, and k is a scalar quantity.

A limited amount of vector algebra is used in this book to simplify notation, where feasible. (A review of elementary vector algebra is provided in Appendix B.) A general point (x, y, z) in three-dimensional space is specified by the position vector

$$\mathbf{r} = x\,\mathbf{x_0} + y\,\mathbf{y_0} + z\,\mathbf{z_0}.$$

The phase of the wave at this point is given by the dot product of $\mathbf{k}$ and $\mathbf{r}$

$$\mathbf{k} \cdot \mathbf{r} = k_1 x + k_2 y + k_3 z, \qquad (1\text{-}12)$$

which is a linear function of x, y, and z. The wave function

$$u(x,y,z,t) = Ae^{j(\omega t - \mathbf{k} \cdot \mathbf{r})} \qquad (1\text{-}13)$$

is a generalization of Eq. (1-2) and is called a plane wave function. The plane wave propagates in the direction of $\mathbf{k}$, which is perpendicular to the planes of constant phase (see Appendix B)

$$k_1 x + k_2 y + k_3 z = \text{constant}.$$

The relationships expressed by Eqs. (1-5), (1-6), and (1-7) hold for plane waves when identifying the wave number, k, of Eq. (1-6) with the length of the wave vector $\mathbf{k}$ in Eq. (1-11). If the time factor is allowed to remain implicit, the plane wave functions have the forms

$$u_+ = Ae^{-j\mathbf{k}\cdot\mathbf{r}}$$

and

$$u_- = Be^{j\mathbf{k}\cdot\mathbf{r}}, \qquad (1\text{-}14)$$

which are useful generalizations of the forms given in Eqs. (1-8) and (1-9). Indeed, if selecting the particular wave vector $\mathbf{k} = k\mathbf{x_0}$, then Eq. (1-14) reduces to Eqs. (1-8) or (1-9). For this special case, the planes of constant phase are parallel to the y-z plane.

1.2 SPHERICAL WAVES AND ENERGY TRANSPORT

The plane waves (introduced in the previous section) represent a useful idealization under certain conditions. However, pure plane waves do not exist in the real world. To visualize how real waves behave, consider the water waves that spread out on the surface of a pond when a stone is dropped into still water. The waves form a circular or ring pattern that diminishes in amplitude (vertical displacement from the undisturbed level) as the ring radius increases. This simple observation is consistent with the law of energy conservation.

The wave function, u, may represent the deflection or local velocity of a vibrating string, voltage, or current at a point along a transmission line or a component of the electromagnetic field. It is known from elementary physics that energy is a quadratic function of the amplitude. For example, the potential energy stored in a spring is proportional to x^2, where x is the deflection; the kinetic energy of a body is proportional to u^2, where u is the velocity; and the potential energy of a charged capacitor is proportional to e^2, where e is the voltage. Waves transport energy along the trajectories of their propagation, and the local density of transported energy is proportional to the square of the wave amplitude. In the first example, the stone that dropped into the pond imparted a certain constant amount of energy to the water as manifested

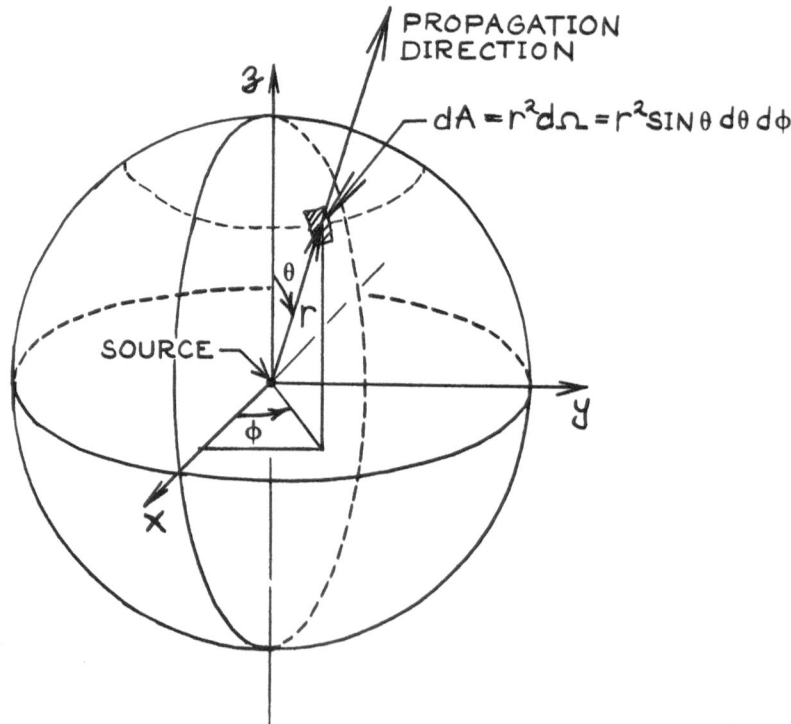

Figure 1-2.   Propagation geometry for point source of spherical waves.

by the surface waves. As the energy spreads out in circles of increasing radii, its density, and hence the wave amplitude, decreases.

In the three-dimensional coordinate system shown in Figure 1-2, r is the distance from the origin. By arguments similar to those of the previous section, a function of the form

$$u(r, \theta, \phi, t) = \frac{A}{r} e^{j(\omega t - kr)} \qquad (1\text{-}15)$$

represents a three-dimensional wave that propagates in a radially outward direction, with phase velocity $c_0 = \omega/k$. The $1/r$ amplitude factor is required to account for spreading of the wave's energy in three-dimensional space. This divergence factor conserves the energy of the source. The amplitude, A, which is in general complex, may be a function of direction, but not of r, that is, $A = A(\theta,\phi)$.

The quantity

$$uu^* = |A(\theta,\phi)|^2/r^2 \qquad (1\text{-}16)$$

is proportional to the energy that flows through a unit of area in a unit of time at the point (r,θ,ϕ). The time average of the energy that crosses a spherical boundary of radius, r, is found by integration of uu^* over the boundary

$$\iint_{\text{sphere}} uu^* \, ds = \int_0^{2\pi} d\phi \int_0^{\pi} d\theta \, \frac{|A(\theta,\phi)|^2}{r^2} \, r^2 \sin\theta$$

$$= \int_0^{2\pi} d\phi \int_0^{\pi} d\theta \sin\theta |A(\theta,\phi)|^2, \qquad (1\text{-}17)$$

which is a constant. In particular, when $A(\theta,\phi) = A_0$ is constant, the wave is *isotropic*, which means independent of direction (i.e., the same in all directions). For an isotropic wave

$$\iint_{\text{sphere}} uu^* ds = A_0^2 \int_0^{2\pi} d\phi \int_0^{\pi} \sin\theta d\theta = 4\pi A_0^2. \qquad (1\text{-}18)$$

If no energy is generated or dissipated anywhere (except at the source) the total outward energy flow (averaged in time) is constant and equal to the power of the source.

In the context of integration over a sphere, familiarity is necessary with the concept of solid angle Ω, which is related to the area on a unit sphere by

$$d\Omega = \sin\theta d\phi d\theta = ds/r^2.$$

Several of the problems at the end of this chapter deal with solid angle concepts.

If an observer is sufficiently far from a source of spherical waves, the sphere of constant phase is nearly planar, and the spatial divergence is, consequently, also very small at the local level. Under such circumstances, the wave function may be described fairly accurately by a plane wave. This useful simplification is explored in the next section.

1.3 RELATIONS BETWEEN PLANE AND SPHERICAL WAVES— THE FAR FIELD

When the source and observation point are separated by a large enough distance to permit the spherical waves to be locally represented by plane waves, the observation point is in the *far field* of the source. The far field region does not begin abruptly; the accuracy of the plane wave approximation always improves as the separation increases.

The local flatness of a spherical phase front depends on both the wavelength and the size of the measuring device (antenna) that samples the phase front. The measuring device itself defines the range of local space over which flatness is considered. Figure 1-3 schematically displays a device, or antenna, of length d, tangent to the phase front of a point source at 0. The phase difference between points C and E can be expressed in terms of the number of wavelengths contained in the length Δ, which is given by

$$\Delta = \overline{EO} - \overline{OC} = \sqrt{r^2 + (d/2)^2} - r.$$

If $d/r \ll 1$,

$$\Delta = r[\sqrt{1 + (d/2r)^2} - 1] \cong d^2/8r. \tag{1-19}$$

The common definition or criterion for the far field requires that Δ be less than 1/16 of a wavelength or 22.5° (one wavelength corresponds to a phase change of 2π radians or 360°). More explicitly, if $\Delta = d^2/8r \leq \lambda/16$, then

$$\boxed{r \geq 2\, d^2/\lambda}\qquad\qquad\textbf{(1-20)}$$

is a useful expression for the far field region. As noted, the far field criterion depends on the size of the measuring device, d, as well as the wavelength, λ. For example, to place a microwave antenna of diameter $d = 3$ m in the far field of a point source requires a range of at least 60 m when the frequency is 1 GHz (L-band, $\lambda = 30$ cm). However, at 10 GHz (X-band, $\lambda = 3$ cm) the required far field range is 600 m, because the criterion varies linearly with frequency (inversely with wavelength).

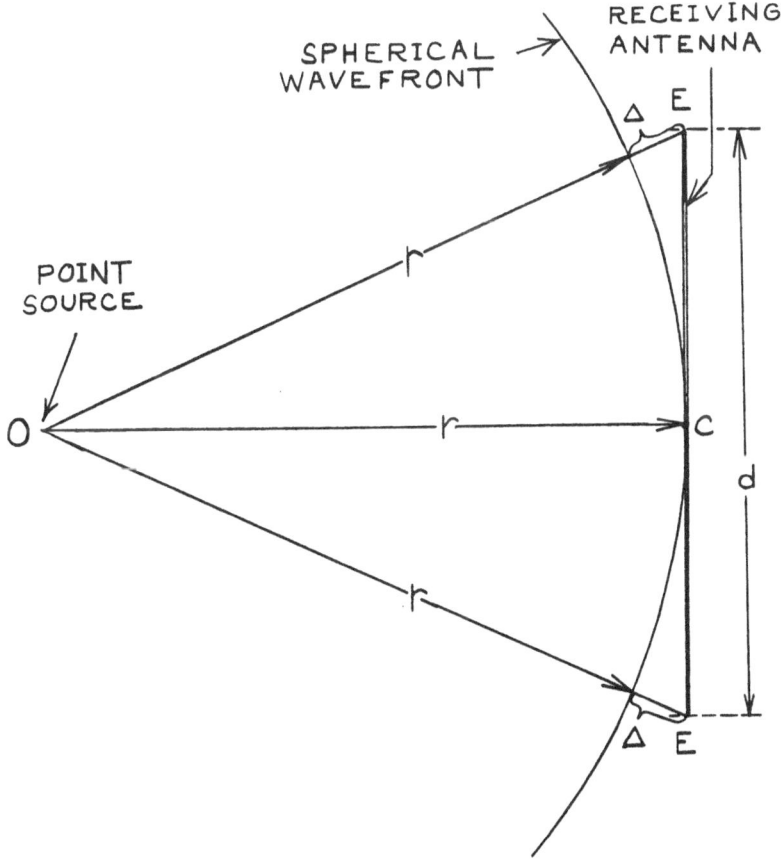

Figure 1-3.   Construction for far field criterion.

The argument can also be applied to a source of finite size, such as an antenna, and to a measuring device (or field probe) of negligible size (point probe). For the far field of the antenna, the numerical value of the criterion is the same. For example, the far field of the 3-m antenna begins at a distance of 60 m, at 1GHz. If both the source and measuring device are of finite size, the larger object's dimension, d, must be used in the criterion of Eq. (1-20).

If the distance between the source and the measuring device is not sufficient, the collected waves will not add up in phase, and the receiving device is said to be defocused with respect to the source.

1.4 WAVE INTERFERENCE

Waves may interact by combining or interfering with each other. This effect is apparent if two stones are dropped simultaneously at different places in a pond. When the ring shaped waves on the water's surface cross each other, there are places where they seem to add to produce a larger water displacement and other places where they appear to cancel. A simple example of interference is two one-dimensional waves that have the same frequency but travel in opposite directions. The wave functions may be expressed

$$u_1(x,t) = A_1 e^{-j(kx - \omega t)}$$

and

$$u_2(x,t) = A_2 e^{j(kx + \omega t)} \tag{1-21}$$

in which A_1 and A_2 are complex amplitudes:

$$A_1 = a_1 e^{-j\alpha_1}, \qquad A_2 = a_2 e^{-j\alpha_2}. \tag{1-22}$$

The sum of the two waves, S, may be written

$$S = u_1 + u_2 = e^{j\omega t}[a_1 e^{-j\psi_1} + a_2 e^{-j\psi_2}], \tag{1-23}$$

in which $\psi_1 = kx + \alpha_1$ and $\psi_2 = -kx + \alpha_2$. The phasor terms in the brackets add constructively to produce a peak in the total wave if ψ_2 and ψ_1 differ by an integer multiple of 2π radians, that is, if

$$(kx + \alpha_1) - (-kx + \alpha_2) = 2\pi m \qquad (m = 0, \pm 1, \pm 2, \ldots).$$

Solving for the positions of the peaks,

$$x_{max} = (2m\pi + \alpha_2 - \alpha_1)/2k = (\lambda/2)\left(m + \frac{\alpha_2 - \alpha_1}{2\pi}\right), \quad \textbf{(1-24)}$$

in which Eq. (1-6) has been used in the last step. The positions at which maxima occur are fixed (time independent) and evenly spaced at intervals of $\lambda/2$. The maxima of the combined waves are equal to $a_1 + a_2$.

Similarly, the phasor terms in Eq. (1-23) are diametrically opposed when ψ_2 and ψ_1 differ by $\pi(2m + 1)$, which leads to

$$x_{min} = (\lambda/2)\left(m + \frac{1}{2} + \frac{\alpha_2 - \alpha_1}{2\pi}\right). \quad \textbf{(1-25)}$$

Clearly, the combined wave has minima of amplitude $a_1 - a_2$ at the positions x_{min}, which are fixed and separated by intervals of $\lambda/2$. The envelope of the total wave function is illustrated in Figure 1-4. Note that the minima and maxima have $\lambda/4$ spacing (i.e., the minima and maxima are uniformly interlaced).

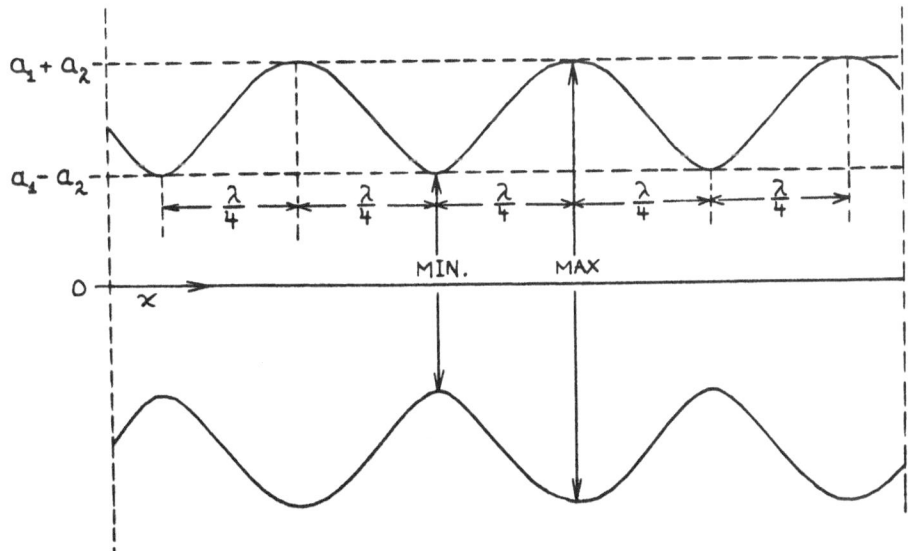

Figure 1-4. Standing wave envelope.

The ratio of the maximum to the minimum amplitude

$$\rho = \left| \frac{a_1 + a_2}{a_1 - a_2} \right| = \left| \frac{a_1/a_2 + 1}{a_1/a_2 - 1} \right| \qquad (1\text{-}26)$$

is called the standing wave ratio (SWR). If either a_1 or a_2 is equal to zero, SWR is unity. However, if $a_1 = a_2$, SWR is infinite. For a given SWR ($1 < \rho < \infty$), the amplitude ratio of the two interfering waves can be computed. Assuming, without loss of generality, that $a_1 > a_2 \geq 0$, then from Eq. (1-26) the obtained result is

$$a_1/a_2 = (\rho + 1)/(\rho - 1). \qquad (1\text{-}27)$$

The interference of two plane waves that have the same frequency but travel in different directions is an interesting example that arises in connection with waveguide theory. In considering two interfering waves that have wave vectors $\mathbf{k}_1$ and $\mathbf{k}_2$, respectively, clearly, $k_1 = k_2$, because the frequencies and phase velocities are equal. For mathematical convenience it can be assumed that the two wave vectors lie in the x-z plane, such that the z axis bisects the angle, α, between them, as shown in Figure 1-5. Using Eqs. (1-12) and (1-13), the following are derived.

$$\mathbf{k}_1 = k[\mathbf{z}_0 \cos(\alpha/2) + \mathbf{x}_0 \sin(\alpha/2)],$$

$$\mathbf{k}_2 = k[\mathbf{z}_0 \cos(\alpha/2) - \mathbf{x}_0 \sin(\alpha/2)],$$

$$\mathbf{k}_1 \cdot \mathbf{r} = k[z \cos(\alpha/2) + x \sin(\alpha/2)],$$

$$\mathbf{k}_2 \cdot \mathbf{r} = k[z \cos(\alpha/2) - x \sin(\alpha/2)],$$

in which $k = 2\pi/\lambda$, and (x,z) is a point in the x-z plane. For the sake of simplicity, it is assumed that the waves have equal amplitudes, which are denoted by A. The total wave function is then given by

$$S = u_1 + u_2 = A[e^{-j\mathbf{k}_1 \cdot \mathbf{r}} + e^{-j\mathbf{k}_2 \cdot \mathbf{r}}]$$

$$= Ae^{-jkz \cos(\alpha/2)}[e^{-jkx \sin(\alpha/2)} + e^{jkx \sin(\alpha/2)}]$$

$$= 2Ae^{-jkz \cos(\alpha/2)} \cos[kx \sin (\alpha/2)] \qquad (1\text{-}28)$$

in which the time factor, $e^{j\omega t}$, is implicit. The Euler-DeMoivre identity was used to obtain the final form of Eq. (1-28),

$$\boxed{e^{jt} = \cos t + j \sin t} \qquad (1\text{-}29)$$

Equation (1-28) represents a wave propagating along the z axis and modulated along the x axis direction. Clearly, when

$$kx \sin(\alpha/2) = \pm\pi\left(m + \frac{1}{2}\right) \qquad (m = 0, 1, 2, \ldots)$$

or

$$x = \pm\frac{\lambda(m + \frac{1}{2})}{2\sin(\alpha/2)}, \tag{1-30}$$

the total wave function is zero, that is, vanishes along a line of nodal points. Similarly, when $x = m\lambda/[2\sin(\alpha/2)]$, the total wave function has twice the amplitude of either one alone. Viewing the wave along the z axis, the distance between peaks (the wavelength) has increased to $\lambda/\cos(\alpha/2)$. Because the frequency of the wave is still ω, and propagation occurs along the z axis, the

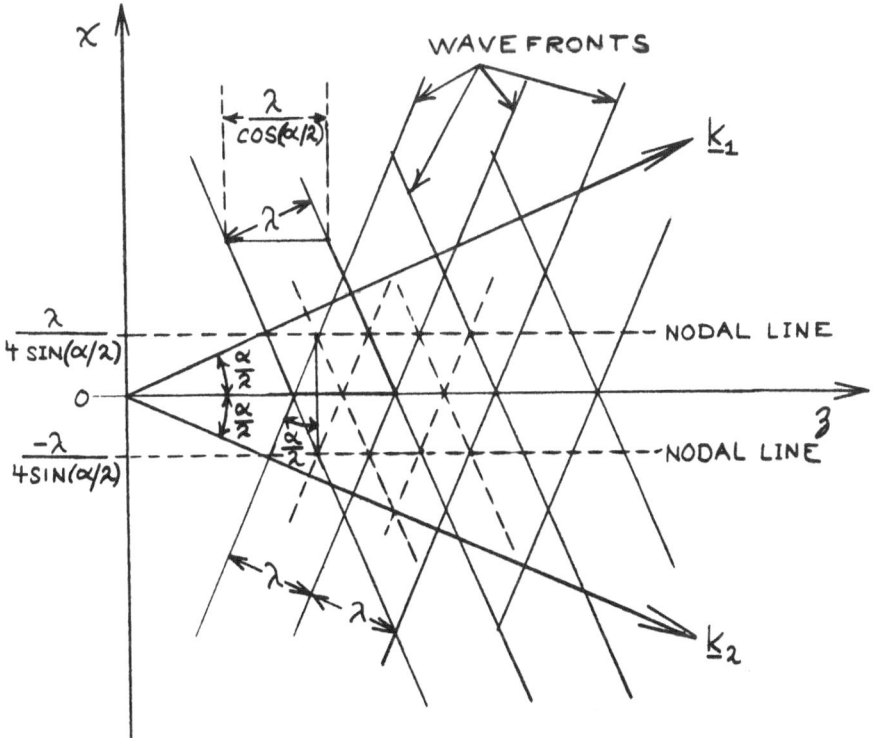

Figure 1-5.   Interference of waves in the x-z plane.

phase velocity of the total wave must be greater than that of either of its components, in accordance with Eq. (1-7).

The interference of spherical waves is somewhat more complicated, but deserves attention because of its major role in antenna and array theory. Two sources of spherical waves of identical frequency separated by a distance 2a are illustrated in Figure 1-6. The sum of the waves at an arbitrary point in space can be expressed

$$S = u_1 + u_2 = A_1 e^{-jkr_1}/r_1 + A_2 e^{-jkr_2}/r_2. \qquad \textbf{(1-31)}$$

This is a general form for the sum. Useful approximations can be made when the observation point lies far from the sources relative to their separation:

$$r \gg a, \quad \text{or} \quad \frac{a}{r} \ll 1. \qquad \textbf{(1-32)}$$

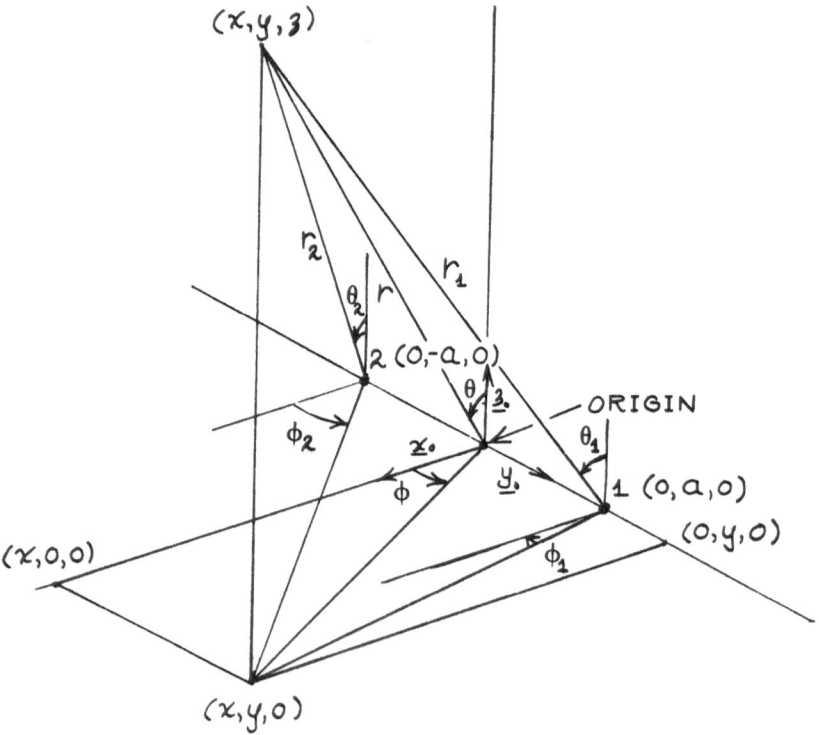

Figure 1-6. Construction for sum of spherical waves.

The position vectors are given by the exact expressions (minus sign refers to r_1):

$$r_{1,2} = \sqrt{x^2 + (y \mp a)^2 + z^2}$$

$$= r\sqrt{1 \mp 2\left(\frac{a}{r}\right)\sin\theta\sin\phi + \left(\frac{a}{r}\right)^2}, \quad \textbf{(1-33)}$$

where, from Figure 1-2

$$r = \sqrt{x^2 + y^2 + z^2}; \quad y = r\sin\theta\sin\phi.$$

The Taylor series expansion is used:*

$$f(x) = (1 + 2bx + x^2)^{p/2}$$

$$= 1 + pbx + \frac{p}{2}[1 + (p - 2)b^2]x^2 + \ldots \quad \textbf{(1-34)}$$

together with

$$x = \frac{a}{r}, \quad b = \pm\sin\theta\sin\phi, \quad p = 1,$$

to obtain

$$r_{1,2} = r \mp a\sin\theta\sin\phi + \frac{a^2}{2r}(1 - \sin^2\theta\sin^2\phi) + \ldots$$

and

$$e^{-jkr_{1,2}} \cong (e^{-jkr})(e^{\pm jka\sin\theta\sin\phi})(e^{jk(a^2/2r)\sin^2\theta\sin^2\phi}). \ldots$$

The third and higher factors of the result are close to unity if

$$\frac{ka^2}{2r} = \frac{\pi a^2}{\lambda r} \ll 1,$$

---

*If not familiar with Taylor series expansions, simply use the approximation $\sqrt{1 + 2bx + x^2}$ $\approx 1 + bx + (1 - b^2)x^2/2$, which holds for $|x| \ll 1$ and $|b| \leq 1$.

or

$$r \gg \frac{\pi}{4} \frac{(2a)^2}{\lambda}. \tag{1-36}$$

This condition resembles Eq. (1-20) and requires that the observation point lie in the far field with respect to the spherical wave sources.

Using Eq. (1-34) with $p = -1$ yields:*

$$\frac{1}{r_{1,2}} = \frac{1}{r}\left[1 \mp \frac{a}{r} \sin\theta \sin\phi + \text{terms of order} \left(\frac{a}{r}\right)^2\right] \tag{1-37}$$

In accordance with condition (1-32), only the first term is retained, and the approximate wave function becomes:

$$e^{-jkr_{1,2}}/r_{1,2} \cong \frac{e^{-jkr}}{r} e^{\pm jka \sin\theta \sin\phi}. \tag{1-38}$$

If the sources have complex amplitudes A_1 and A_2 as given by Eq. (1-22), the sum of Eq. (1-31) is approximately:

$$S \cong \frac{e^{-jkr}}{r} e^{-j(\alpha_1 + \alpha_2)/2} (a_1 e^{j\psi} + a_2 e^{-j\psi}), \tag{1-39}$$

in which

$$\psi = ka \sin\theta \sin\phi - \left(\frac{\alpha_1 - \alpha_2}{2}\right). \tag{1-40}$$

Thus, when the condition is satisfied (in the far field), the superposition of two spherical wave sources appears like a single spherical wave source located at the origin, as in Eq. (1-15). If the two sources have equal strengths, $a_1 = a_2 = a_0$, Euler's identity,

$$\cos x = (e^{jx} + e^{-jx})/2,$$

_____

*Similar to the approximation used for Eq. (1-34), use here

$$\frac{1}{\sqrt{1 + 2bx + x^2}} \cong 1 - bx - (1 - 3b^2)x^2/2,$$

which leads to Eq. (1-37).

can be used and the sum becomes

$$S \cong 2 \, a_0 \, A(\theta,\phi) e^{-j(\alpha_1 \, + \, \alpha_2)/2} \left(\frac{e^{-jkr}}{r} \right)$$

$$A(\theta,\phi) = \cos \psi \qquad \qquad \textbf{(1-41)}$$

in which ψ is given by Eq. (1-40). For certain directions in space, $\psi = m\pi$, so that $|\cos \psi| = 1$ and reinforcement occurs:

$$|S| = 2a_0.$$

For other directions, $\psi = (m + \frac{1}{2})\pi$ is satisfied, causing cancellation. The function $A(\theta,\phi)$ can be visualized as a three-dimensional surface whose magnitude is a function of θ, ϕ, and the normalized source separation $2\pi a/\lambda = ka$. In antenna theory, $A(\theta,\phi)$ is identified with the radiation pattern.

The far field approximation of the fields from only two sources can be extended to many sources (for example, in array theory, as discussed in Chap. 8).

PROBLEMS

PROBLEM 1-1. Figure 1-7 is employed in the following exercises, which involve the calculation of solid angular areas.

a. Consider the spherical cap of area Ω_1, which is bounded by:

$$0 \leqslant \phi < 2\pi; \qquad 0 \leqslant \theta \leqslant \alpha/2.$$

Show that

$$\Omega_1 = 2\pi[1 - \cos(\alpha/2)].$$

(Hint: Recall that $d\Omega = \sin \theta d \theta d\phi$.)

b. Consider the spherical ring shaped area Ω_2, which is bounded by

$$0 \leqslant \phi < 2\pi; \; \pi/2 - \beta/2 \leqslant \theta \leqslant \pi/2 + \beta/2.$$

Show that

$$\Omega_2 = 4\pi \sin(\beta/2).$$

c. Consider a section, Ω_3, of the ring of part b. Points in Ω_3 must satisfy:

$$\phi' \leq \phi \leq \phi''; \; \pi/2 - \beta/2 \leq \theta \leq \pi/2 + \beta/2.$$

Show that if $\Delta\phi = \phi'' - \phi'$, then

$$\Omega_3 = 2(\Delta\phi) \sin (\beta/2).$$

PROBLEM 1-2. Consider a source of sinusoidal spherical waves whose radiation is distributed uniformly within a spherical cap of area Ω_1 (see Fig. 1-7). This condition may be expressed:

$$A(\theta,\phi) = \begin{cases} 1 & 0 \leq \phi < 2\pi; \; 0 \leq \theta \leq \alpha/2 \\ 0 & \text{Elsewhere} \end{cases}$$

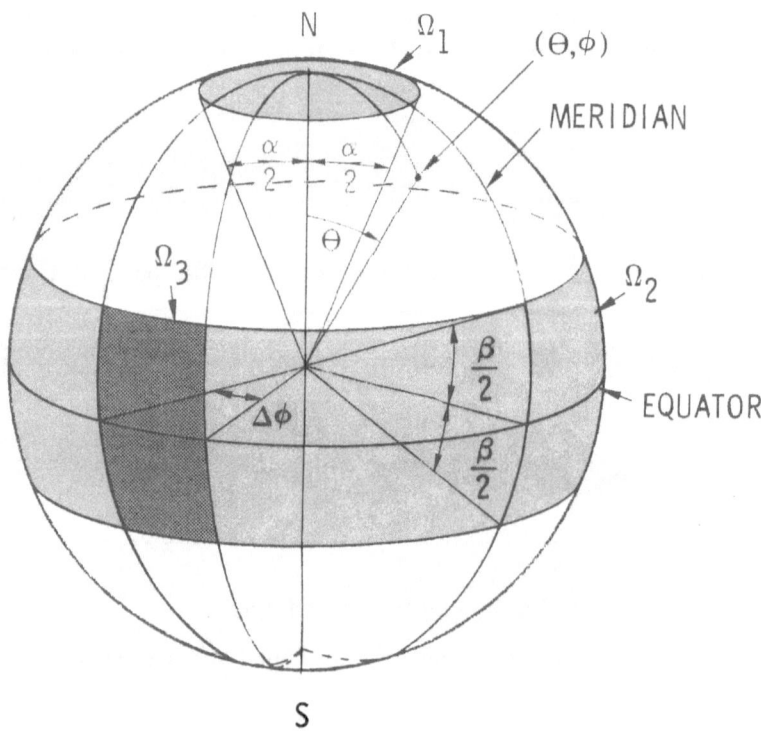

Figure 1-7. A unit sphere.

By what factor does the radiation flux within Ω_1 exceed that of an isotropic source of the same total radiated power if $\alpha = 30°$? Express your result in decibels (dB).

PROBLEM 1-3. Consider a source of sinusoidal spherical waves whose radiation is uniformly distributed within a spherical ring of area Ω_2 (see Fig. 1-7). If all radiation is confined to the ring, this condition may be expressed:

$$A(\theta,\phi) = \begin{cases} 1 & 0 \leq \phi \leq 2\pi; \ \pi/2 - \beta/2 \leq \theta \leq \pi/2 + \beta/2 \\ 0 & \text{Elsewhere} \end{cases}$$

By what factor (in decibels) does the radiation flux within Ω_2 exceed that of an isotropic source of the same total radiated power if $\beta = 30°$? How does this result compare to that of problem 1.2?

PROBLEM 1-4. Calculate the percentage error incurred by the approximation

$$\sqrt{1 + 2bx + x^2} \approx 1 + bx + (1 - b^2)\, x^2/2,$$

If $b = 0.5$ and $|x| < 0.2$

PROBLEM 1-5. Consider two point sources located at $(0, a, 0)$ and $(0, -a, 0)$, with the scalar fields

$$U_1 = A_1 e^{-jkr_1}/r_1, \ U_2 = A_2 e^{-jkr_2}/r_2,$$

respectively. The far field expression for the sum is $(A_1 = A_2 = 1)$

$$S = U_1 + U_2 = \frac{2e^{jkr}}{r} \cos[ka \sin \theta \sin \phi]$$

(see Fig. 1-6 for angle conventions).

Draw the shape of the radiation pattern in the x-y plane for $a = \lambda/4$, λ, and 4λ. How do these patterns compare to those that lie in the y-z plane? Did you expect this result on the basis of symmetry?

PROBLEM 1-6.* Given four point scores U_1, U_2, U_3, and U_4 with positions $(b, a, 0)$, $(b, -a, 0)$, $(-b, a, 0)$, and $(-b, -a, 0)$:

---

*Advanced problem.

a. Derive a far field formula for $S = U_1 + U_2 + U_3 + U_4$;
b. Determine the conditions that must hold for this formula to be a valid approximation; and
c. Work out in detail the special case for equal amplitudes $(A_1 = A_2 = A_3 = A_4 = 1)$, and show that for $b = 0$ the formula for two point sources that appears in problem 1.5 is recovered.

PROBLEM 1-7.* By use of the conservation of energy principle described in this chapter, show that the amplitude of water waves that arise from a point source on the surface of a pond diminish as $1/\sqrt{r}$, where r is the distance along the surface, as measured from the source.

_____

*Advanced problem.

CHAPTER 2

Transmission Lines

The discussion of general wave characteristics included those of one-dimensional waves. A uniform transmission line is a fine example of a structure that supports one-dimensional waves. Transmission lines are of practical importance in general and of great importance to antennas in particular. Indeed, antennas can be viewed as transducers: in the transmit mode they transform a one-dimensional wave from a transmission line feed into a three-dimensional radiation field; in the receive mode they transform a three-dimensional field arriving from a distant source into a one-dimensional wave that propagates along a transmission line to a receiver.

The review presented here is of limited scope and assumes some familiarity with transmission lines. For example, Smith Chart theory and practice are omitted. The Annotated Bibliography at the end of the book provides additional sources for studying or reviewing this subject.

2.1 BASIC RELATIONSHIPS OF VOLTAGES AND CURRENTS

The radiation effects of AC currents (discussed in the next chapter) can be neglected when a circuit's typical dimensions are small compared to a wavelength. An example is to view a transmission line as an AC circuit in which one dimension may be large relative to the wavelength, and waves propagate along that dimension. Figure 2-1 shows a simple lossless transmission line model and the corresponding equivalent lumped parameter network. Assume that the conducting wires are straight and parallel with a separation, d, which is small compared to the wavelength. Capacitance, c, and inductance, ℓ, per unit length are measured in farads per meter and henrys per meter, respectively. At any position along the line, the conductors carry currents equal in magnitude and opposite in direction. With AC circuit theory, the relationship between voltages, u, and currents, i, is derived in the transmission line segment of length Δx, and can be summarized as follows:

$$i_1 - i_2 = i_3,$$

$$u_1 - u_2 = u_3,$$

$$i_3 \cong c\Delta x \frac{\partial u_2}{\partial t},$$

$$u_3 \cong \ell\Delta x \frac{\partial i_1}{\partial t}. \tag{2-1}$$

Partial derivatives are used, because the voltages and currents are functions of both x and t. (When taking a partial derivative with respect to one variable, the other variable is assumed to be a constant.) The current and voltage differences may be expressed

$$i_2 - i_1 \cong \frac{\partial i_1}{\partial x} \Delta x,$$

TRANSMISSION LINE   LUMPED EQUIVALENT NETWORK

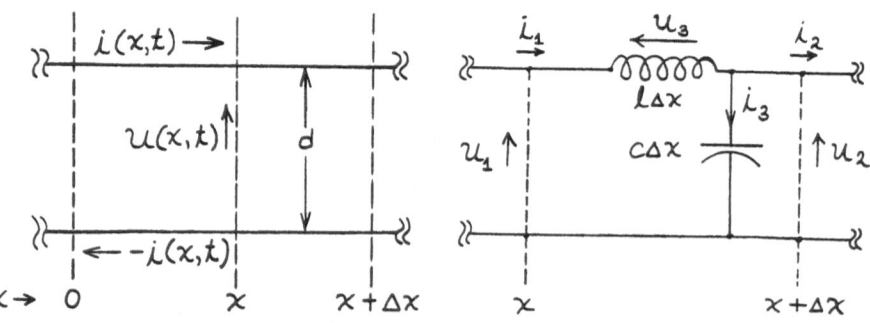

Figure 2-1. Schematic of transmission line and its equivalent network.

and

$$u_2 - u_1 \cong \frac{\partial u_1}{\partial x}\Delta x. \qquad (2\text{-}2)$$

When differentiating the second of these expressions with respect to time,

$$\frac{\partial u_2}{\partial t} \cong \frac{\partial u_1}{\partial t} + \frac{\partial^2 u_1}{\partial x \partial t}\Delta x. \qquad (2\text{-}3)$$

Using Eqs. (2-2) and (2-3) in Eq. (2-1),

$$\frac{\partial i_i}{\partial x}\Delta x \cong -c\Delta x \left(\frac{\partial u_1}{\partial t} + \frac{\partial^2 u_1}{\partial x \partial t}\Delta x, \right),$$

$$\frac{\partial u_1}{\partial x}\Delta x \cong -\ell\Delta x \frac{\partial i_i}{\partial t}. \qquad (2\text{-}4)$$

In the limit as $\Delta x \to 0$, the subscript 1 can be omitted, and

$$\boxed{\begin{aligned} \frac{\partial i}{\partial x} &= -c\,\frac{\partial u}{\partial t} \\[1mm] \frac{\partial u}{\partial x} &= -\ell\frac{\partial i}{\partial t} \end{aligned}} \qquad (2\text{-}5)$$

which are the differential equations for a lossless transmission line. To obtain entirely equivalent second order differential equations for the current and voltage, second derivatives can be taken of the above expressions with respect to x or t and by eliminating the i or u dependence by substitution from one equation to the other. The result is:

$$\frac{\partial^2 u}{\partial x^2} - \ell c\,\frac{\partial^2 u}{\partial t^2} = 0,$$

and

$$\frac{\partial^2 i}{\partial x^2} - \ell c\,\frac{\partial^2 i}{\partial t^2} = 0. \qquad (2\text{-}6)$$

The expressions of Eq. (2-6) have the form of the *wave equation*

$$\frac{\partial^2 y}{\partial x^2} - \frac{1}{v^2}\frac{\partial^2 y}{\partial t^2} = 0,$$

in which v is the propagation velocity. The velocity of propagation along the transmission line is, by analogy

$$c_0 = 1/\sqrt{\ell c}. \tag{2-7}$$

The solutions of Eqs. (2-5) or (2-6) may be separated in the x and t variables, if sinusoidal time dependence (steady state) is assumed:

$$u(x,t) = U(x)e^{j\omega t},$$

and

$$i(x,t) = I(x)e^{j\omega t}. \tag{2-8}$$

Substitution of these trial forms into Eqs (2-5) yields

$$\frac{dU}{dx} = -j\omega\ell\, I,$$

$$\frac{dI}{dx} = -j\omega c\, U. \tag{2-9}$$

Substitution of Eq. (2-8) into (2-6) produces

$$\frac{d^2 U}{dx^2} + \left(\frac{\omega}{c_0}\right)^2 U = 0,$$

and

$$\frac{d^2 I}{dx^2} + \left(\frac{\omega}{c_0}\right)^2 I = 0. \tag{2-10}$$

The general solution of the first equation is

$$U(x) = Ae^{-j(\omega/c_0)x} + Be^{j(\omega/c_0)x}, \tag{2-11}$$

in which A and B are arbitrary complex amplitudes. This is done by substituting Eq. (2-11) and its second derivative into Eq. (2-10). If Eq. (2-11) is differentiated with respect to x and substituted into the first of Eq. (2-9), the expression for the current is

$$I(x) = \frac{1}{Z_0} [Ae^{-j(\omega/c_0)x} - Be^{(j\omega/c_0)x}], \qquad (2\text{-}12)$$

in which

$$Z_0 = \frac{k}{\omega l} = \sqrt{l/c} \qquad (2\text{-}13)$$

is referred to as the transmission line's *characteristic impedance*. From Eq. (1-7), the quotient of the angular frequency and propagation speed is the wave number

$$\frac{\omega}{c_0} = k = \frac{2\pi}{\lambda} = \frac{2\pi f}{c_0}. \qquad (2\text{-}14)$$

Insertion of Eqs. (2-11) and (2-12) into Eq. (2-8) yields the complete steady-state solution for the voltage and current along the lossless transmission line:

$$u(x,t) = Ae^{-j(kx-\omega t)} + Be^{j(kx+\omega t)},$$

$$i(x,t) = \frac{1}{Z_0} [Ae^{-j(kx-\omega t)} - Be^{j(kx+\omega t)}]. \qquad (2\text{-}15)$$

2.2 REFLECTIONS AT DISCONTINUITIES

The solution of Eq. (2-15) represents waves of arbitrary amplitude traveling in both the $+x$ and $-x$ directions. Equations (2-15) show that the voltage wave is always accompanied by a current wave and vice versa. The general solution's arbitrary constants, A and B, can be determined, if the appropriate boundary conditions are specified. For example, if a line is terminated by a load with impedance Z_L at $x = 0$, that is,

$$U(0) = Z_L I(0), \qquad (2\text{-}16)$$

as seen in part a of Figure 2-2, the ratio of the amplitude of the reflected wave (traveling away from the load), B, to the incident wave (traveling toward the load), A can be obtained. If Eqs. (2-11) and (2-12) are substituted in Eq. (2-16), the amplitudes evidently obey:

$$A + B = \frac{Z_L}{Z_0} (A - B). \qquad (2\text{-}17)$$

The amplitude ratio is defined as the reflection coefficient of the load:

$$\boxed{\Gamma_L = \frac{B}{A}}$$

(2-18)

(a) TERMINATION

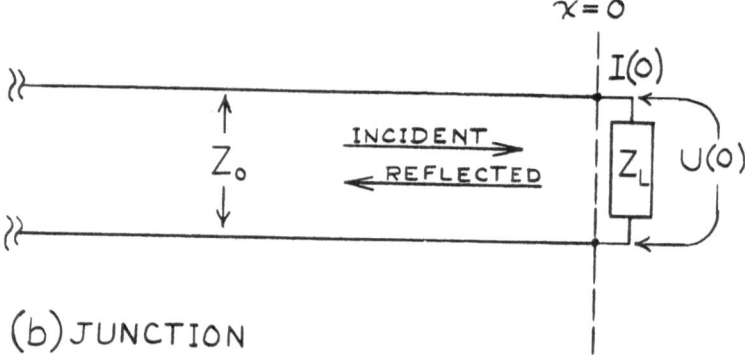

(b) JUNCTION

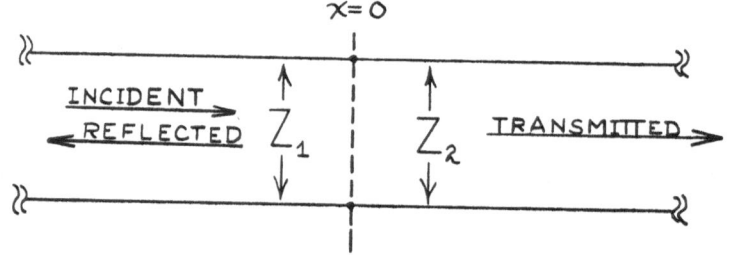

(c) INPUT IMPEDANCE

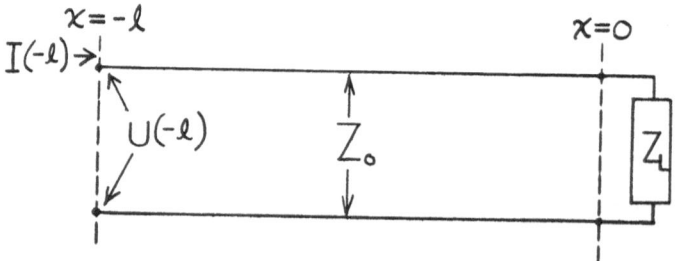

Figure 2-2. Transmission line configurations.

Division of Eq. (2-17) by A, and use of the definition given in Eq. (2-18) yields

$$\Gamma_L = \frac{Z_L - Z_0}{Z_L + Z_0}. \qquad (2\text{-}19)$$

For the line terminated at $x = 0$ with the load Z_L, the solutions for the voltage and current may now be expressed:

$$U(x) = A(e^{-jkx} + \Gamma_L e^{jkx}),$$

and

$$I(x) = \frac{A}{Z_0}(e^{-jkx} - \Gamma_L e^{jkx}). \qquad (2\text{-}20)$$

(The $e^{j\omega t}$ time dependence is implicit.)

Standing waves describe both voltage and current. The *voltage standing wave ratio* (VSWR) is the transmission line analog of the *standing wave ratio* defined in Eq. (1-26):

$$\rho = \text{VSWR} = \frac{U_{max}}{U_{min}} = \frac{I_{max}}{I_{min}} = \frac{1 + |\Gamma_L|}{1 - |\Gamma_L|}. \qquad (2\text{-}21)$$

The transmission line solution can be interpreted as interference of an incident and reflected wave. Accordingly, maxima and minima of the standing wave pattern for the current or voltage are spaced one-quarter wavelength apart (see Fig. 1-4). When the termination is a short circuit ($Z_L = 0$) or open circuit ($Z_L = \infty$), the reflection coefficient is -1 or 1, respectively. The minima (zeros) of the current standing wave lie at the maxima of the voltage standing wave whenever Z_L is real.

A second example is a wave incident from the left on a junction of two lines at $x = 0$, as illustrated in Figure 2-2b. The line on the left has a characteristic impedance Z_1, while that on the right has a characteristic impedance Z_2. Boundary conditions for this case arise from the required continuity of the current and voltage at the boundary, which is the junction $x = 0$. The general solutions are

$$U(x) = \begin{cases} Ae^{-jk_1 x} + Be^{jk_1 x} & x < 0 \\ Ce^{-jk_2 x} & x > 0 \end{cases} \qquad (2\text{-}22)$$

and

$$
I(x) = \begin{cases}
\dfrac{A}{Z_1} e^{-jk_1 x} - \dfrac{B}{Z_1} e^{jk_1 x} & x < 0 \\[1em]
\dfrac{C}{Z_2} e^{-jk_2 x} & x > 0
\end{cases}
$$

$$(2\text{-}23)$$

The assumption of both Eqs. (2-22) and (2-23) is that on the right ($x > 0$) there is only an outward-going wave (i.e., that nothing is returning from $x = \infty$). Continuity of voltage and current at $x = 0$ places the following constraints on the coefficients:

$$A + B = C,$$

$$A - B = \frac{Z_1}{Z_2} C. \qquad (2\text{-}24)$$

A transmission coefficient represents the amplitude of the wave that propagates to the right along the semi-infinite line characterized by the characteristic impedance Z_2:

$$T = C/A \qquad (2\text{-}25)$$

and the reflection coefficient is, as before,

$$\Gamma = B/A. \qquad (2\text{-}26)$$

Simultaneous solution of Eq. (2-24) yields

$$
\boxed{\begin{aligned}
T &= 2Z_2/(Z_1 + Z_2) \\[0.5em]
\Gamma &= T + 1 = \frac{Z_2 - Z_1}{Z_2 + Z_1}
\end{aligned}}
\qquad (2\text{-}27)
$$

The second equation of Eq. (2-27), and Eq. (2-19) indicate that the semi-infinite line on the right appears as a load that is numerically equal to its characteristic impedance, Z_2. Thus, the input impedance of a semi-infinite line is equal to its characteristic impedance. Reflections and associated "ringing" in electrical cables can be eliminated by termination with a load that is equal to the cable's characteristic impedance. Finite length lines then behave as though they were semi-infinite.

Consider a line of length, l, and characteristic impedance Z_0 terminated by a load of impedance Z_L, as illustrated in Figure 2-2c. The input impedance is defined:

$$Z_{in} \equiv \frac{U(-l)}{I(-l)}.$$ (2-28)

The equations of Eq. (2-20) yield the expression

$$Z_{in} = Z_0 \left(\frac{e^{jkl} + \Gamma_L e^{-jkl}}{e^{jkl} - \Gamma_L e^{-jkl}} \right).$$ (2-29)

Using the Euler-DeMoivre identity, $e^{\pm jkl} = \cos(kl) \pm j \sin(kl)$, Eq. (2-29) may be reduced to the more convenient form:

$$Z_{in} = Z_0 \left[\frac{Z_L \cos kl + j Z_0 \sin kl}{Z_0 \cos kl + j Z_L \sin kl} \right]$$ (2-30)

Several special cases serve to further illustrate these points.

CASE 2-1   If $Z_L = 0$, the Figure 2-2c configuration is a *shorted stub*. From Eq. (2-30), the input impedance is simply

$$Z_{in} = j Z_0 \tan(kl).$$ (2-31)

Because the impedance is purely imaginary, it must be capacitive for those values of the product kl that cause the tangent to be negative:

$$(2n - 1)\frac{\pi}{2} < kl < n\pi \qquad n = 1, 2, 3 \ldots$$

or

$$(2n - 1)\frac{\lambda}{4} < l < \frac{n\lambda}{2}.$$

On the other hand, when l satisfies the relations

$$\frac{n\lambda}{2} < l < \frac{\lambda}{4}(2n + 1) \qquad n = 0,1,2\ldots$$

the tangent is positive, and the shorted stub is therefore inductive,

CASE 2-2 If $Z_L \to \infty$, the Figure 2-2c configuration is an open line, and the input impedance is given by

$$Z_{in} = -jZ_0 \cot(kl). \qquad \textbf{(2-32)}$$

Because the cotangent is the reciprocal of the tangent as in Eq. (2-31), the open line is capacitive for those lengths that cause a shorted stub to be inductive and vice versa. In contrast to the shorted stub, an open line is difficult to achieve in practice, because some radiation occurs at the open end, and the condition $Z_L \to \infty$ is unrealistic.

CASE 2-3 If Z_L has an arbitrary value and the line length is $\lambda/4$, the configuration of Figure 2-2c is a *quarter wave transformer*. By Eq. (2-30), the input impedance is

$$Z_{in} = Z_0^2/Z_L. \qquad \textbf{(2-33)}$$

Shorted stubs and quarter wave transformers are used to match loads to lines (i.e., eliminate reflections from the load). Figure 2-3 illustrates two common applications. A shorted stub matches a load Z_L to a line of characteristic impedance, $Z_0 \neq Z_L$, by the proper selection of the lengths, l_1 and l_2, as shown in Figure 2-3a. A graphic aid, the Smith Chart,* facilitates length determination.

Figure 2-3b shows load-to-line matching using a quarter wave transformer. The characteristic impedance of the quarter wave line segment is selected so that

$$Z_1 = \sqrt{Z_0 Z_L}. \qquad \textbf{(2-34)}$$

---

*For a description of the Smith Chart and its applications see the Annotated Bibliography at the end of the book.

Input impedance at the terminals A_1 and A_2 is, by Eq. (2-33),

$$Z_{in} = \frac{Z_1^2}{Z_L} = Z_0,$$

which represents the desired impedance match to the line.
Several special cases illustrate the results.

CASE 2-4 Let $Z_0 = 50 \ \Omega$ and $Z_L = 100 + j60 \ \Omega$ for the case illustrated in Figure 2-2a.

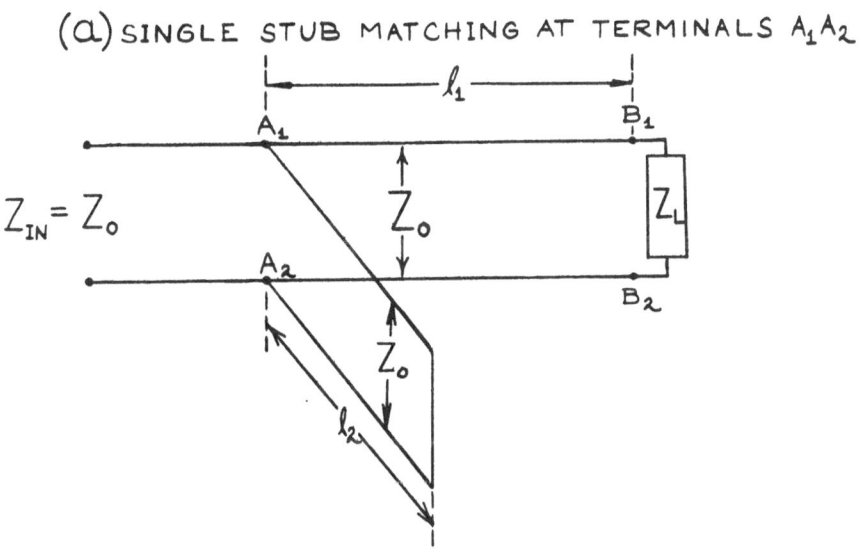

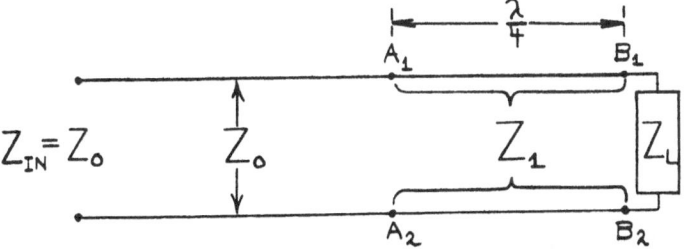

Figure 2-3. Impedance matching methods.

From Eq. (2-19), the reflection coefficient is $\Gamma_L = (100 + j60 - 50)/(100 + j60 + 50) = 0.425 + j0.230$ and has a magnitude

$$|\Gamma_L| = (0.425^2 + 0.230^2)^{1/2} = 0.483.$$

Note that the reflection coefficient is dimensionless.

The reflected power can be expressed relative to the incident power in decibels:

$$10 \log|\Gamma_L|^2 = 20 \log|\Gamma_L| = -6.32 \text{ dB}.$$

If the load were matched to the line, the reflection coefficient would be zero, giving a relative reflected power of $-\infty$ dB.

The voltage standing wave ratio ρ is given by Eq. (2-21). This result can also be expressed in decibels: $20 \log \rho = 9.16$ dB. When the load and line are matched, $\rho = 1$ or 0 dB.

CASE 2-5   The load can be matched to the line with a two-step procedure.

For step one, at some distance (l) along the line from the load, the imped-ance is purely real. From Eq. (2-30).

$$Z_{in} = 50 \left[\frac{10 + j(6 + 5 \tan kl)}{(5 - 6 \tan kl) + j10 \tan kl} \right].$$

Denoting $\tan kl = t$ for convenience, the form is rewritten:

$$Z_{in} = 50 \left[\frac{(10^2 + (6 + 5t)^2)^{\frac{1}{2}} \exp\left(j\tan^{-1}\left(\frac{6 + 5t}{10}\right)\right)}{(100t^2 + (5 - 6t)^2)^{\frac{1}{2}} \exp\left(j\tan^{-1}\left(\frac{10t}{5 - 6t}\right)\right)} \right].$$

The impedance is purely real if the exponential in the numerator's argument equals the exponential in the denominator's argument:

$$\frac{6 + 5t}{10} = \frac{10t}{5 - 6t}.$$

The result is a quadratic equation in t, with the solutions $t = 0.253, -3.953$. Using the positive root, $\tan kl = 0.253$, implies $l = 0.0394\lambda$, which is the shortest distance from the load at which the input impedance is real. The input impedance is given at this point by

$$Z_{in} = 50 \sqrt{\frac{10^2 + (6 + 5t)^2}{100t^2 + (5 - 6t)^2}} = 143.6 \ \Omega.$$

For step 2, the characteristic impedance of the quarter wave transformer needed to match the 143.6 Ω load (Z_L plus 0.0394λ line segment), is then, from Eq. (2-34)

$$Z_1 = \sqrt{(50)(143.6)} = 84.7 \ \Omega.$$

2.3 VARIOUS TYPES OF TRANSMISSION LINES

Figure 2-4 illustrates four common types of transmission lines.

a. The familiar dual conductor line with a plastic spacer is often used for television antennas or in other applications that allow low-power handling, high-characteristic impedance, and high losses.
b. The frequently encountered coaxial line, shown in Figure 2-4b, may be either rigid or flexible. This type of line is completely shielded and does not radiate, can be built to handle high power, and has low-characteristic impedance. However, at very high frequencies (X-band and above) dielectric losses are considerable.
c. Microstrip, easily fabricated into complicated shapes, integrates with other circuit components to produce truly compact assemblies. It radiates at bends, junctions, and terminations.
d. Stripline combines some of the advantages of coaxial line and microstripline; it is shielded (therefore, does not radiate) and is easily fabricated.

Stripline and microstripline usually cannot handle very high power and become increasingly lossy with increasing frequency. Transmission of high-power, high-frequency electromagnetic (EM) waves with low losses requires waveguides (described in Chap. 6).

Thus far, voltage and current waves propagating only along the transmission line have been assumed. Another (implicit) assumption was that the electric, **E,** and magnetic, **H,** fields that accompany those voltages and currents are completely transverse to the line's direction and that they propagate along the line, rather than in directions normal to the line. In practice, such

(a) DUAL CONDUCTOR LINE

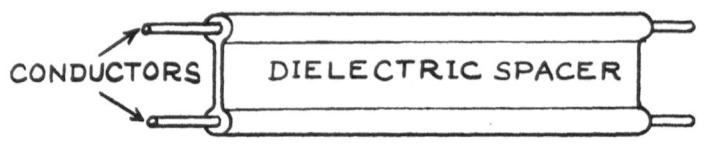

CONDUCTORS | DIELECTRIC SPACER

(b) COAXIAL LINE

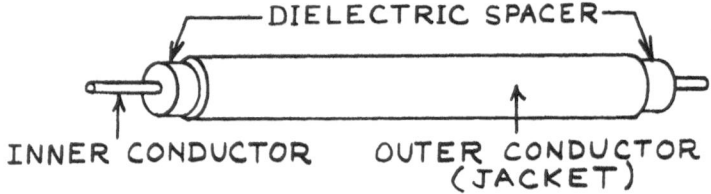

DIELECTRIC SPACER

INNER CONDUCTOR   OUTER CONDUCTOR
                        (JACKET)

(c) MICROSTRIP

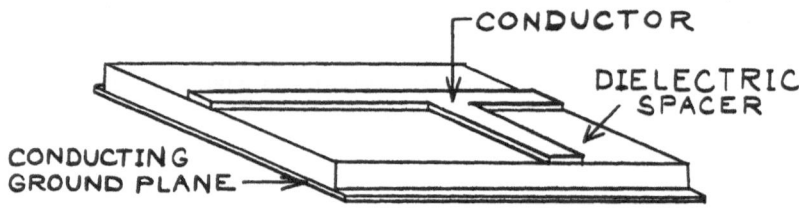

CONDUCTOR

DIELECTRIC
    SPACER

CONDUCTING
GROUND PLANE

(d) STRIPLINE

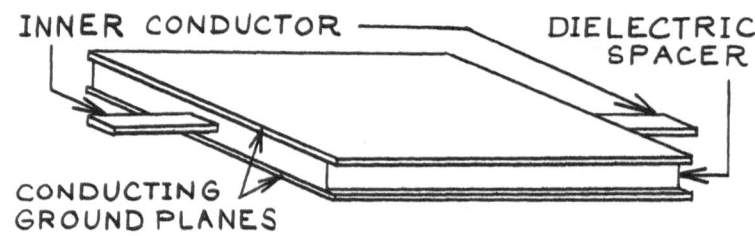

INNER CONDUCTOR —————   DIELECTRIC
                             SPACER

CONDUCTING
GROUND PLANES

Figure 2-4. Common types of transmission line.

an ideal situation can only be approached, because radiation and other effects occur at discontinuities, such as bends and terminations.

A dual conductor transmission line terminated with an open end bent into a "tee" shape is shown in Figure 2-5a. The arrows and dotted lines indicate the instantaneous current direction and magnitude, respectively, along the line. Along most of the line the currents are of equal magnitude and opposite direction and produce only negligible fields at a distance from the line. However, along the top of the tee, the currents are collinear and, therefore, reinforce one another. Configured this way, the termination constitutes a dipole antenna of length 2*l*, which is fed by the line and radiates EM energy in practically all directions. As shown in Figure 2-5b, the antenna is equivalent to a load at the end of a transmission line, fed by a generator. The real part of load impedance is defined as the *radiation resistance*. (This concept is

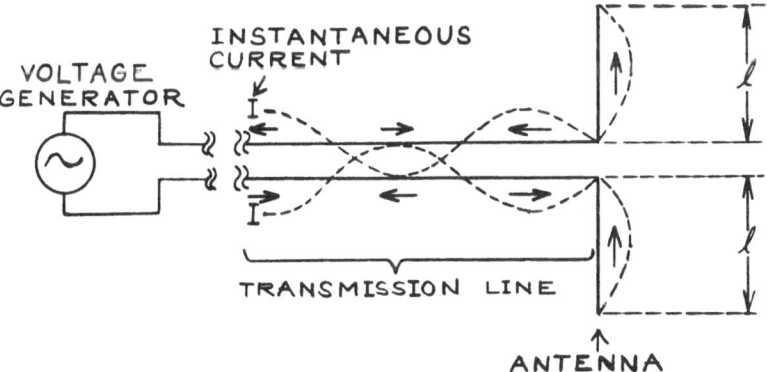

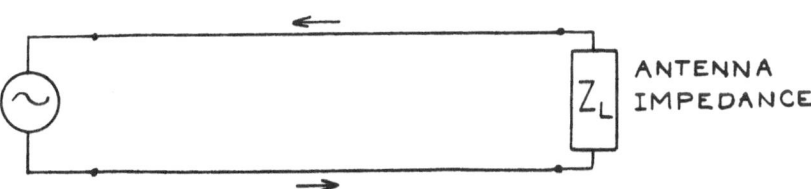

Figure 2-5. Primitive antenna formed by bending ends of transmission line.

described in Chap. 4.) This illustrates how an antenna acts as a transducer, which converts the one-dimensional wave on the transmission line into three-dimensional space wave.

PROBLEMS

PROBLEM 2-1.   Complete the derivation of the formulas provided in Eqs. (2-10), (2-12), (2-15), (2-19), (2-21), (2-27), (2-29), (2-30), and (2-33).

PROBLEM 2-2.   A transmission line with a characteristic impedance $Z_0 = 50\ \Omega$ is terminated by a load $Z_L = R_L + jX_L$ and the VSWR must be less than 3.5 dB.

  a. Find the limits of the reflection coefficient Γ_L.
  b. Can Z_L be purely reactive (i.e., can $Z_L = jX_L$ represent a solution)?
  c. Find the limits of Z_L if the load is purely ohmic (i.e., $Z_L = R_L$).
  d. Find the limits of X_L when $R_L = 50\ \Omega$.
  e.* Find the locus of points in the complex plane that bound the allowed values of the load impedance, Z_L.

PROBLEM 2-3.*At the junction of two transmission lines (see Fig. 2-2b) the voltage and current must be continuous (i.e., at $x = 0$, $U_1 = U_2$ and $I_1 = I_2$). These boundary conditions lead to Eq. (2-24). A more general solution for the transmission line junction includes a wave returning from the right, so that, in analogy with Eqs. (2-22) and (2-23):

$$U = \begin{cases} A_1 e^{-j\beta_1 x} + B_1 e^{j\beta_1 x} & x < 0 \\ A_2 e^{-j\beta_2 x} + B_2 e^{j\beta_2 x} & x > 0 \end{cases}$$

$$I = \begin{cases} Y_1(A_1 e^{-j\beta_1 x} - B_1 e^{j\beta_1 x}) & x < 0 \\ Y_2(A_2 e^{-j\beta_2 x} - B_2 e^{j\beta_2 x}) & x > 0 \end{cases}$$

with $Y_1 = 1/Z_1$ and $Y_2 = 1/Z_2$ being the characteristic admittances. Evidently, continuity at $x = 0$ requires that the above coefficients satisfy

$$A_1 + B_1 = A_2 + B_2,$$

$$Y_1 (A_1 - B_1) = Y_2 (A_2 - B_2).$$

---

*Advanced problem.

a. Use these results to determine the elements of the matrix, M, that relates A_1 and B_1 to A_2 and B_2 (i.e., find M_{ij}) in

$$\begin{pmatrix} A_2 \\ B_2 \end{pmatrix} = \begin{pmatrix} M_{11} & M_{12} \\ M_{21} & M_{22} \end{pmatrix} \begin{pmatrix} A_1 \\ B_1 \end{pmatrix}$$

b. Express the reflection and transmission coefficients at the junction in terms of the matrix elements.

PROBLEM 2-4.*Consider two points along a uniform transmission line, x and $x + d$. If the characteristic impedance is Z, and the wave number is β, determine the matrix elements that connect the incident and reflected waves at x and at $x + d$ (see Prob. 2-3).

---

*Advanced problem.

CHAPTER 3

The Electromagnetic Field

Antennas are used to launch electromagnetic (EM) waves into space or, in the reverse process, to capture energy from an incident EM field. Some antennas are quite elaborate; however, others are merely pieces of wire. How does a wire radiate and what makes a wire an antenna? To answer these basic questions requires electromagnetic theory. Described here are only some properties of the EM field (without derivations) relevant to antenna studies.

EM energy propagates through space as EM waves. The propagation velocity of EM waves (including visible light) in free space is one of the basic constants of nature, with the approximate value

$$c_0 = 2.998 \times 10^8 \text{ m/s} \qquad \textbf{(3-1)}$$

EM fields are closely related to charges and currents, thus the basic relationships between currents, fields, and matter are reviewed.

43

3.1 THE ELECTROMAGNETIC FIELD VECTORS AND BASIC CONSTANTS

Electric charge is a basic quantity of the physical world. Particles, such as electrons, protons, and neutrons, are characterized by their electric charge (or lack of it). A total charge of approximately 6.242×10^{18} electrons equals one coulomb. Currents arise from the motion of charge and are measured in amperes (1 A = 1 C/s). Charge motion on the atomic scale, such as occurs in atomic orbitals, must be described by quantum mechanics, the modern theory of matter. Quantum theory accounts for many material bulk properties like conductivity, specific heat, and magnetic susceptibility. Motion of electrons in the vacuum of a cathode ray (television) tube represents microscopic charge motion across macroscopic distances. Macroscopic motion of charges can also occur inside conductors (wires) and semiconductors. These phenomena are more complicated, however, because the charges interact with the material's microstructure.

To study antennas, requires only an understanding of two macroscopic material phenomena—conduction and dielectric polarization.* Most antenna materials are either good conductors, meaning charge moves freely within them, or good dielectrics (insulators), which do not support free charge flow. Microwave technology also uses magnetic phenomena and magnetic effects, but these are not commonly exploited in antennas. Nevertheless, magnetic materials, as in the case of ferrite phase shifters, are used in phased arrays (discussed in Chap. 9).

Charges exert forces on each other, as is formulated in Coulomb's Law. The force per unit charge "sensed" by a point charge q, is defined as the electric field, **E**, which is a vector quantity. The electrostatic force on a charge q is given by

$$\mathbf{F}_e = q\mathbf{E}. \tag{3-2}$$

The force between two point charges is inversely proportional to the square of the distance between the charges. The static electric field is therefore a relatively *short-range* phenomenon.

Electric currents in materials can be imagined as the motion of a swarm of charged particles. Particles move rapidly but continually scatter off the matrix of atoms. This interaction impedes particle flow and results in a relatively slow average drift velocity along the direction of the current. The charge density per unit volume is defined as a function of position and time,

$$\rho(x,y,z,t), \quad (C/m^3)$$

---

*This is different from wave polarization, which is covered in Chapter 5.

and the corresponding velocity vector,

$$\mathbf{v}(x,y,z,t), \qquad (\text{m/s})$$

describes the local drift (average motion) of the charge. The current density vector, $\mathbf{j}$, is the product of the velocity and the charge density.

$$\mathbf{j} = \rho\mathbf{v}. \qquad \textbf{(3-3)}$$

The vector magnitude equals the amount of charge that passes through a unit area normal to $\mathbf{v}$ in a unit of time. The unit of the current density is therefore

$$(\text{C/m}^3) \times (\text{m/s}) = (\text{C/s})/\text{m}^2 = \text{A/m}^2.$$

When charges flow through a *tube* (a wire, for example), the current, i, is equal to the product of the current density and the tube's cross sectional area.

   In a vacuum, charged particles are accelerated by the electric field in accordance with Newton's second law of motion,

$$\mathbf{F} = m\mathbf{a} \qquad \textbf{(3-4)}$$

in which m is the particle mass. Inside materials, however, the atomic matrix scatters the charged particles, randomizing their motion and impeding their progress along the field direction. This process creates resistance and leads to an average uniform (unaccelerated) particle motion along the field direction. Under these circumstances the current density vector is proportional to the electric field vector

$$\mathbf{j} = \sigma\mathbf{E} \qquad \textbf{(3-5)}$$

in which the proportionality constant, σ, is the conductivity. Eq. (3-5) is a microscopic analog of Ohm's law,

$$V = iR. \qquad \textbf{(3-6)}$$

   The uniform average motion of a swarm of charged particles represents a steady current. Ampere's law describes the force that exists between steady currents—the magnetostatic force. The force experienced by an element of filamentary current, i, that flows in a straight line segment of length $\mathbf{dl}$ defines the magnetic field vector $\mathbf{B}$. The magnetostatic force is

$$d\mathbf{F}_m = i\mathbf{dl} \times \mathbf{B}, \qquad \textbf{(3-7)}$$

in which the cross product indicates that the force is perpendicular to both the magnetic vector, **B** and the current element, **dl**. The magnetostatic force, like the electrostatic force, is inversely proportional to the square of the distance between two interacting current elements and is therefore a short-range force.

Electric and magnetic fields can exist inside material media and are modified by interactions with charges and atomic scale currents in the media. A detailed description of these phenomena is not given here, but some relationships are reviewed.

When considering EM fields in materials, it is useful to define two additional EM field vectors that are closely related to **E** and **B**. In free space, these are

$$\boxed{\begin{aligned} \mathbf{D} &= \varepsilon_0 \mathbf{E} \\ \mathbf{H} &= \mathbf{B}/\mu_0 \end{aligned}} \qquad \textbf{(3-8)}$$

in which ε_0 is the electric susceptibility and μ_0 is the magnetic permeability of free space. These fundamental constants have the numerical values

$$\boxed{\begin{aligned} \varepsilon_0 &= 8.8542 \times 10^{-12} \ (\text{F/m}) \\ \mu_0 &= 4\pi \times 10^{-7} = 1.2566 \times 10^{-6} \ (\text{H/m}) \end{aligned}} \qquad \textbf{(3-9)}$$

and are related to another fundamental constant, the speed of light, c_0, by

$$\boxed{c_0 = 1/\sqrt{\mu_0 \, \varepsilon_0}} \qquad \textbf{(3-10)}$$

as can be verified by comparing Eqs. (3-10) and (3-1).

In material media, Eqs. (3-8) must be modified, so that the fields are now related by

$$\boxed{\begin{aligned} \mathbf{D} &= \varepsilon_0 \, \varepsilon \mathbf{E} \\ \mathbf{B} &= \mu_0 \, \mu \mathbf{H} \end{aligned}} \qquad \textbf{(3-11)}$$

in which ε and μ are dimensionless quantities referred to as the relative dielectric susceptibility and the relative magnetic permeability of the medium, respectively. With the exception of magnetic materials, μ is very close to unity, so that

$$\boxed{\mu = 1} \qquad (3\text{-}12)$$

can be assumed for all of the materials considered here. The relative dielectric constant of air is also very close to unity ($\varepsilon_{air} = 1$) and the speed of light in air is therefore quite close to c_0. For natural dielectrics such as glass, plastics, paper, and mineral oil, ε is a complex number and can be expressed as

$$\varepsilon = \varepsilon'(1 - j \tan \delta) = \varepsilon' - j\varepsilon''. \qquad (3\text{-}13)$$

Although ε', the real part of ε is larger than unity, the imaginary part, ε'', is usually much smaller than unity. The factor $\tan\delta$ is often called the *loss tangent* or the *dissipation factor,* and is described later.

It is important to realize that static charges and steady currents are idealizations that do not actually exist in nature. Real charged particles are incessantly accelerated and decelerated (i.e., do not move uniformly). In the theory of electric circuits, the concepts of electro- and magnetostatics are extended to electric currents that change in time. Coulomb's and Ampere's laws must be supplemented by Faraday's law of induction to obtain a satisfactory physical description of electric circuits. This law expresses the relationship between rate of change of the magnetic flux and the induced voltage (or *electromotive force*) as follows:

$$\boxed{u = -d\Phi/dt} \qquad (3\text{-}14)$$

In Eq. (3-14) u is an electric voltage induced along a closed loop and Φ is the magnetic flux that passes through any closed surface bounded by that loop. The magnetic flux is often visualized as the number of lines of the magnetic field that pass through a specified area bounded by a closed curve.

3.2 MAXWELL'S FIELD EQUATIONS AND ELECTROMAGNETIC WAVES

The British physicist James Clerk Maxwell (1831–1879) realized that the laws of Coulomb, Ampere, and Faraday, which were experimentally discovered, contained certain contradictions, and therefore could not represent a complete and self-consistent physical theory. By making the theory consistent, Maxwell unified all electromagnetic and optical phenomena. Maxwell modified Ampere's law through addition of the so called "displacement" term.

The unified theory, as concisely expressed by Maxwell's equation, predicted the existence of electromagnetic waves, which propagate in free space with the velocity of light and which transport energy (and thus can exert forces) to great distances. The components of the dynamic EM field fall off as the reciprocal of the distance from the source, in contrast with the electrostatic and magnetostatic forces, which fall off as the inverse square of the distance. Thus, the EM field represents a long-range force, and EM energy in the form of light and radio waves can travel to earth from unimaginably distant sources in the universe.

Maxwell's organization and completion of the field equations is comparable in scientific significance to Newton's formulation of the laws of motion. Several years after Maxwell's death, his prediction of the existence of electromagnetic waves was experimentally confirmed by the German physicist Heinrich Rudolf Hertz (1857–1894). This discovery paved the way for the practical development of radio by Guglielmo Marconi (1874–1937) and for the theoretical development of the theory of relativity by Albert Einstein (1879–1955).

Maxwell's equations provide exact relationships between the electric and magnetic field vectors and the charge and current densities. In differential (microscopic) form they are*

$$
\begin{aligned}
\nabla \times \mathbf{E} &= -\frac{\partial \mathbf{B}}{\partial t}, \\
\nabla \times \mathbf{H} &= \frac{\partial \mathbf{D}}{\partial t} + \mathbf{j}, \\
\nabla \cdot \mathbf{D} &= \rho, \\
\nabla \cdot \mathbf{B} &= 0.
\end{aligned}
$$

(3-15)

---

*These equations are not required in the remainder of this book.

The first equation represents Faraday's law, and the second equation represents Ampere's law modified by the addition of the displacement term. The third equation results from Coulomb's law, and the last equation maintains that magnetic field lines are always closed loops (i.e., that magnetic point sources analogous to electric charges are not known to exist). Maxwell's field equations do not prove the nonexistence of magnetic point sources or of magnetic currents; they merely state that such sources have not been observed. Some physicists, believing that magnetic monopoles (particles analogous to electrons) do exist, are performing experiments to prove their existence. In texts more advanced than this book, extensive use is made of "equivalent magnetic currents," which help simplify certain antenna calculations.

The interdependent electric and magnetic field vectors characterize the dynamic EM field, which carries both energy and momentum. The energy flux (power per unit area) is expressed by the Poynting vector (named after John Henry Poynting [1852–1914]).

$$\boxed{\mathbf{P} = \mathbf{E} \times \mathbf{H}; \qquad (\text{W/m}^2)} \tag{3-16}$$

If the fields are in a steady state with sinusoidal time dependence, the average energy flux is given by

$$\mathbf{P} = \frac{1}{2} \text{Re}(\mathbf{E} \times \mathbf{H}^*) \tag{3-17}$$

in which $\mathbf{E}$ and $\mathbf{H}$ are the field's complex vector amplitudes at the point of interest. The path of energy propagation, under certain conditions considered later, can be described by rays like those of geometrical optics. Indeed, geometrical optics provides a *zero order* approximate solution to the EM field equations (discussed in greater detail in Chap. 5).

Under steady-state conditions, Eqs. (3-5) and (3-11) can be substituted into the right-hand side of the second equation in Eq. (3-15) to yield

$$\mathbf{j} + \frac{\partial \mathbf{D}}{\partial t} = j\omega\varepsilon_0\left(\varepsilon + \frac{\sigma}{j\omega\varepsilon_0}\right)\mathbf{E}. \tag{3-18}$$

This form suggests the complex dielectric constant, which is defined as

$$\boxed{\varepsilon = \varepsilon' - j\frac{\sigma}{\omega\varepsilon_0}} \tag{3-19}$$

For "ideal" dielectrics the conductivity is zero, and ε is real. The quantity,

$$\tan \delta \equiv \frac{\sigma}{\omega\varepsilon_0\varepsilon'} \qquad (3\text{-}20)$$

(the loss tangent) represents the magnitude of the material's ohmic losses when an alternating field is present. For good dielectrics, $\tan \delta$ is small, however, the frequency dependence of the loss tangent should be recognized. Table 3-1 lists loss tangents for some common materials.

For a good conductor $\sigma \gg \omega\varepsilon_0$ and ε is almost purely imaginary. On the inside of a good conductor, the dynamic EM field decays exponentially with distance from the surface, and fields and currents are primarily confined to a thin boundary layer characterized by a *skin depth* parameter,

$$\boxed{d = 1/\sqrt{\pi f \mu_0 \sigma}} \qquad (3\text{-}21)$$

in which f is the frequency $(f = \omega/2\pi)$. Fields decay exponentially like $\exp(-dx)$ with distance x from the conductor's surface. Fields and currents are

Table 3-1. Dielectric Properties of Common Materials

Material	Dielectric Constant	Loss Tangent
E-Glass	6.2	0.006
Quartz	3.8	0.0002
Porcelaine (dry)	5.0	0.01
Aluminum oxide	9.0	0.0003
Bakelite	4.5	0.07
Plexiglass	2.7	0.01
Polyethylene	2.2	0.0002
Epoxy resin	3.3	0.02
E-Glass epoxy laminate	4.3	0.012
Urethane foam (10lb/ft³)	1.16	0.0024
Mineral oil	2.2	0.0004
Paper (impregnated)	3.0	0.07
Distilled water	81	0.005
Ice	3.2	0.0008
Air (atmos. pressure)	1.0006	~0

Note: Dielectric properties vary widely with frequency, temperature, humidity, and the presence of impurities. These values are typical averages, at room temperature and at 1 GHz.

entirely excluded from a perfect conductor's interior ($\sigma \to \infty$). However, a surface current density $\mathbf{j}_s$ is permitted in an "infinitely thin" boundary layer. Tangential components of the EM field at an ideal conductor's surface satisfy

$$\boxed{\begin{aligned} \mathbf{n} \times \mathbf{E} &= 0 \\ \mathbf{n} \times \mathbf{H} &= \mathbf{j}_s \end{aligned}} \tag{3-22}$$

in which $\mathbf{n}$ is the local unit normal to the surface (see Fig. 3-1). Physically, Eqs. (3-22) require that there be no tangential component of $\mathbf{E}$ at the conductor's surface. This means that $\mathbf{E}$ must lie along $\mathbf{n}$, and the magnetic field's tangential component is numerically equal to and spatially perpendicular to the surface current density $\mathbf{j}_s$.

The tangential components of $\mathbf{E}$ and $\mathbf{H}$ are continuous at the interfaces between materials. These boundary conditions are basic and can be derived directly from Maxwell's equations (3-15). Eqs. (3-22) pertain to the special case in which $\sigma \to \infty$ in one medium.

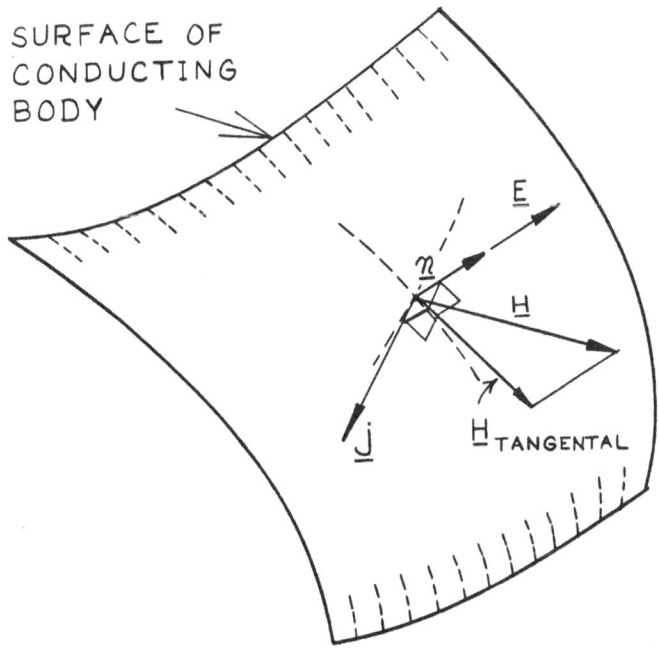

Figure 3-1. EM field vectors at the surface of a conducting body.

The speed of EM energy propagation in matter is always less than its speed in a vacuum, c_0, and for an ideal dielectric material is given by

$$c = 1/\sqrt{\mu_0\,\varepsilon_0\varepsilon} = \frac{c_0}{\sqrt{\varepsilon}} \qquad (3\text{-}23)$$

The refractive index, n, is defined:

$$n \equiv \sqrt{\varepsilon} = c_0/c. \qquad (3\text{-}24)$$

For a nonideal or lossy dielectric medium, the dielectric constant is complex. Eq. (3-24) implies that the propagation velocity is complex, which is clearly nonphysical. Although the interpretation does not hold for lossy dielectrics, if the loss tangent is very small, the propagation velocity is approximately

$$c \cong c_0/Re[\sqrt{\varepsilon}]. \qquad (3\text{-}25)$$

3.3 THE RELATION BETWEEN ELECTROMAGNETISM AND CIRCUIT THEORIES

Maxwell's field equations imply that any time-varying electric current radiates electromagnetic waves. The amount of energy radiated depends on the ratio of the EM wavelength to the size of the circuit supporting the current. Referring to the relationships among frequency, wavelength, and propagation velocity in Chapter 1, Eq. (1-7), with $c_0 = 3 \times 10^8$ m/s, most man-made circuits (except long-distance power lines) are very small compared to a wavelength at power and audio frequencies ($f < 30$ kHz). Thus, alternating current (AC) theory, which neglects radiation, provides a good description of electrical circuits at low frequencies. Radiation becomes an important factor when circuits are comparable in size to the EM wavelength. When this happens, a mere wire may become an antenna. Transmission line theory is used to illustrate the relationship between the quasi-static approach of AC circuit theory and the full treatment of electrodynamics. Figure 2-1 illustrated an ideal uniform parallel wire transmission line as a two-port L,C network, with parameters $\ell\Delta x$ and $c\Delta x$. In Figure 3-2(a), a uniform transmission line segment of length d, fed by a sinusoidal voltage source and terminated by a load, R, is shown. The line is characterized by the parameters ℓ and c, which determine its phase velocity and characteristic impedance. In Figure 3-2(b), a two-port network with parameters L and C, and a source load R, similar to that of part a, is shown. If $L = \ell d$ and $C = cd$, the two configurations "look similar."

Using AC circuit theory, it is not difficult to show that the input impedance of the circuit of Figure 3-2b is

$$Z_{in} = \frac{U}{I} = \frac{R}{1 + (\omega CR)^2} + j/\omega \left[L - \frac{R^2 C}{1 + (\omega CR)^2} \right]. \qquad \textbf{(3-26)}$$

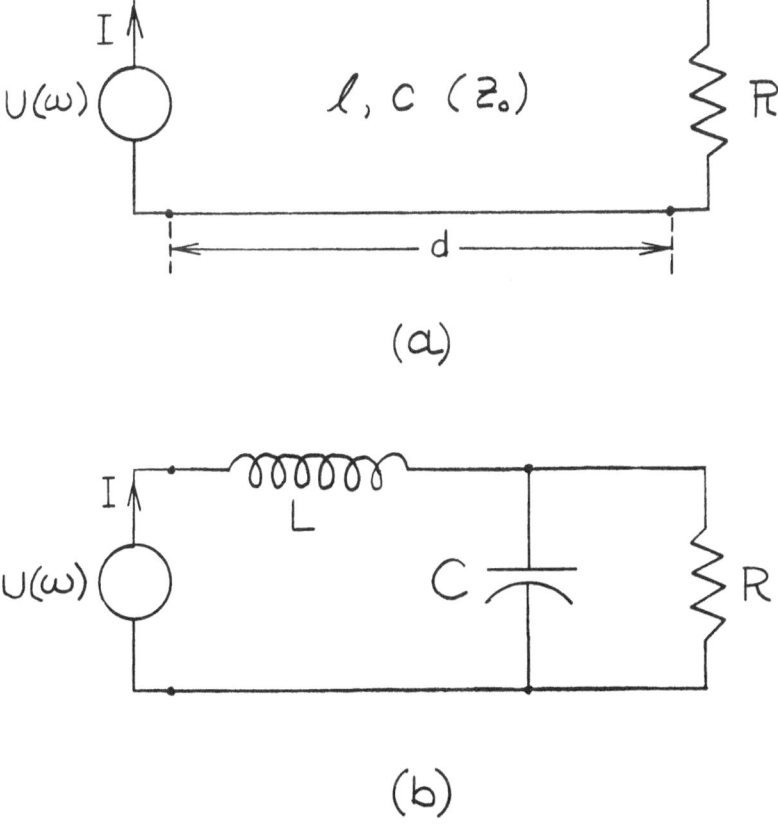

Figure 3-2. Comparison of transmission line section and AC circuit.

On the other hand, using Eq. (2-30), the input impedance of the line of part a is derived,

$$Z_{in} = \frac{U}{I} = Z_0 \left(\frac{R\cos\beta d + jZ_0\sin\beta d}{Z_0\cos\beta d + jR\sin\beta d} \right), \qquad (3\text{-}27)$$

in which $Z_0 = \sqrt{\ell/c}$ and $\beta = 2\pi/\lambda = \omega\sqrt{\ell c}$. Clearly, the results of Eqs. (3-26) and (3-27) are quite different and can not be reconciled. These conflicting results are surprising, especially with AC circuit theory being used to derive the transmission line equations and, therefore Eq. (3-27)! Indeed, there is no error. The discrepancy arises from a difference in the basic assumptions that underlie the two cases. In deriving AC circuit theory and Eq. (3-26), the assumption is that the circuit is vanishingly small compared to a wavelength; therefore, its size does not even appear as a parameter. Under these circumstances, Eq. (3-26) is correct. The assumptions in deriving the equations for a transmission line are that AC circuit theory can be used to model an infinitesmally short segment of the line and also, implicitly, that the separation of the parallel wires is very small compared to λ. The length, d, of the line is unrestricted. Without these assumptions, the AC model of the transmission line would lead to incorrect results. If the two wires are close to each other, but are arbitrarily long, electromagnetic waves propagate only in the long dimension, and Eq. (3-27) is valid.

It can be accepted, as a law of nature, that accelerated charges radiate. Therefore any wire that supports a time-varying current radiates. In certain cases, however, radiation is negligible.

PROBLEMS

PROBLEM 3-1.   Consider an electromagnetic field with

$$\mathbf{E} = \mathbf{z}_0\, E_0\, e^{-jky};$$
$$\mathbf{H} = \mathbf{x}_0\, \frac{E_0}{\eta}\, e^{-jky}$$

in which $k = 2\pi/\lambda = \omega\sqrt{\varepsilon_0\,\mu_0}$, $\eta = \sqrt{\mu_0/\varepsilon_0}$ (characteristic impedance of five space), and E_0 is a constant (scalar).

   a. Use Eq. (3-17) to determine the magnitude and direction of the energy flux.
   b. What is the field's energy density?
   c. Interpret ky as the scalar (Dot) product of a wave vector $\mathbf{k}$ and position vector $\mathbf{r}$, as in Eq. (1-10) to (1-12). What is the direction of $\mathbf{k}$?

PROBLEM 3-2. Assume that the electromagnetic field defined in the previous problem is advancing toward a perfectly conducting wall given by $y = y_0$. Use the boundary conditions of Eq. (3-22).

    a. Obtain the reflected EM field.

    b. Obtain the directions of the currents that are induced in the wall.

PROBLEM 3-3.* The conducting wall of the previous problem is replaced by a half space, by $y > y_0$, filled with a medium characterized by $\mu = 1$, $\varepsilon > 1$, $\sigma > 0$. Calculate both the reflected fields for $y < y_0$ and transmitted field in $y > y_0$.

PROBLEM 3-4.* As a continuation of the previous problem, determine the flux of energy (in watts per square meter) that enters the half space $y > y_0$. What happens to this power?

PROBLEM 3-5. Below is a list of indices of refraction for some common substances at a wavelength of 5.893×10^{-7}m (free space, sodium light)

Dry air (at 15°C, sea level)	1.0002765
Pure water (at 20°C)	1.333
Ethyl alcohol (at 20°C)	1.360
Light flint glass	1.575

    a. Determine the velocity of light in the materials.

    b. Determine the relative dielectric constants of the materials.

    c. What is the wavelength of sodium light in these media?

---

*Advanced problem.

CHAPTER 4

Wire Antennas and Basic Antenna Theory

Alternating currents radiate electromagnetic (EM) energy as electromagnetic waves. The transmission line, discussed in Chapter 2, is a very special circuit. Ideally, waves propagate only in the line's direction because adjacent currents are equal in magnitude and opposite in direction at each point along the line.

In a general circuit, as illustrated in Figure 4-1, a source causes a current to flow in the wire. If the circuit size is not very small compared to a wavelength, the current complex amplitude, I, is not constant, but rather a function of the position vector, $\mathbf{r}'$. An observer at position $\mathbf{r}$ receives an EM field contribution from the current at position $\mathbf{r}'$. The field strength is a function of the separation, R, between current and observer, the current's strength at $\mathbf{r}'$, and the element length, Δl. If the circuit is subdivided into a series of elements, the total field at $\mathbf{r}$ is determined by summing the elemental contributions. To perform this calculation, the current distribution, $I(\mathbf{r}')$, along the wire, and the field due to an isolated element of length, Δl, must be known.

The basic "building block," radiation from a small current element, is described in this chapter followed by analyses of practical wire antennas. Also introduced are the definitions of a number of important general antenna concepts, such as radiation resistance, directivity, gain, sidelobes, bandwidth, and polarization.

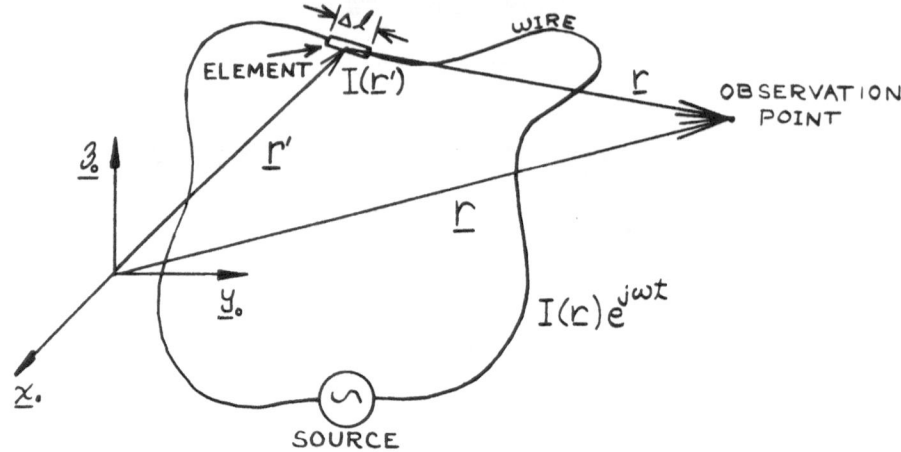

Figure 4-1.   General geometry for calculation of circuit field.

4.1 RADIATION BY AN ELECTRIC DIPOLE

The fields due to a short filamentary current element* can be derived rigorously from Maxwell's equations but are given here without proof. Consider a current element of length Δl and constant amplitude I_0, oriented along the z axis at the origin of a spherical coordinate system, as depicted in Figure 4-2. At a point with spherical coordinates (r, ϕ, θ), the EM field's components are oriented as shown in the figure and have magnitudes given by

$$H_\phi = jc\left(1 - \frac{1}{jkr}\right) u(r)\sin\theta,$$

$$E_\theta = j\eta c\left(1 + \frac{1}{jkr} - \frac{1}{k^2 r^2}\right) u(r)\sin\theta,$$

$$E_r = 2j\eta c\left(\frac{1}{jkr} - \frac{1}{k^2 r^2}\right) u(r)\cos\theta. \qquad \textbf{(4-1)}$$

In Eqs. (4-1)

$$k = \omega\sqrt{\varepsilon_0 \mu_0} = 2\pi/\lambda$$

---

*The current element is also referred to as an electric dipole, a small dipole, and a current dipole.

is the free space wave number,

$$\eta = \sqrt{\frac{\mu_o}{\varepsilon_o}} = 377\Omega$$

is the characteristic impedance of free space,

$$c = I_0\left(\frac{k\Delta l}{4\pi}\right) = I_0\frac{\Delta l}{2\lambda}$$

is a constant (not to be confused with the velocity of light, c_o), and $u(r) = e^{-jkr}/r$. The field components, H_ϕ and E_θ, which are tangent to a sphere of radius, r, centered on the origin, decrease less rapidly than E_r with increasing distance. At a great distance from the dipole, where $kr \gg 1$, all the terms of order $1/kr$ and $1/(kr)^2$ in the parentheses of Eqs. (4-1) are neglected, yielding

$$H_\phi \cong jc\,\frac{e^{-jkr}\sin\theta}{r}$$

$$E_\theta \cong \eta H_\phi$$

$$E_r \cong 0 \qquad\qquad \textbf{(4-2)}$$

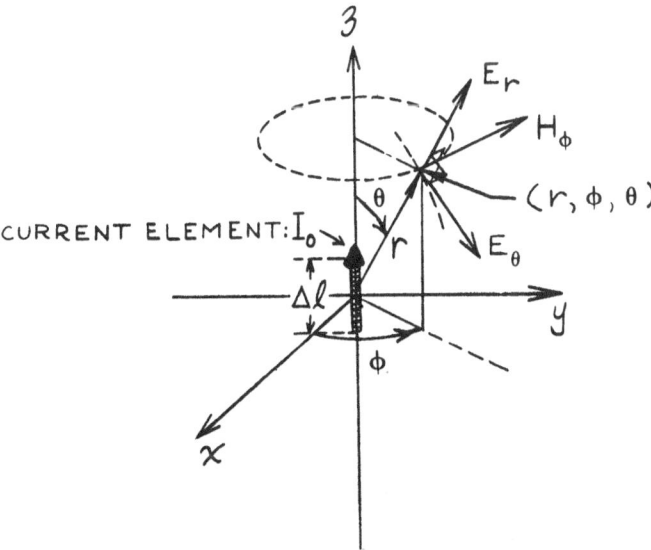

Figure 4-2. Fields of a single current element.

In the current element's far field, the radial component, E_r, is vanishingly small, and the tangential components are proportional to each other and have the form of spherical waves as in Eq. (1-15).

The EM field's energy flux is obtained from the Poynting vector of Eq. (3-16). For the special case of the current dipole, the Poynting vector's radial component is given by

$$P_r = \frac{1}{2} E_\theta H_\phi^* = \frac{1}{2}\eta c^2 (\sin \theta / r)^2. \tag{4-3}$$

The tangential components, P_θ and P_ϕ, become vanishingly small in the far field. Energy travels outward (radiates) from the elemental source. Chapter 1 described how the total power radiated by a source of spherical waves is found, as for example in Eq. (1-17). By integrating P_r over a sphere of radius r the total radiated electromagnetic power W due to the current element is found to be

$$W = \iint_{\substack{\text{sphere} \\ \text{surface}}} P_r dA = \int_0^{2\pi} d\phi \int_0^{\pi} d\theta\, r^2 \sin\theta \left(\frac{\eta c^2 \sin^2\theta}{2r^2}\right)$$

$$= \pi\eta c^2 \int_0^{\pi} \sin^3\theta\, d\theta = \frac{\pi}{3}\eta I_0^2 \left(\frac{\Delta l}{\lambda}\right)^2. \tag{4-4}$$

With the approximation

$$\frac{\pi\eta}{3} = 394.8\Omega \cong 400\Omega,$$

Eq. (4-4) may be expressed in the approximate form

$$\boxed{W = 400\, I_0^2 \left(\frac{\Delta l}{\lambda}\right)^2 \text{(watts)}} \tag{4-5}$$

4.2 RADIATION RESISTANCE AND DIRECTIVITY

The current element's total radiated power equals the total power dissipated if the same current, I_0, flowed through a resistance, R, such that

$$W = 400\, I_0{}^2 \left(\frac{\Delta l}{\lambda}\right)^2 = \tfrac{1}{2} I_0^2 R.$$

The equivalent resistance,

$$R \cong 800 \left(\frac{\Delta l}{\lambda}\right)^2 \Omega, \tag{4-6}$$

is referred to as the current element's radiation resistance and measures the element's radiating capability. It was assumed earlier that Δl is small compared to λ. So small, in fact, that the current was considered constant over the element's entire length. If, for example, $\Delta l = \lambda/30$ then $R = 0.9\Omega$. Many straight wire antennas are not short relative to a wavelength, and the current cannot be considered constant over the antenna's length. In these cases Eq. (4-6) no longer applies. Yet, the concept of radiation resistance applies to all wire antennas.

In Eq. (4-3) the energy flux is *anisotropic* (i.e., not the same in all directions). Due to the rotational symmetry, no azimuthal variation is expected. However, in elevation there is a $\sin^2\theta$ dependence. Figure 4-3 shows the power radiation pattern of the current element's far field. Although the pattern appears as a figure "8" shaped curve in this view, rotational symmetry about the current dipole's axis reveals that the full pattern is a torus (a donut, with an infinitely small center hole). The dashed circle represents the spherical pattern of an isotropic radiator of identical total radiated power. The energy flux at r from the isotropic radiator is given by

$$P_{\text{ave}} = W/4\pi r^2 \ (\text{W/m}^2). \tag{4-7}$$

Directivity or directional gain, an important antenna parameter, is a dimensionless quantity given by the ratio of the power density in a given direction, to the average (isotropic) radiated power:

$$g_d(\theta,\phi) = \frac{P_r(r,\theta,\phi)}{P_{ave}} = \frac{4 \pi r^2 P_r(r,\theta,\phi)}{\displaystyle\iint_{sphere} P_r dA} \qquad \textbf{(4-8)}$$

For a current element aligned along the z axis,

$$g_d(\theta) = \frac{3}{2}\sin^2\theta, \qquad \textbf{(4-9)}$$

which is obtained by using Eq. (4-3) and (4-4) in Eq. (4-8). Along the equatorial plane of Figure 4-3, the directivity is the greatest. When $\theta = 90°$,

$$g_d(90°) = g_{dmax} = 1.5. \qquad \textbf{(4-10)}$$

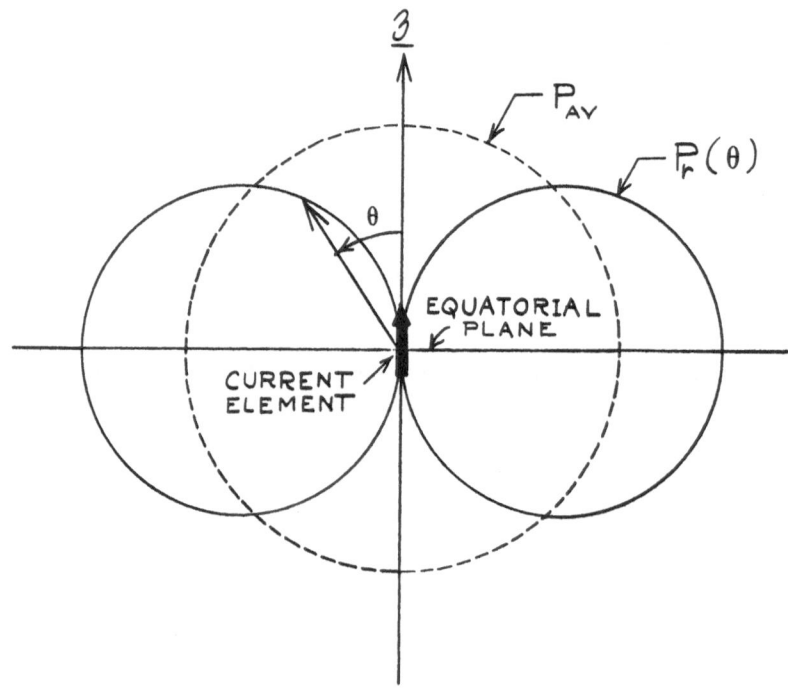

Figure 4-3.   Power radiation pattern of current element.

As is common practice, the directional gain is expressed in decibels:

$$G_d = 10 \log(g_{d\max}) \text{ (dB)}.$$

For the current element $G_d = 1.76$ dB.

The directivity or directional gain of an antenna should be distinguished clearly from the overall gain. Directivity is a measure of an antenna's ability to concentrate *radiated power*—the power that flows out into space. The overall gain or simply *antenna gain* is a measure of an antenna's ability to concentrate the power accepted at its input terminals. Because accepted power is always greater than the radiated power due to a variety of losses, directivity is always greater than gain.

4.3 STRAIGHT WIRE ANTENNAS—STANDING WAVE EXCITATION

The characteristics of a straight wire antenna of general length can be determined using the results derived above. For convenience, it is assumed that the wire is oriented along z axis of the coordinate system illustrated in Figure 4-4. The wire lies between z_1 and z_2 and supports a current of complex amplitude $I(z')$. The contribution at (r,θ,ϕ) in the far field, due to the element of current $\Delta z'$ at position z', using Eq. (4-2), is

$$\Delta E \cong \frac{j\eta k}{4\pi} \frac{e^{-jkR}}{R} \sin\psi \, I(z')\Delta z'. \tag{4-11}$$

With suitable approximations, the field contributions are added from all the wire's current elements. This procedure is similar to the one used for point sources in Chapter 1. The distance from the element at position z' to the point of observation may be expressed

$$R = \sqrt{r^2 + z'^2 - 2r z' \cos\theta} = r \sqrt{1 - 2\frac{z'}{r} \cos\theta + \left(\frac{z'}{r}\right)^2},$$

which can be expanded in a Taylor series as in Eq. (1-34), to yield

$$R \cong r - z' \cos\theta.$$

If z/r is small (in the far field), the first order approximation then gives

$$\frac{e^{-jkR}}{R} \cong \frac{e^{-jkr}}{r} \, e^{jkz'\cos\theta},$$ (4-12)

which is like the results of Eq. (1-38).

From examination of Figure 4-4,

$$\sin\alpha = \frac{z'}{r}\sin\psi,$$

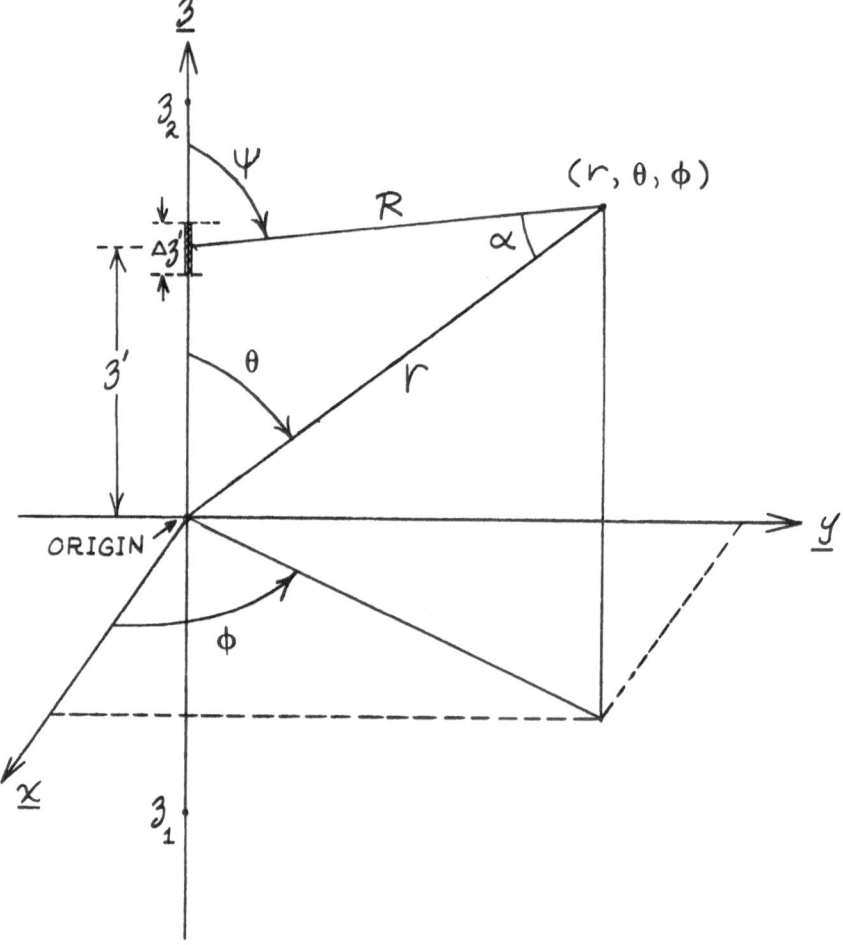

Figure 4-4.   Straight wire antenna geometry.

and because $|\sin\psi| \leqslant 1$, α must be of order z'/r, which is assumed to be very small, thus

$$\psi = \theta + \alpha = \theta + \text{term order of } \left(\frac{z'}{r}\right) \qquad \textbf{(4-13)}$$

and to the same accuracy as expression (4-12).

$$\sin\psi \cong \sin\theta. \qquad \textbf{(4-14)}$$

The results of Eqs. (4-11) through (4-14) mathematically express an antenna-to-observer separation that is very large relative to antenna size and wavelength. This causes r and R in Figure 4-4 to be nearly parallel.

If Eqs. (4-12) and (4-14) are used in the expression for the elemental field contribution of Eq. (4-11), the total field may be calculated by integration:

$$E = \frac{j\eta k}{4\pi}\frac{e^{-jkr}\sin\theta}{r}\int_{z_1}^{z_2} I(z')e^{jkz'\cos\theta}dz'. \qquad \textbf{(4-15)}$$

Then, by Eq. (4-2),

$$H_\phi = \frac{E_\theta}{\eta}. \qquad \textbf{(4-16)}$$

In comparing Eqs. (4-15), and (4-16), and (4-2), for an antenna of arbitrary length, the constant current, I_0, of the elemental section is replaced by

$$F(I) = \int_L I(z')\,e^{jkz'\cos\theta}dz', \qquad \textbf{(4-17)}$$

which is closely related to the Fourier transform* of the current distribution. This relation between the far field and the Fourier transform of the source distribution is important to remember because it is encountered often.

Field calculations for wire antennas are straightforward for a known current distribution, $I(z')$, because (in principle) it is not difficult to evaluate integrals as in Eq. (4-17). The current distribution must be known, however, to determine the field. Rigorous methods for finding the current distribution can only be found in very special cases, and good approximations require com-

---

*A brief review of Fourier transforms appears in Appendix C.

plicated computations. Nevertheless, by assuming reasonable current distributions, Eqs. (4-15) and (4-16) can produce useful information. A few such cases follow.

CASE 4-1.   An antenna of length L supports a constant current, such that

$$I(z') = I_0. \qquad \text{(4-18)}$$

This is an oversimplification, because realistically the current must vanish at the ends of the antenna. Yet, a constant current presents a simple calculation and is a sensible approximation for certain situations. For example, if a plane EM wave illuminates a long wire with the electric field vector, **E,** aligned with the wire, the induced current will be approximately constant over most of the wire's length. The wire will reradiate (or scatter) like an antenna with a constant distribution. Combining Eqs. (4-18) and (4-17) yields:

$$\int_L I(z')e^{jkz'\cos\theta}dz' = I_0 \int_{-L/2}^{L/2} e^{jkz'\cos\theta}dz'$$

$$= I_0 \left. \frac{e^{jkz'\cos\theta}}{jk\cos\theta} \right|_{-L/2}^{L/2} = LI_0 \left(\frac{\sin\left(\dfrac{kL\cos\theta}{2}\right)}{\dfrac{kL\cos\theta}{2}} \right)$$

$$= LI_0 \left(\frac{\sin u(\theta)}{u(\theta)} \right); \qquad u(\theta) = \frac{\pi L\cos\theta}{\lambda}. \qquad \text{(4-19)}$$

This result may be substituted into Eq. (4-15) to obtain

$$E_\theta = \left(\frac{j\eta kI_0 L}{4\pi} \frac{e^{-jkr}}{r} \sin\theta \right) \frac{\sin\mu(\theta)}{\mu(\theta)}. \qquad \text{(4-20)}$$

The expression in the parentheses corresponds to the field of a current element of length L as verified by Eq. (4-2). The second factor in the radiation pattern depends on the ratio L/λ and attains a maximum value of unity when $\theta = 90°$ (in the equatorial plane of the polar coordinate system in Fig. 4-4). When the antenna is long compared to the wavelength (a large L/λ), the sin u/u function oscillates rapidly. The power radiation pattern is proportional to

$$f(\theta) = \sin^2\theta \left(\frac{\sin u(\theta)}{u(\theta)} \right)^2 \qquad \text{(4-21)}$$

and is plotted in decibels in Figure 4-5 for two values for L/λ. The figure shows a *main beam* and side lobes that decrease in size with increasing separation from the main beam. The patterns are symmetrical about $\theta = 90°$. As the ratio of antenna length to wavelength increases, the main beam narrows, and the maximum directional gain, from Eq. (4-8)

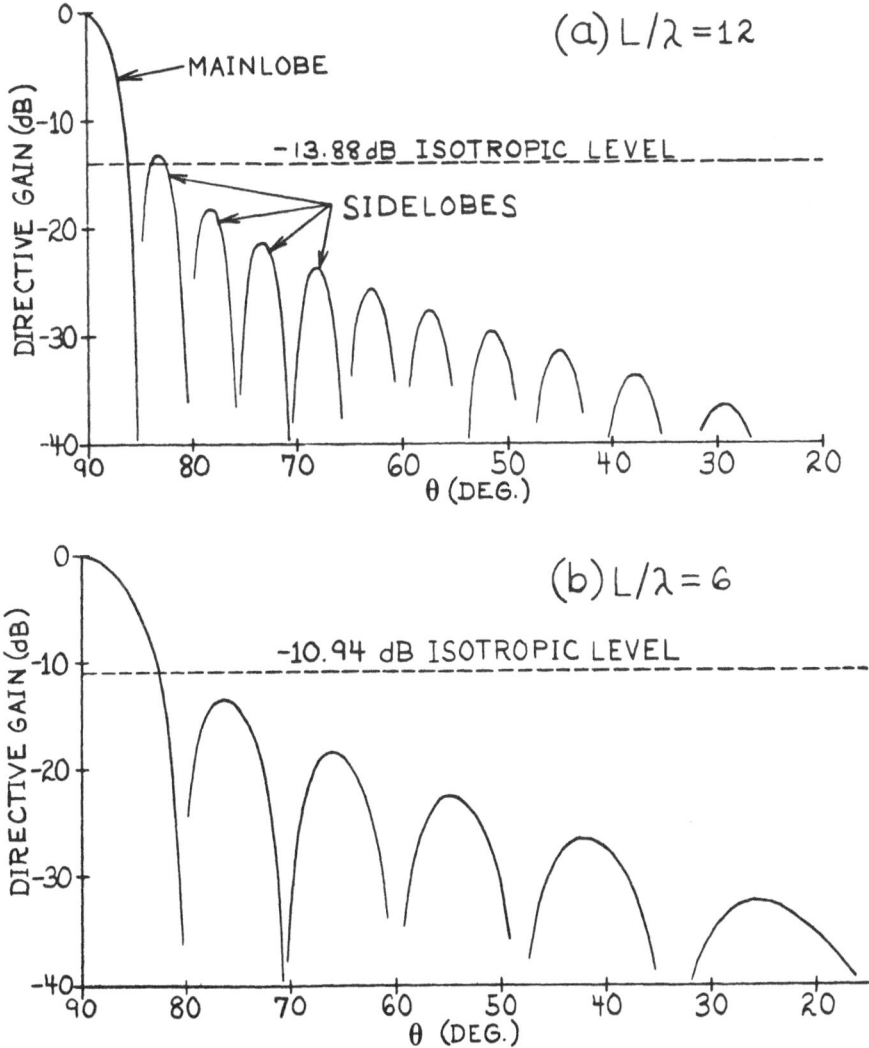

Figure 4-5. Power patterns for idealized straight wire antenna.

$$g_{dmax} = \frac{4\pi f(90°)}{\displaystyle\iint_{unit\ sphere} f(\theta)\,d\omega} \qquad \textbf{(4-22)}$$

increases. Note, however, that the envelope of the side lobe peaks stays about the same.

CASE **4-2.**   Figure 4-6 illustrates another simple wire antenna configuration. The ends of a transmission line are bent out at right angles to form an antenna of overall length $2l$. If it is assumed that the current distribution along the wires approximately remains as it was before bending the ends, the following expressions may be used:

$$I(z') = \begin{cases} I_0 \sin[k(l - z')] & l \geq z' > 0 \\[2mm] I_0 \sin[k(l + z')] & 0 > z' \geq -l. \end{cases} \qquad \textbf{(4-23)}$$

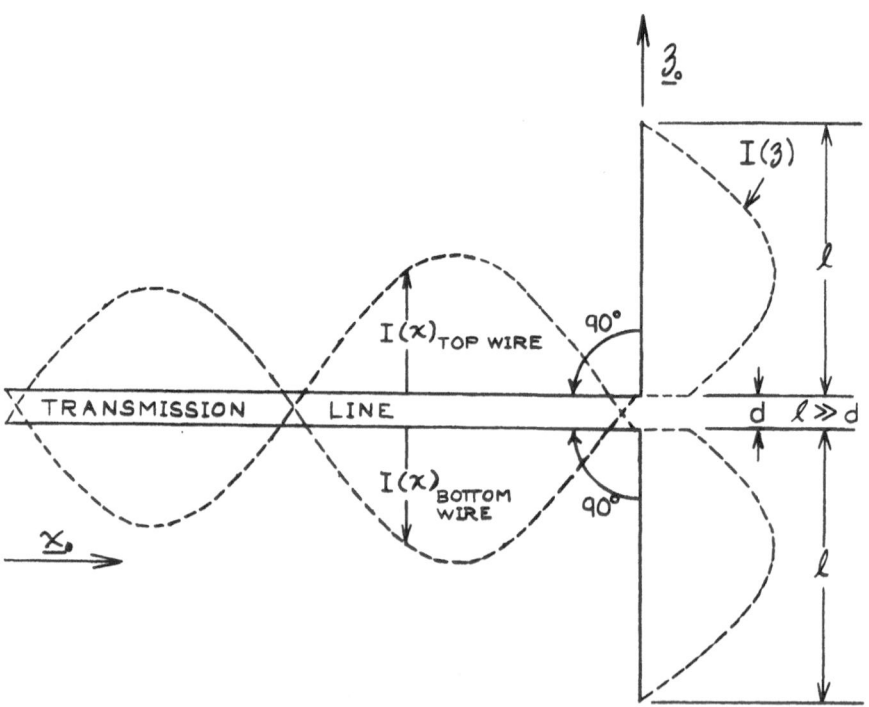

Figure 4-6.   Transmission line derived wire antenna.

Eq. (4-15) is then used to find the total field at positions far from the antenna. After some algebraic manipulation,

$$E_\theta = \frac{j\eta I_0}{2\pi} \frac{e^{-jkr}}{r} \left[\frac{\cos(kl\cos\theta) - \cos(kl)}{\sin\theta} \right]. \tag{4-24}$$

A superficial look at this result might suggest that the field is infinite along the antenna axis ($\sin\theta = 0$), however, application of L'Hospital's rule reveals that

$$\lim_{\theta \to 0} \left[\frac{\cos(kl\cos\theta) - \cos(kl)}{\sin\theta} \right] = 0.$$

This means that the antenna does not radiate in the direction of the wire, as would be expected, because none of the elemental currents radiate in this direction. If the antenna is long compared to the wavelength, there are nulls and lobes in the power pattern similar to those of the previous example. The *half-wave dipole* is a special case. The antenna's length is half a wavelength, $l = \lambda/4$,

$$kl = 2\pi \frac{l}{\lambda} = \frac{\pi}{2}$$

which implies $\cos kl = 0$, and the equation for the field becomes

$$E_\theta = \frac{j\eta I_0}{2\pi} \frac{e^{-jkr}}{r} \frac{\cos\left(\frac{\pi}{2}\cos\theta\right)}{\sin\theta}. \tag{4-25}$$

As shown before,

$$E_\theta = H_\phi/\eta, \qquad P_r = E_\theta H_\phi^*/2.$$

Thus, calculating the gain for the half-wave dipole requires an integral of the form

$$\int_0^\pi \frac{\cos^2\left(\frac{\pi}{2}\cos\theta\right)}{\sin\theta} \, d\theta.$$

This expression cannot be evaluated in closed form, but can be reduced to accurately computed tabulated functions. Important results derived this way include $R \cong 73\Omega$ for the radiation resistance, and

$$g_{dmax} \cong 1.64; \qquad G_{dmax} \cong 2.15 \ dB.$$

The radiation pattern is qualitatively similar to the current element's pattern (Fig. 4-3) but is somewhat compressed in the z direction.

4.4 STRAIGHT WIRE ANTENNAS—TRAVELING WAVE EXCITATION

A wave that travels along a single wire in one direction is also a radiation source. A wave may be launched, for example, as illustrated in Figure 4-7a. The wave has a phase velocity, c, that is generally not equal to c_0, the free space speed of light. In the configuration shown in Figure 4-7a, the phase velocity is less than c_0; however, other traveling wave arrangements may have phase velocities in excess of c_0.*

The relationships among phase velocity, frequency, and wave number were established in Eq. (2-4) as

$$k = \omega/c_0. \tag{4-26}$$

In the discussion of dielectrics in Chapter 3, the propagation velocity was shown to be related to the refraction index via Eq. (3-24)

$$n = c_0/c. \tag{4-27}$$

To preserve the Eq. (4-26) relationship, a modified wave number, kn, is adopted such that

$$kn = n\frac{\omega}{c_0} = \frac{\omega}{c}. \tag{4-28}$$

---

*A phase velocity greater than c_0 does not violate laws of the special theory of relativity, which stipulate that matter or energy cannot be transported at speeds in excess of c_0. Phase velocity alone does not imply energy transport. In this book, phase velocities that exceed the speed of light are included in the discussion of propagation in waveguides. The related concept of "group velocity" can be used to account for matter and energy transport. It is the group velocity that cannot exceed c_0.

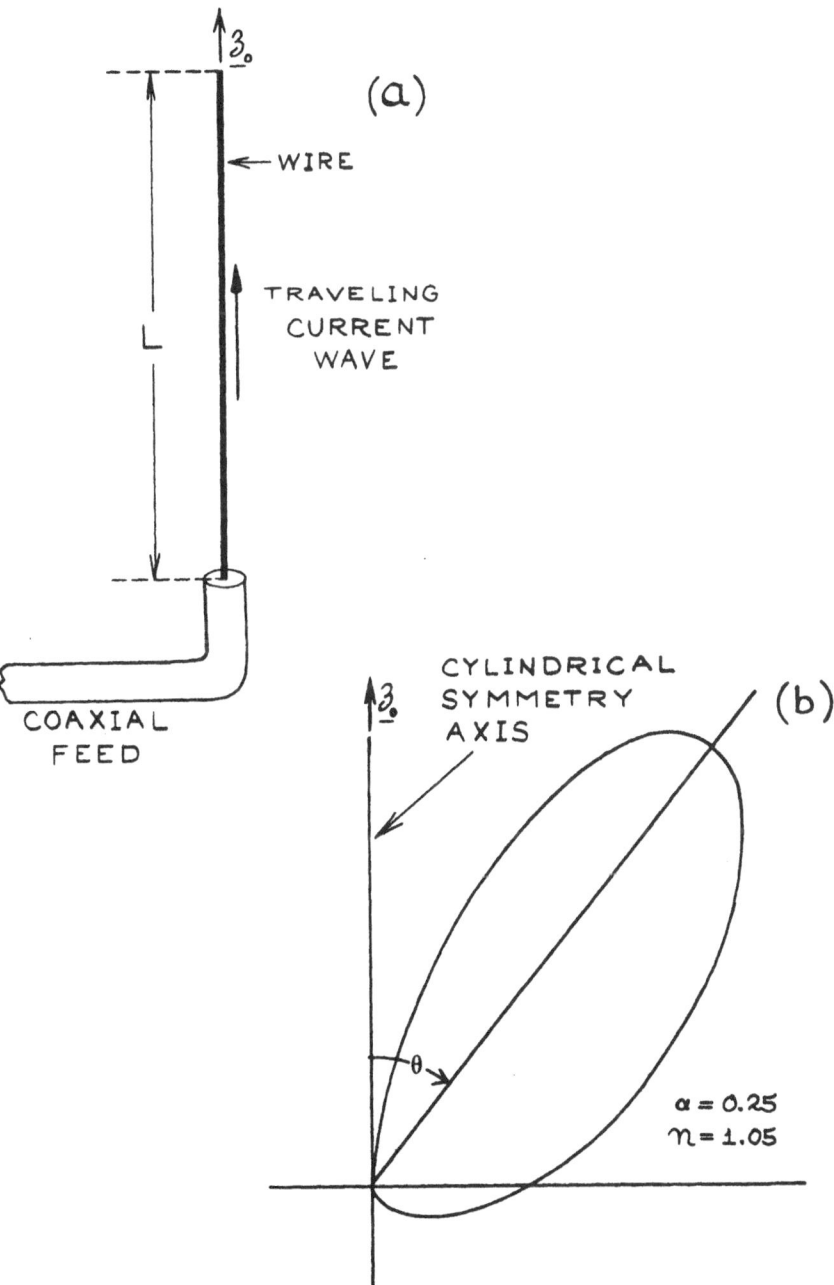

Figure 4-7. Traveling wave wire antenna and power pattern.

As the current wave propagates along the wire of Figure 4-7a, its amplitude is progressively reduced. The wire is assumed to be very long. As an additional precaution, reflections may be eliminated by dissipating any remaining energy in a suitable termination. The current may be expressed as

$$I(z') = I_0 \, e^{-k\alpha z'} \, e^{-jknz'}, \tag{4-29}$$

in which α is a normalized attenuation constant. Substitution of Eq. (4-29) into Eq. (4-17) gives

$$F(I) = \int_L I(z') \, e^{jkz'\cos\theta} \, dz' = I_0 \int_0^L e^{-k[\alpha + j(n-\cos\theta)]z'} \, dz'$$

$$= \frac{1 - e^{-kL[\alpha + j(n-\cos\theta)]}}{k[\alpha + j(n - \cos\theta)]}.$$

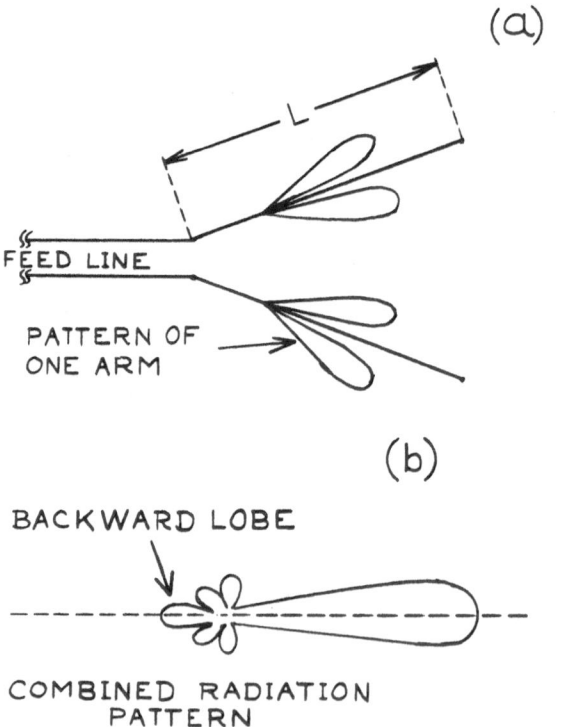

(a)

FEED LINE

PATTERN OF
ONE ARM

(b)

BACKWARD LOBE

COMBINED RADIATION
PATTERN

Figure 4-8.   The "vee" antenna and its power pattern.

The exponential term can be discarded if the wire is long enough so that

$$F(I) \cong 1/k[\alpha + j(n - \cos\theta)]. \qquad \textbf{(4-30)}$$

The power radiation pattern, which is the angle dependent part of P_r, is then proportional to $\sin^2\theta F(I) \ F^*(I)$, with the result

$$f(\theta) = \frac{\sin^2\theta}{\alpha^2 + (n - \cos\theta)^2}. \qquad \textbf{(4-31)}$$

For a *slow wave*, $c \leq c_0$ (or $n \geq 1$), the power pattern will look like the example shown in Figure 4-7b.

A simple traveling wave antenna of higher directivity is formed when the ends of a transmission line make a "vee" as shown in Figure 4-8a. At the divergent section of the line, the currents on each arm radiate in a pattern as shown in Figure 4-8a. Vectorial addition of the two fields results in a radiation pattern with a strong forward lobe. The gain of this lobe can be as high as 10 dB. If the current waves are not sufficiently attenuated over the arms, a strong backward reflection at the open end can produce backward waves and a resultant backward lobe, as shown in Figure 4-8b. Proper loading of the open ends reduces this lobe by suppressing reflections.

4.5 HELIX AND LOOP

Figure 4-9a shows a helix. This is a practical example of a slow wave structure. A current wave traveling along the helix, with a phase velocity near c_0, has a velocity component along the helix axis reduced by a factor that is the sine of the thread angle or $s/2\pi d$ as illustrated in Figure 4-9a. None of the current elements along the helix point in the axial direction, therefore, unlike previous examples, the radiated field is strong along the helix's axis. A qualitative sketch of the helix radiation pattern appears in Figure 4-9b. To produce such a mode of radiation, the helix circumference, measured in a plane perpendicular to its axis, should be comparable to a wavelength $2\pi d \approx \lambda$. When this condition is satisfied, the waves traveling along the helix reinforce one another as viewed from a vantage point along the axis. This reinforcement produces high gain compared to a wire antenna. Also, as the wave proceeds around the helix, the field vectors associated with its current elements rotate, producing a circularly polarized field of frequency ω, as shown in Figure 4-9b. The helix's traveling wave can be roughly seen as rotating current elements. Circular polarization is discussed in more detail in the next chapter in connection with plane electromagnetic waves.

The flat loop is another basic wire antenna configuration. Figure 4-10 shows a loop of radius a, and the geometric parameters used to integrate individual current element fields. Although the loop geometry causes integration to be somewhat complicated, for a small loop ($a \ll \lambda$) with a current of constant amplitude and phase, the resultant fields are remarkably similar to those obtained for the elemental current dipole in Eq. (4-2) and are given by

$$H_\theta \cong -\left(\frac{ka}{2}\right)^2 I_0 \frac{e^{-jkr}}{r} \sin\theta$$

$$E_\phi = -\eta H_\theta$$

$$H_r = \text{order of } (kr)^{-2} \to 0. \tag{4-32}$$

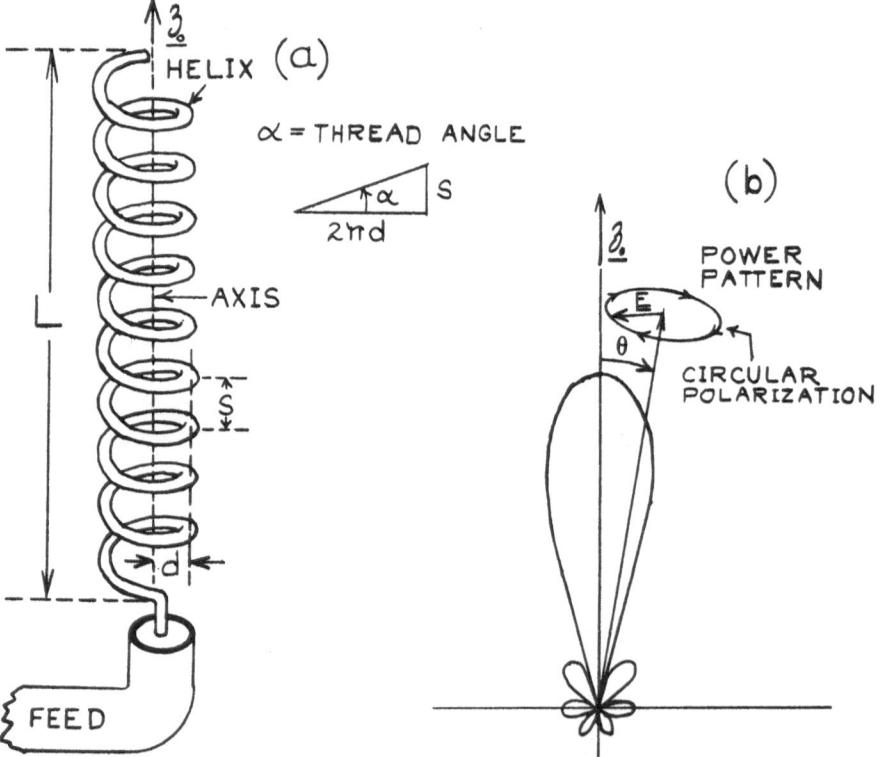

Figure 4-9. Helical antenna and its power pattern.

The last equation indicates that H_r is of order $1/(kr)^2$ and, therefore, does not contribute to the far field. Comparison of Figures 4-2 and 4-10 show that the small loop and the current element fields are related by a 90° rotation about the propagation direction (apart from multiplicative constants).

The power and radiation resistance of the small single loop are given by

$$W = \eta\frac{\pi}{3}\left(\frac{ka}{2}\right)^4 I_0^2 = \eta\frac{\pi}{12}\left(\frac{p}{\lambda}\right)^4 I_0^2,$$

$$R \cong 20\pi^2\left(\frac{p}{\lambda}\right)^4, \tag{4-33}$$

in which p is the loop's circumference, and

$$ka = p/\lambda. \tag{4-34}$$

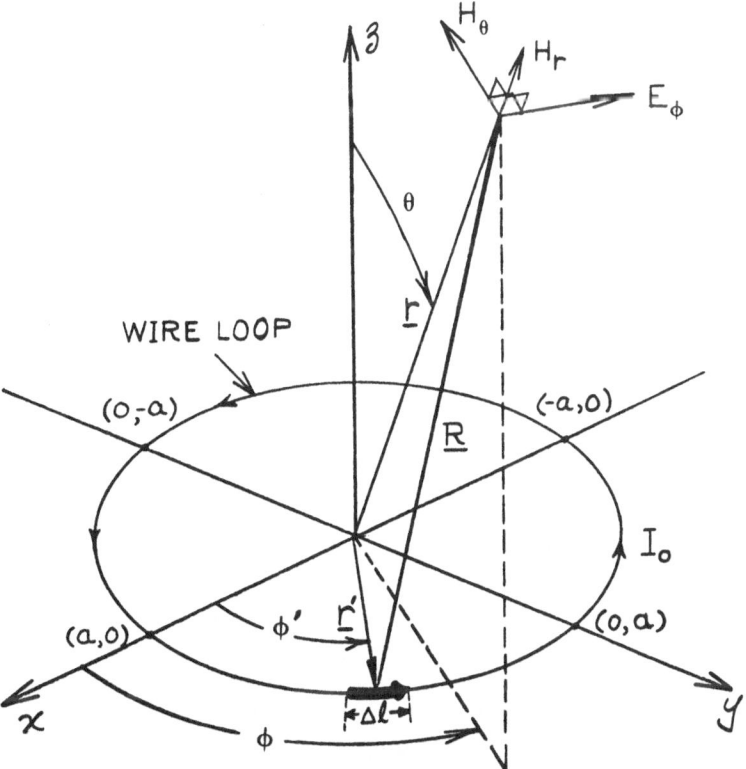

Figure 4-10. Construction for fields of a wire loop.

Thus, for a loop with a circumference measuring a quarter wavelength ($p = \lambda/4$), the resistance is less than 0.8Ω. Because the small loop and current element have the same radiation pattern, they both have a maximum directive gain of 1.5 or 1.76 dB (Eq. (4-10)).

Because the small loop resistance decreases so rapidly with increasing wavelength (inverse fourth power dependence), an isolated loop is not a very efficient radiator and multiwinding loops are sometimes used. For large loops, field calculations become more involved because the current is not constant.

The helical antenna considered earlier is basically a distended N-winding loop. However, to understand how it operates requires understanding a progressive current wave that propagates along the helix. This feature of the helical antenna sets it apart from loop antennas. For example, a helical antenna is relatively insensitive to frequency change—a property referred to as broad bandedness. The helix radiation pattern and radiation resistance are reasonably constant over a fairly wide bandwidth. If f_L, f_o, and f_H denote lowest, middle, and highest operational frequencies, respectively, then proportional bandwidth is defined as

$$\Delta f/f_o = (f_H - f_L)/f_o.$$

Helical antennas may have proportional bandwidths greater than 50%. It is difficult to calculate the helical antenna's parameters, because the current distribution along the helix is not known. However, semiempirical expressions for the maximum directivity and radiation resistance have been determined:

$$g_{dmax} \cong 15 \, N\left(\frac{p}{\lambda}\right)^3 \tan\alpha, \tag{4-35}$$

$$R \cong 140 \, p/\lambda. \tag{4-36}$$

N is the number of windings, α is the thread angle (Fig. 4-9a), and p/λ is the normalized circumference.

4.6 ANTENNAS ABOVE A FLAT CONDUCTING GROUND

The antennas discussed thus far are seen as isolated objects in free space, which is unrealistic. An antenna's free space performance is modified by the presence of any material bodies as a result of electromagnetic coupling. Although a general analysis is too complex to consider here, the effects of a flat ground plane are described as a simple yet important example. The ground is modeled as a perfectly conducting, infinite plane that lies below an antenna.

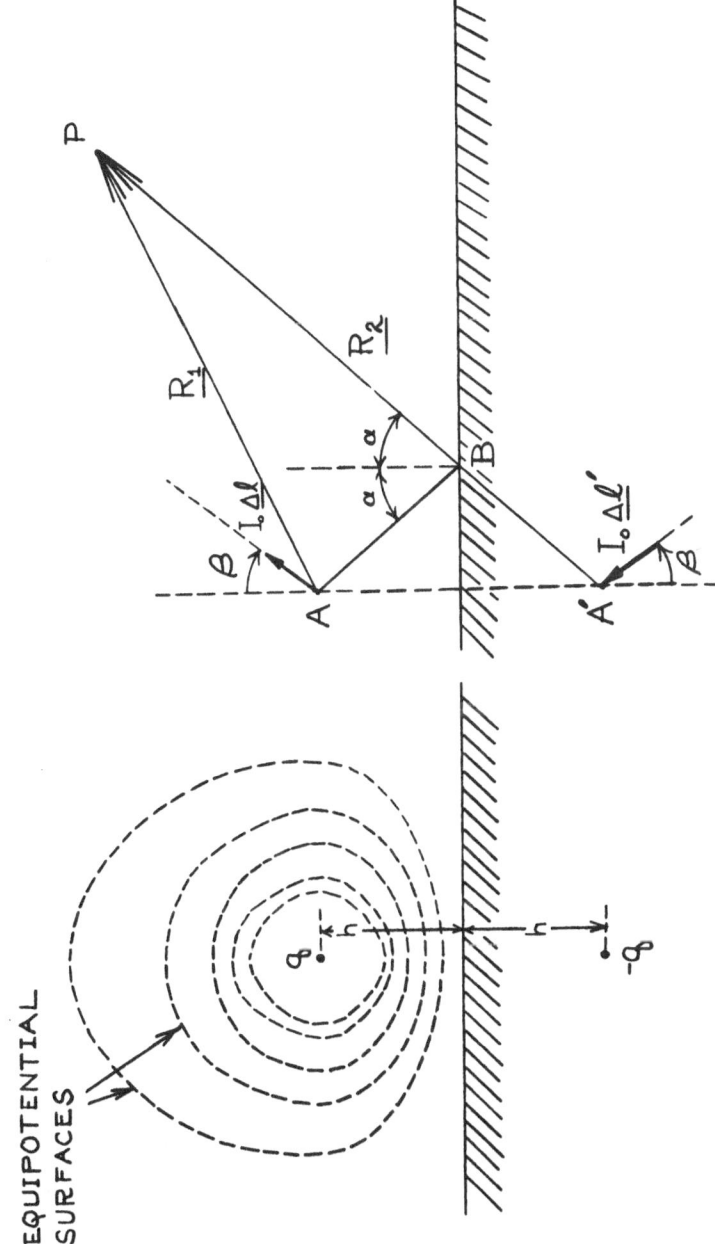

Figure 4-11. Method of images for charge and currents.

The effect of this idealized ground is conveniently described by the *method of images* illustrated in Figure 4-11. The figure shows the classic electrostatic problem of a charge q located above a perfectly conducting plane. Because the surface is a perfect conductor, it must be an equipotential surface. If a charge $-q$, located in the image position replaces the lower half-space, the middle plane is still an equipotential plane. This equivalence allows accounting for the effects of the conducting plane in the region above it by placing an equal charge of opposite sign in the image position. An observer in the upper half-space, who can only measure the electrostatic field, would not be able to distinguish the two-charge from the charge-over-conducting-plane configuration.

This method can be generalized to include charges in motion (currents). Because the image of q is $-q$, the image of a current I_0, flowing along an elemental directed line segment $\Delta \mathbf{l}$, is a current I_0 flowing along the segment $\Delta \mathbf{l}'$, also shown in Figure 4-11. The EM field at a point P above the plane is given by the vector addition of the field due to the real current element and its image. Because the normal to the ground at point B bisects the angle ABP, the image's field is regarded as the reflection of the real current element's field. Note also that the reflection reverses the direction of the tangential components of E (this is analogous to the image charge's sign change). Alternatively, the image current may be determined by use of the boundary conditions of Eq. 3-22. The field at P has two components: a direct field that propagates along AP, and a reflected or indirect field that propagates along ABP. Later the validity of this interpretation is demonstrated by use of geometrical optics.

Because the effect of the ground on the field of a current element can be treated by the method of images, the field of any aboveground wire antenna can be determined by taking into account its properly placed image. A horizontal wire above a conducting plane has an image that consists of a horizontal wire with an opposite current situated an equal distance below the plane. A single wire above a ground plane forms a transmission line as shown in Figure 4-12a. This configuration is similar to the microstripline of Figure 2-4c.

In Figure 4-12b, a horizontal antenna is shown above the ground. Because the image is opposite, very little radiation is produced by this configuration when the ground–antenna separation is small. In contrast, the vertical aboveground antenna's field is enhanced by its image, as illustrated in Figure 4-12c. Similarly, the vertical quarter-wave monopole shown in Figure 4-12d produces the field of a half-wave dipole, relative to an observer in the upper half-space.

Calculation of the radiation resistance of an antenna above a ground plane is tricky. For example, Figure 4-13 shows a small current element above ground and its equivalent in free space. Careless application of Eq. (4-6)

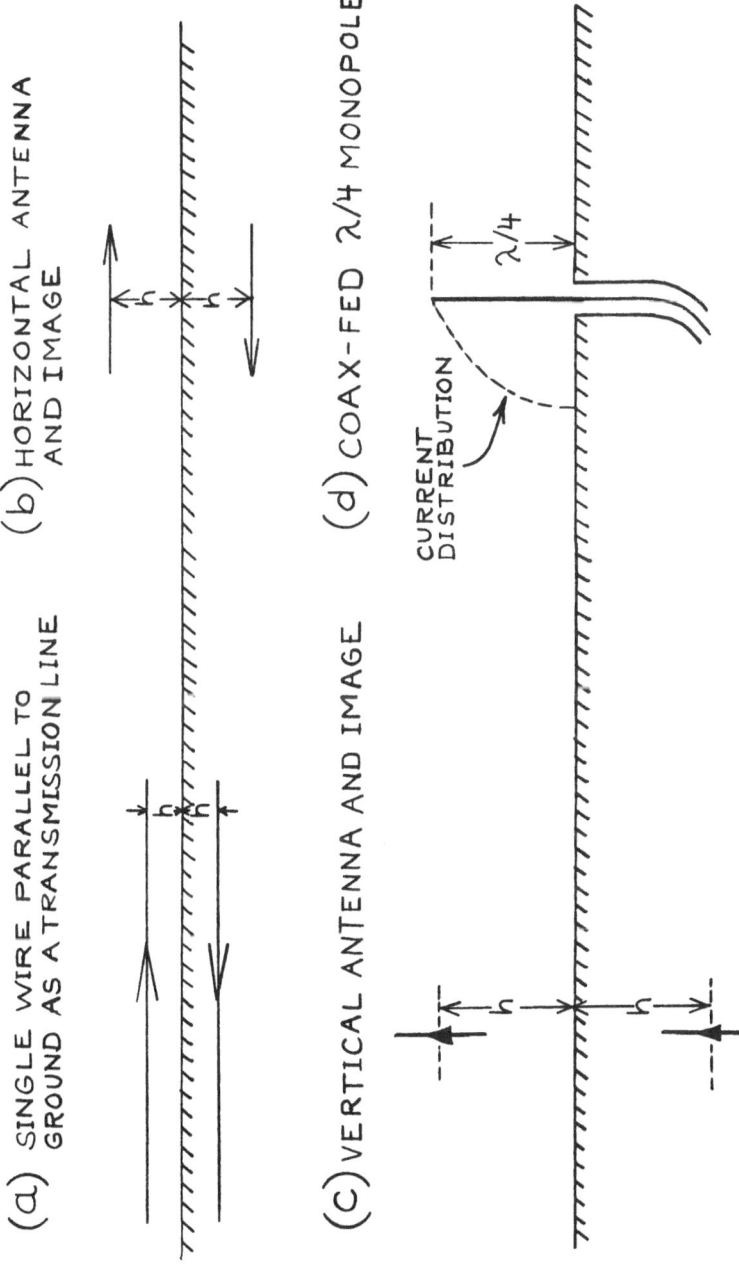

Figure 4-12. Method of images applied to antennas.

Figure 4-13.   Above-ground current element and its equivalent.

indicates that the short current element's radiation resistance over the ground plane is four times that of the same element in free space. However, the element over the ground plane can radiate into only half of all space (2π steradians), and therefore, the integration limits on θ used in Eq. (4-4), from $0 \to 180°$, must be changed to $0 \to 90°$. This means that the radiated power and radiation resistance of the vertical, aboveground current element are only twice that of the same element in free space. For a current element over a ground plane,

$$W \cong 800 \, I_0^2 \left(\frac{\Delta l}{\lambda}\right)^2 = \frac{1}{2} R I_0^2 \qquad \textbf{(4-37)}$$

and

$$R \cong 1600 \left(\frac{\Delta l}{\lambda}\right)^2. \qquad \textbf{(4-38)}$$

Despite such cautions, it is easy to assume, for example, that the monopole of Figure 4-12d has twice the radiation resistance of a half-wave dipole (146Ω)— its free space equivalent. However, Eq. (4-38) actually states that the radiation resistance of the quarter-wave monopole above a ground plane is twice that of a hypothetical quarter-wave monopole in free space—not its equivalent, the half-wave dipole. Because the half-wave dipole is the equivalent antenna, the quarter-wave monopole must have only half the dipoles' radiation resistance ($73/2 = 36.5$) because the monopole's ground plane cuts off half of space, meaning half of the integration volume for the radiation of power.

The method of images may also be applied to problems in spherical wave interference. Figure 1-6 shows the geometry of spherical wave interference associated with two sources. If a conducting surface is positioned at the plane $y = 0$, and the source at $y = a$ is a z_0 oriented wire antenna, an image antenna at the $y = -a$ can be envisioned. Because the real antenna is parallel to the conducting surface, the configuration resembles the one in Figure 4-12b, and the image source phase differs from that of the real source by 180° or π radians. Eqs. (1-39) through (1-44) may be used to obtain the radiation pattern, with $\alpha_1 = 0$ and $\alpha_2 = \pi$ to account for the phase opposition of the sources. In the equatorial or $\theta = 90°$ plane, the far field pattern is therefore given by

$$|S(\phi)| = 2a_0 \sin(ka \sin\phi), \qquad (4\text{-}39)$$

in which a_0 is the amplitude of the source and a is the distance of the source from the conducting surface. At the conducting surface, $\phi = 0$, the field vanishes, because the image field always cancels that of the real source. If $a = \lambda/4$, the field pattern becomes

$$|S(\phi)| = 2a_0 \sin\left(\frac{\pi}{2}\sin\phi\right), \qquad (4\text{-}40)$$

which has a maximum at $\phi = 90°$: $|S(90°)|^2 = 4a_0^2$. Thus, this configuration has a power gain of a factor of 4 (6 dB) in the y direction with respect to the gain of a single source in free space. This result can be predicted without using Eq. (4-40), because the source and image are separated by half a wavelength along the y direction, which compensates for their phase opposition and produces reinforcement.

The ground effect is important in antenna design and deployment. Because the real ground is neither a plane nor a perfect conductor, the simple method of images is often only a crude approximation. To make the electromagnetic environment more predictable, a wire mesh is sometimes placed on the ground below an antenna, extending as far out as practical, to simulate a perfectly conducting ground plane.

PROBLEMS

PROBLEM 4.1.   Let the angular dependence of the electric field of the current element be $|f(\theta,\phi)| = $ (const) $\sin\theta$. Compare this with Eq. (4-2).

   a. Show that $r = a \sin\theta$ is the equation of a circle that is tangent to the origin. (Hint: you may convert to Cartesian coordinates.)
   b. Sketch the field pattern characterized by $|f(\theta,\phi)| = $ (const) $\sin^2\theta$. Does this pattern have a higher or lower peak directive gain than that of the current element of Eq. (4-10)? Explain your answer.
   c. Calculate the peak directive gain of the hypothetical power pattern of part b.

PROBLEM 4.2   Draw the power radiation pattern and calculate the level of the first side lobe (in dB below the main beam peak) for each of the following two cases:

    a. An antenna with a uniform current distribution as in Eq. (4-20) with $L = 3\ \lambda$.

    b. An antenna with a sinusoidal current distribution as in Eq. (4-24) with $L = 2l = 3\lambda$.

PROBLEM 4.3* Consider a straight wire antenna of length L. Using the far field criterion described in Chapter 1, find the far field region if $\lambda = 1\text{m}$ and $L = 3\lambda$; $L = 10\lambda$; $L = 30\lambda$. What are the corresponding far field regions for a loop antenna of circumference L?

PROBLEM 4.4 Use Eqs. (4-28) through (4-31) to compute and graph the radiation pattern of a traveling wave antenna characterized by $L = 4\lambda$. $\alpha = 0.2$, and $C = 0.98\ C_0$.

PROBLEM 4.5 The method of images can be used to account for the effect of a corner reflector on the pattern of a wire antenna. Consider a straight wire antenna that is equidistant from two infinite perpendicular planes.

    a. Apply the method of images to show, in principle, how the pattern of the wire antenna in the quarter space can be obtained.

    b. Generalize the method to include planes that form a dihedral angle β such that $\beta = 180°/m$ $(m = 2, 3, 4, \ldots)$

PROBLEM 4.6* A linear current distribution is given by:

$$I(z) = I_0 \left[(1 - b)\cos^2 \frac{\pi z}{L} + b\right]; \ -L/2 < z < L/2$$

Find "b" such that the first two side lobes are of equal strength. Calculate the side lobe level with respect to the main beam.

---

*Advanced problem.

CHAPTER 5

Plane Waves, Rays, and Geometrical Optics

Both waves in general and spherical electromagnetic (EM) waves radiated by wires have been discussed. The resemblance of spherical waves to plane waves at observation points sufficiently far from the source (i.e., in the far field) has also been shown. At great distances from the source, the EM field components are perpendicular to each other and also perpendicular to the propagation direction, which, for spherical waves, is the radial direction. This perpendicularity is characteristic of plane EM waves, which will be described here. Exact plane waves do not exist in the real world. Nevertheless, they represent a convenient means of describing many EM phenomena, as a result of their mathematical simplicity.

The striking analogy between the laws of geometrical optics (GO) and the laws governing EM plane waves indicates that GO is an approximate method for solving EM problems. GO methods have many applications in antenna theory and design, as discussed later in this and subsequent chapters.

5.1 PLANE ELECTROMAGNETIC WAVES

The properties of a homogeneous, isotropic and nonmagnetic medium can be described by the parameters

$$\mu = 1$$

and

$$\varepsilon = n^2. \tag{5-1}$$

(See notation and discussion in Chap. 3.) EM field vectors of the form

$$\mathbf{E} = \mathbf{E}_0 \exp(-j\mathbf{k} \cdot \mathbf{r})$$

and

$$\mathbf{H} = \mathbf{H}_0 \exp(-j\mathbf{k} \cdot \mathbf{r}) \tag{5-2}$$

are solutions to Maxwell's equations if the following conditions are satisfied:

$\mathbf{E}_0$, $\mathbf{H}_0$, and $\mathbf{k}$ are constant, mutually perpendicular vectors

$$\mathbf{H}_0 = \frac{\mathbf{k} \times \mathbf{E}_0}{\omega \mu_0}, \ \mathbf{E}_0 = \frac{\mathbf{k} \times \mathbf{H}_0}{-\omega \varepsilon_0 \varepsilon} \quad \text{(see Fig. 5-1), and}$$

$$|k| = \sqrt{(k_1^2 + k_2^2 + k_3^2)} = \omega \sqrt{\mu_0 \varepsilon_0 \varepsilon} = kn, \tag{5-3}$$

in which $k = \omega \sqrt{\mu_0 \varepsilon_0} = \omega/c_0 = 2\pi/\lambda$ and $n = \sqrt{\varepsilon}$. Using these conditions, the following can be derived,

$$|\mathbf{E}_0| = \eta |\mathbf{H}_0|,$$

where $\eta = \sqrt{\mu_0/\varepsilon_0 \varepsilon}$ is called the *space impedance*.

Recall from Eq. (1-12):

$$\mathbf{k} \cdot \mathbf{r} = k_1 x + k_2 y + k_3 z.$$

Thus, the derivatives of Eqs. (5-2) are readily obtained. For example, $\partial \mathbf{E}/\partial x = -jk_1 \mathbf{E}$, $\partial^2 \mathbf{E}/\partial y^2 = -k_2^2 \mathbf{E}$, and so on. In this way it is easy to show that the fields of Eqs. (5-2) are solutions of the three-dimensional wave equations:

$$\frac{\partial^2 \mathbf{E}}{\partial x^2} + \frac{\partial^2 \mathbf{E}}{\partial y^2} + \frac{\partial^2 \mathbf{E}}{\partial z^2} + k^2 n^2 \mathbf{E} = 0$$

and

$$\frac{\partial^2 \mathbf{H}}{\partial x^2} + \frac{\partial^2 \mathbf{H}}{\partial y^2} + \frac{\partial^2 \mathbf{H}}{\partial z^2} + k^2 n^2 \mathbf{H} = 0. \tag{5-4}$$

These results can be verified as an exercise. Eqs. (5-4) are the three-dimensional analogues of the one-dimensional wave Eqs. (2-10), which de-

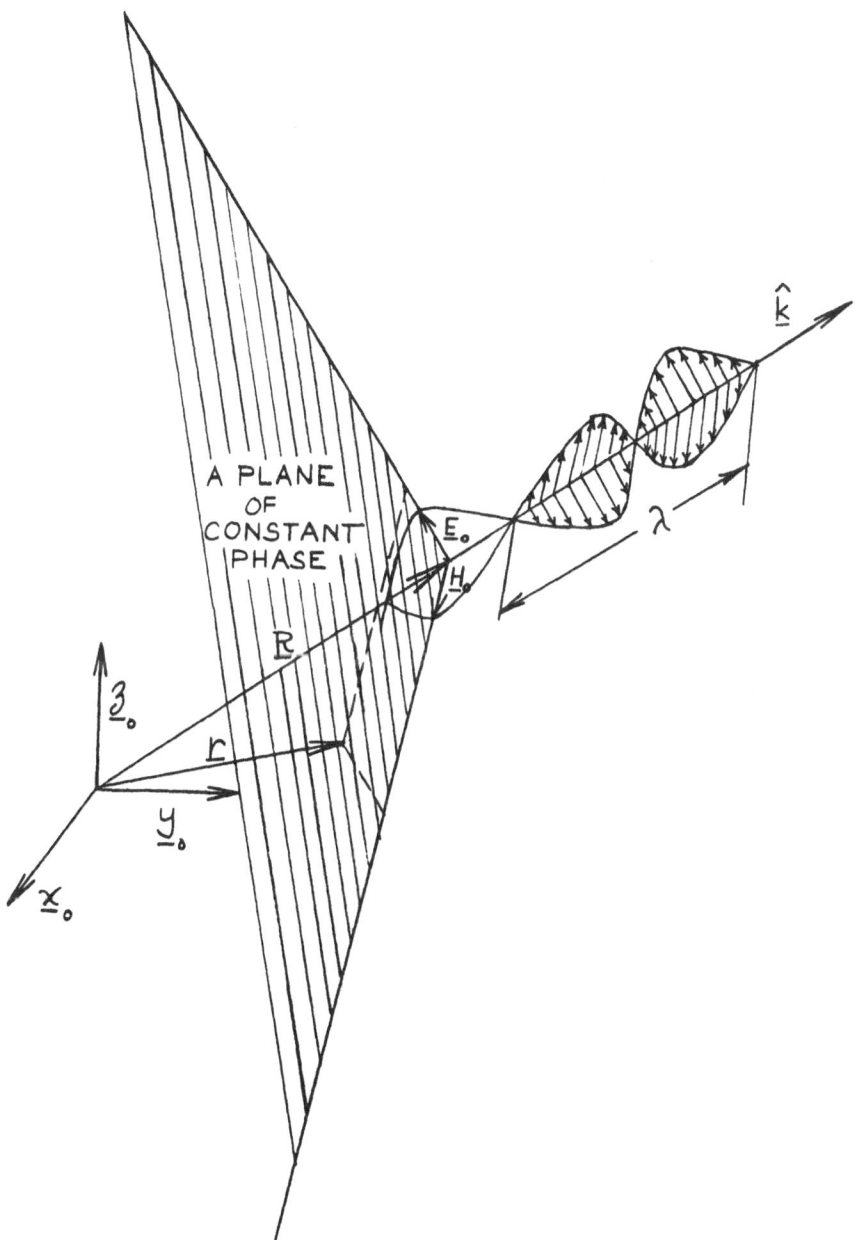

Figure 5-1. Plane wave geometry.

scribe the voltage and current along a transmission line. Indeed, it is often helpful to think of **E** and **H** as the analogues of voltage and current, respectively. Like voltage and current along a transmission line, the electric field is always accompanied by a magnetic field.

According to Eqs. (3-16), (3-17), and (5-2), the plane EM wave conveys energy in the direction **P**, given by,

$$\mathbf{P} = \tfrac{1}{2} \operatorname{Re}(\mathbf{E} \times \mathbf{H}^*) = \frac{1}{2\omega\mu_0} \mathbf{E}_0 \times (\mathbf{k} \times \mathbf{E}_0).$$

This may be reexpressed by the use of the vector identity:

$$\mathbf{E}_0 \times (\mathbf{k} \times \mathbf{E}_0) = \mathbf{k}(\mathbf{E}_0 \cdot \mathbf{E}_0) - \mathbf{E}_0(\mathbf{k} \cdot \mathbf{E}).$$

Because **k** and **E**$_0$ are perpendicular, as in the first condition of Eq. (5-3), their dot product is zero, and the remaining term gives the result,

$$\mathbf{P} = \frac{|\mathbf{E}_0|^2}{2\eta}\mathbf{k}_0, \qquad [\text{W/m}^2] \qquad \textbf{(5-5)}$$

in which $\eta = \sqrt{\mu_0/\varepsilon_0\varepsilon} = \eta_o/n$, (space impedance) and **k**$_0$ is a unit vector along **k**. According to Eq. (5-5), energy flows in the **k**$_0$ direction, with an intensity equal to $|\mathbf{E}_0|^2/2\eta$, which is analogous to $P = V^2/2R$ from circuit theory. The spatial relationship between the vectors **E**$_0$, **H**$_0$, **k**, and **P** is shown in Figure 5-1. As stated previously, the vectors **E**$_0$, **H**$_0$, and **P** are constant (i.e., their orientation in space does not change as the wave propagates). The orientation of **E**$_0$ is a property of the plane wave called *polarization*. If **E** always points in the same direction, as is the case for the plane wave of Figure 5-1, the field is said to be linearly polarized.

5.2 POLARIZATION

When two or more waves combine, their fields are added vectorially at each point in space at any specified time. The sinusoidal time variation of the fields is represented by the factor exp $(j\omega t)$. If all waves have the same angular frequency, ω, the time dependence is often suppressed, as in Eqs. (5-2). If the plane waves travel in different directions, a standing wave pattern results, similar to that considered in Chapter 1 (see Fig. 1-5). On the other hand, waves may be codirectional, that is, propagate in the same direction, but have different phases and polarizations. In Figure 5-2, for example, two waves, each with wave vector **k** along **z**$_0$ and frequency ω, have vector amplitudes that are chosen to lie along the x and y axes, respectively. If the field along the y axis lags the field along the x axis by a phase of ϕ degrees, then

$$E_x = Re\ E_1\ e^{j\omega t} = A \cos(\omega t)$$

and

$$E_y = Re\ E_2\ e^{j(\omega t - \phi)} = B \cos(\omega t - \phi),\qquad \textbf{(5-6)}$$

in which A, B, and ϕ are real numbers. In general, the combined vectors produce a total field vector that traces an elliptical path when observed in the x-y plane, as illustrated in Figure 5-3. The following cases are presented before exploring the general case further.

CASE 5-1. Linear polarization: If ϕ is equal to $0°$ or $180°$, the total electric field is confined to a plane that is tilted by an angle, α, with respect to the x-z plane, where

$$|\alpha| = \tan^{-1}(B/A).$$

The tilt angle, α, is positive for $\phi = 0°$ and negative for $\phi = 180°$.

CASE 5-2. Circular polarization: If ϕ is equal to $\pm 90°$, and $A = B$, the total electric field vector is constant in magnitude and rotates in the x-y plane with

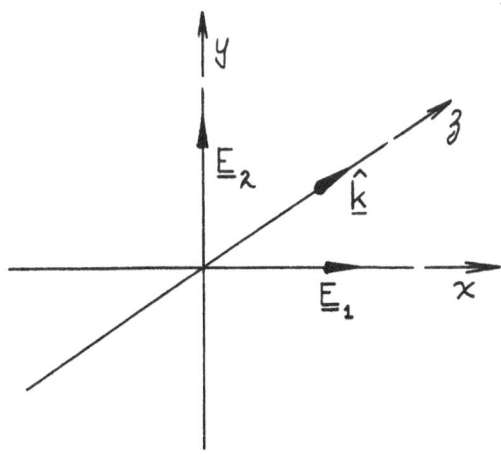

Figure 5-2. Coordinate system for field addition.

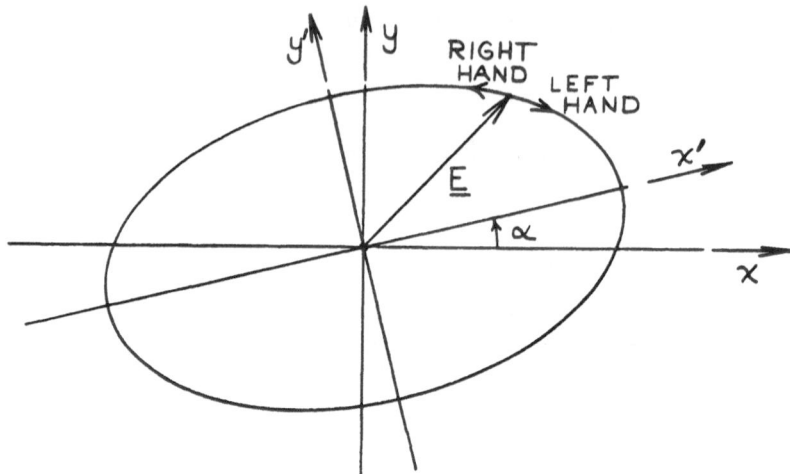

Figure 5-3.   General polarization ellipse.

angular frequency, ω. If $\phi = 90°$, an observer viewing the field along the $-\mathbf{z}_0$ direction will see the total field vector rotate counterclockwise. By Institute of Electronics and Electrical Engineering (IEEE) convention, this polarization is called right-hand circular. To remember this definition, point your right-hand thumb at your face, to represent an approaching wave. Observe that the fingers curl in the direction of field rotation for polarization of the right-hand sense. If $\phi = -90°$, the approaching field rotates clockwise, and the memorization trick requires the use of the left hand. Accordingly, this polarization is referred to as left-hand circular.

CASE 5-3.   Elliptical polarization with zero tilt angle: If $A > B$ and $\phi = \pm 90°$ Eqs. (5-6) become,

$$E_x = A \cos (\omega t)$$

and

$$E_y = \mp B \sin (\omega t), \tag{5-7}$$

which satisfy,

$$\left(\frac{E_x}{A}\right)^2 + \left(\frac{E_y}{B}\right)^2 = 1. \tag{5-8}$$

Clearly, this equation describes an ellipse with major axis A and minor axis B. As in Case 5-2, (circular polarization), the right-handed elliptical polarization sense corresponds to $\phi = 90°$, and the left-handed elliptical sense corresponds to $\phi = -90°$. The tilt angle is zero, because the major axis of the polarization ellipse lies along the x axis.

For the most general polarization, which is elliptical with arbitrary tilt angle (i.e., arbitrary orientation of the ellipse major axis), $A \neq B$, $-180° < \phi < 180°$, and, $-90° < \alpha < 90°$. The right-hand sense corresponds to $0° < \phi < 180°$ and the left-hand sense corresponds to $-180° < \phi < 0°$.

As discussed, antennas may produce either linear, circular, or elliptically polarized fields. The field of a straight wire antenna, for example, is linear when viewed along the radial direction. In other words, the E field has just one component, E_θ, whose direction (in the far field) is independent of time and distance from the source. On the other hand, the helical antenna radiates a field whose E vector rotates as illustrated in Figure 4-9b. As viewed along the z axis, the field is circularly polarized, with a sense that depends on the sense of the helix. When viewed along directions off the z axis ($\theta > 0°$), the field is elliptically polarized, in general. The ratio of the major and minor axes of the polarization ellipse, as A/B for the case of Eq. (5-8), is called the axial ratio, which is a dimensionless number, often expressed in decibels.

In general, an antenna that transmits EM waves of a particular polarization will absorb waves of the same polarization when operated as a receiving antenna. This property of antennas is one manifestation of the *principle of reciprocity* (discussed in Chap. 13). Polarization is a basic antenna characteristic of great practical importance.

5.3 REFLECTION AND REFRACTION

The plane wave in an unbounded, homogeneous, and isotropic medium considered thus far is an idealization that is analogous to the infinitely long transmission line. Wave behavior at the interface or boundary between dissimilar media is discussed in this section.

A simple plane interface $z = 0$ is shown schematically in Figure 5-4. The refractive index in the semi-infinite region for which $z < 0$ is denoted by n_1. For the semi-infinite region on the other side of the boundary ($z > 0$), the refractive index is n_2. A plane wave with wave vector $\mathbf{k_i}$ is incident upon the boundary. The plane of incidence is defined by $\mathbf{k_i}$ and $\mathbf{z_0}$, the unit vector that is perpendicular to the boundary plane $z = 0$. For convenience, and without loss of generality, $\mathbf{k_i}$ is chosen to lie in the y-z plane, which is therefore the plane of incidence, as shown in Figure 5-4. Under these conditions

Eqs. (5-1) and (5-3) stipulate that the incident electric field has the form of Eq. (5-2)

$$\mathbf{E}_i = \mathbf{E}_1 \, e^{-jkn_1(y \, \sin\theta_1' + z \, \cos\theta_1')}, \qquad (5\text{-}9)$$

in which θ_1' is the angle of incidence, as indicated in Figure 5-4. The incident field is completely specified by the constant vectors $\mathbf{E}_1$ and $\mathbf{k}_i$. Chapter 3 explained that the tangential components of $\mathbf{E}$ and $\mathbf{H}$ must be continuous at the interface. To satisfy this general requirement, in the present case, the x

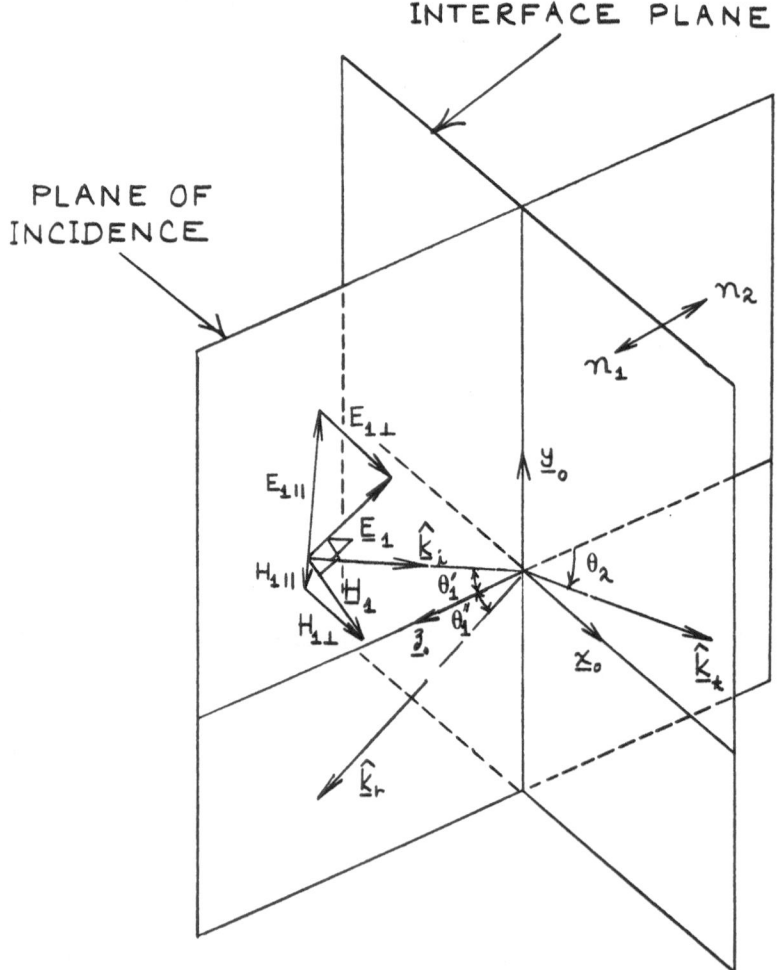

Figure 5-4.   Geometry of plane wave propagation at interface.

and y components of the fields must be continuous across the boundary, and as in the case of the transmission line junction, reflected and transmitted waves arise. The incident field, as specified by Eq. (5-9), is independent of the x coordinate. This property must also hold for the reflected and transmitted waves, so that $\mathbf{k_r}$ and $\mathbf{k_t}$ also lie in the y-z plane, the plane of incidence. Therefore, the reflected and transmitted fields must have the mathematical form

$$\mathbf{E}_{ref} = \mathbf{E}_2 \, e^{-jkn_1(y \, \sin\theta_1'' - z \, \cos\theta_1'')}$$

and

$$\mathbf{E}_{tran} = \mathbf{E}_3 \, e^{-jkn_2(y \, \sin\theta_2 + z \, \cos\theta_2)}. \tag{5-10}$$

The requirement of continuity at $z = 0$ also stipulates that the fields on either side of the interface have the same y dependence. Comparison of Eqs. (5-9) and (5-10) indicates that the solutions for the transmitted and reflected fields are constrained by the conditions

$$\boxed{\begin{array}{l} \theta_1' = \theta_1'' = \theta_1 \\ n_1 \sin\theta_1 = n_2 \sin\theta_2 \end{array}} \tag{5-11}$$

The second of these results is known in optics as Snell's law. The geometrical laws of plane wave propagation at the interface can be summarized as follows.

The wave vectors of the incident, reflected, and transmitted waves all lie in the plane that contains both the incident wave vector and the normal vector at the interface.
The angle of incidence is equal to the angle of reflection.
The angles of incidence and transmission obey Snell's law.

The reader may verify that the second and third results can also be obtained by using Fermat's principle of least time. According to this principle, EM energy travels along that path that minimizes the travel time. For example, the shortest distance from a point on one side of an interface, to the interface, and back to another point on the same side of the interface is realized for the *specular path* defined by the first and second results above. Similarly, it is possible to show that Snell's law can be derived from Fermat's principle when the speed of propagation has different values on opposite sides of the interface. (See problem 5-2.)

Calculation of the reflected and transmitted field amplitudes, E_2 and E_3, is straightforward but algebraically cumbersome; therefore, the results are discussed here without derivation.

The most general incident electric field can be expressed as the vector sum of components parallel and perpendicular to the plane of incidence, as shown in Figure 5-4. The perpendicular component, in this case, lies along $\mathbf{x}_0$, and the parallel component lies in the plane of incidence, along the unit vector $\mathbf{y}_0 \cos \theta_1 - \mathbf{z}_0 \sin \theta_1$. The continuity of the tangential **E** and **H** components at $z = 0$ leads to the following results:

$$\left. \begin{array}{l} |E_2| = R_\perp |E_1| \\ |E_3| = T_\perp |E_1| \end{array} \right\} \quad \perp \text{ Polarization}$$

$$\left. \begin{array}{l} |E_2| = R_\parallel |E_1| \\ |E_3| = T_\parallel |E_1| \end{array} \right\} \quad \parallel \text{ Polarization}$$

$$(5\text{-}12)$$

where

$$\boxed{\begin{array}{l} R_\perp = \dfrac{n_1 \cos\theta_1 - n_2 \cos\theta_2}{n_1 \cos\theta_1 + n_2 \cos\theta_2} \\[3mm] R_\parallel = \dfrac{n_1 \cos\theta_2 - n_2 \cos\theta_1}{n_1 \cos\theta_2 + n_2 \cos\theta_1} \\[3mm] T_\perp = 1 + R_\perp, \quad T_\parallel = 1 + R_\parallel \end{array}}$$

$$(5\text{-}13)$$

The quantities $R_\perp$, $R_\parallel$ and $T_\perp$, $T_\parallel$ are the reflection and transmission coefficients, respectively. Their values depend on polarization, incidence angle (θ_1), and the refractive indices n_1 and n_2 (recall that θ_2 can be obtained by Snell's law, given θ_1, n_1, and n_2). Several interesting results that can be derived from Eqs. (5-12) and (5-13) are discussed below. In lossy media, n_1 and n_2 are complex, so that R and T are also complex. This is evident from the definition of the refractive index, as in Eqs. (3-19), (3-23), and (3-24). A lossless dielectric, for which the dielectric constant is real ($\sigma = 0$), has a refractive index that is also real.

If dielectric losses are small, i.e., when in Eq. (3-19), $\sigma/\omega\varepsilon_o \ll \varepsilon'$, the refractive index has a small imaginary part. That part does not significantly affect the reflection and transmission coefficients, but does introduce some attenuation along the propagation path:

$$n = n_r - jn_i,$$

$$e^{-jknz} = e^{-kn_iz}\, e^{-jkn_rz}.$$

The real part of the refractive index, n_r, still represents a ratio of velocities, and the imaginary part, n_i, represents an attenuation factor.

On the other hand, for a good conductor, $\sigma/\omega\varepsilon_o \gg \varepsilon'$, and therefore,

$$\varepsilon \cong -j\,\frac{\sigma}{\omega\varepsilon_o} = e^{j\pi/2}\,\frac{\sigma}{\omega\varepsilon_o},$$

$$n = \sqrt{\varepsilon} = e^{-j\pi/4}\,\sqrt{\frac{\sigma}{\omega\varepsilon_o}} = (1 - j)\,\sqrt{\frac{\sigma}{2\omega\varepsilon_o}}, \qquad \textbf{(5-14)}$$

and the wave is strongly attenuated by the large imaginary component of the refractive index.

At the boundary between a good dielectric, characterized by n_1, and a good conductor, characterized by n_2, Eqs. (5-13) yield,

$$R_\perp \cong R_\| \cong -1$$

$$T_\perp \cong T_\| \cong 0 \qquad \textbf{(5-15)}$$

because $|n_2| \gg |n_1|$. Thus, at the boundary of a good conductor, a plane wave is totally reflected, regardless of its polarization and angle of incidence. The boundary of a conductor is analogous to a short circuit at the end of a transmission line.

When a wave crosses a boundary while entering a region of lower refractive index, the angle of refraction must exceed the angle of incidence, in accordance with Snell's law:

$$\boxed{\sin\theta_2 = \frac{n_1}{n_2}\,\sin\theta_1; \qquad n_1 > n_2.}$$

However, if the angle of incidence exceeds a critical angle

$$\theta_c = \sin^{-1}\!\left(\frac{n_2}{n_1}\right); \qquad n_1 > n_2 \qquad \textbf{(5-16)}$$

the angle of refraction cannot be a real number, because for any real angle the sine is equal to or less than unity. When $\theta_1 > \theta_c$, the wave is said to experience *total internal reflection,* because no propagating solution exists, as shown in Figure 5-5. As a result, there is no power transfer into the lossless n_2 medium, and all the power of the incident plane wave is converted into that of the reflected wave.

Examination of Eqs. (5-13) reveals that $R_{\parallel}$ will vanish if $n_1 \cos\theta_2 = n_2 \cos\theta_1$. With a little mathematical manipulation to solve for θ_1,

$$\theta_1 = \sin^{-1} \sqrt{\frac{n_2^2}{n_1^2 + n_2^2}} = \sin^{-1} \sqrt{\frac{\varepsilon_2}{\varepsilon_1 + \varepsilon_2}}$$

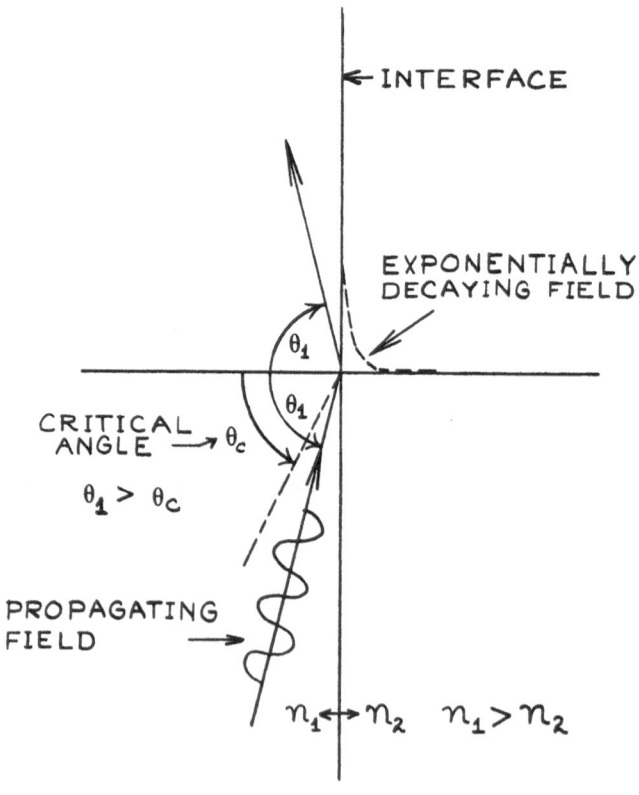

Figure 5-5.   Total internal reflection.

$$\theta_1 \equiv \theta_B = \tan^{-1} \sqrt{\frac{\varepsilon_2}{\varepsilon_1}} \qquad\qquad \textbf{(5-17)}$$

The angle θ_B, for which the reflected parallel polarization component, $R_\|$, vanishes, is known as the *Brewster* angle or *polarizing* angle. A parallel polarized plane wave incident at the Brewster angle is entirely transmitted; however, a perpendicularly polarized plane wave is both transmitted and reflected. Thus, for $\theta_1 = \theta_B$, the reflected field has pure perpendicular polarization; hence, the term polarizing angle. The Brewster angle geometry is illustrated in Figure 5-6. Note that the angles of incidence and refraction add up to 90°. This construction shows that if the reflected parallel polarized field is imagined to arise from the excitation of dipoles distributed along the path of the refracted ray, the polarizing condition occurs when the reflected ray would lie in the direction of the dipoles axes, along which the field is zero, as has been shown in Chapter 4.

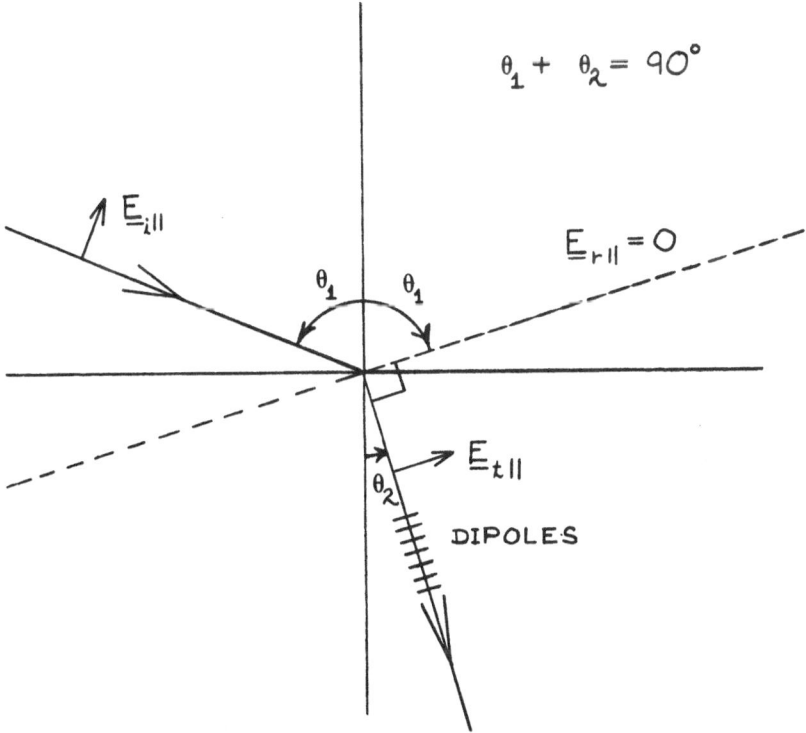

Figure 5-6. Brewster angle condition.

5.4 GEOMETRICAL OPTICS

At this point it is apparent that the laws that govern the reflection and transmission of plane waves are identical to those encountered in geometrical optics. Interestingly, these laws were discovered long before Maxwell showed that visible light is a manifestation of the electromagnetic field. Many years of effort were required to completely reconcile Maxwell's EM theory and optics. It is now realized that geometrical optics represents an approximate solution of Maxwell's equations under certain circumstances. For example, when wavelengths are short relative to typical dimensions that characterize the propagation environment, such as lens diameters, aperture sizes, or radii of curvature (in the case of reflectors), geometrical optics may provide valid results, even when employed for radiations well outside the optical regime. Geometrical optics, due to its simplicity, is a valuable tool for the analysis of EM fields and antennas, despite some limitations, which are discussed below.

Rays represent the trajectories of energy transport (i.e., the direction of Poynting's vector, P). In homogeneous media, for which the refractive index (and hence propagation speed) is constant, rays are straight lines, consistent with the principle of least time. On the other hand, the refractive index may vary with position, as in the case of the atmosphere or ionosphere, as a result of variation in density or charge concentration. Rays in such media curve— the EM energy does not travel along straight lines. In this discussion of antennas, the ray bending effects of inhomogeneous media will not be considered further.

Rays that represent a plane wave are parallel to the propagation vector, k, of Eqs. (5-2) and (5-3) and perpendicular to the planes of constant phase (wave fronts), given by $k \cdot r$ = constant. Similarly, spherical waves are characterized by families of rays that are normal to the spherical surfaces of constant phase, given by r = constant. They represent the radially outward flow of energy from the source. A *tube* of rays may be constructed, as is shown in Figure 5-7. A tube of rays is a small tubeshaped volume, bounded by surfaces that are either parallel or perpendicular to a bunch of close rays. The cross-sectional area of the tube generally will change as the rays that are parallel to the tube change direction. It should be realized that the cross-sectional area of the tube is inversely proportional to the power density within the tube. This follows from the law of energy conservation as discussed in Chapter 1 in connection with spherical waves.

The parallel rays associated with a plane wave remain parallel following reflection and refraction at a plane interface, and therefore, the reflected and refracted waves are also plane waves. It has been seen that the field due to a source over a ground plane can be accounted for by placing an inverted source in the image position. This is the same as considering a spherical wave reflected from a plane surface. The reflected spherical wave appears to originate from an image source. Although general propagation environments are more complex than these simple illustrations suggest, a few rules can be established from geometrical optics that can be used advantageously in antenna work.

If an interface is smooth and gently curved, it can be regarded as locally flat. A smooth surface has no sharp points or edges, so that at each point on the surface a well-defined normal vector can be erected. Curvature is said to be "gentle" if the local radii of curvature are much larger than the wavelength

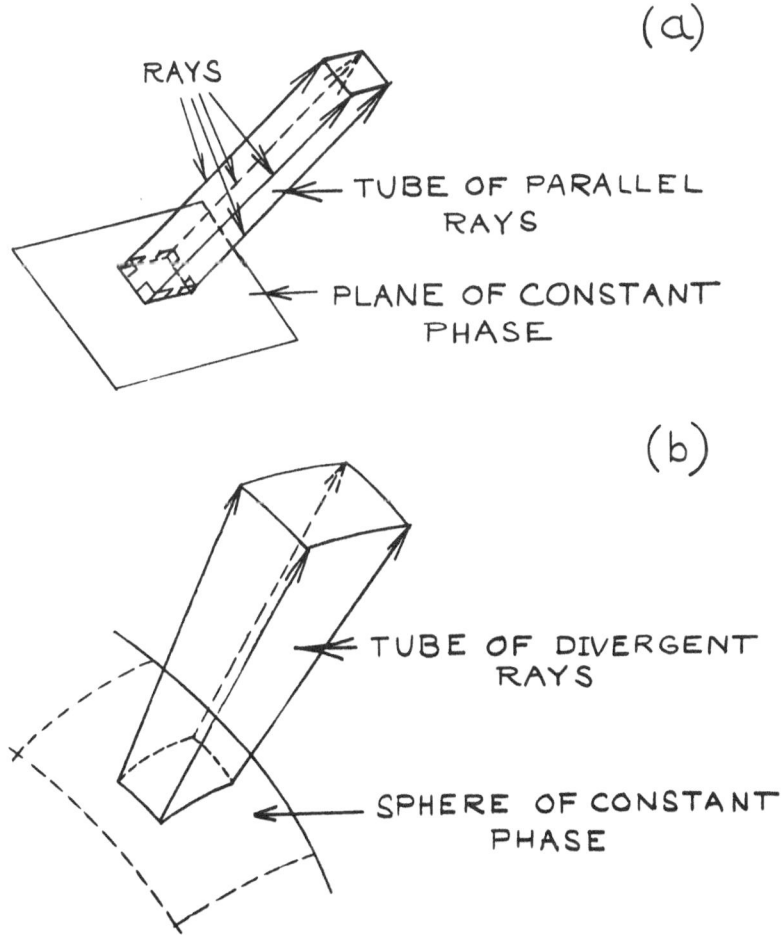

Figure 5-7. Plane wave (a) and spherical wave (b) ray tubes.

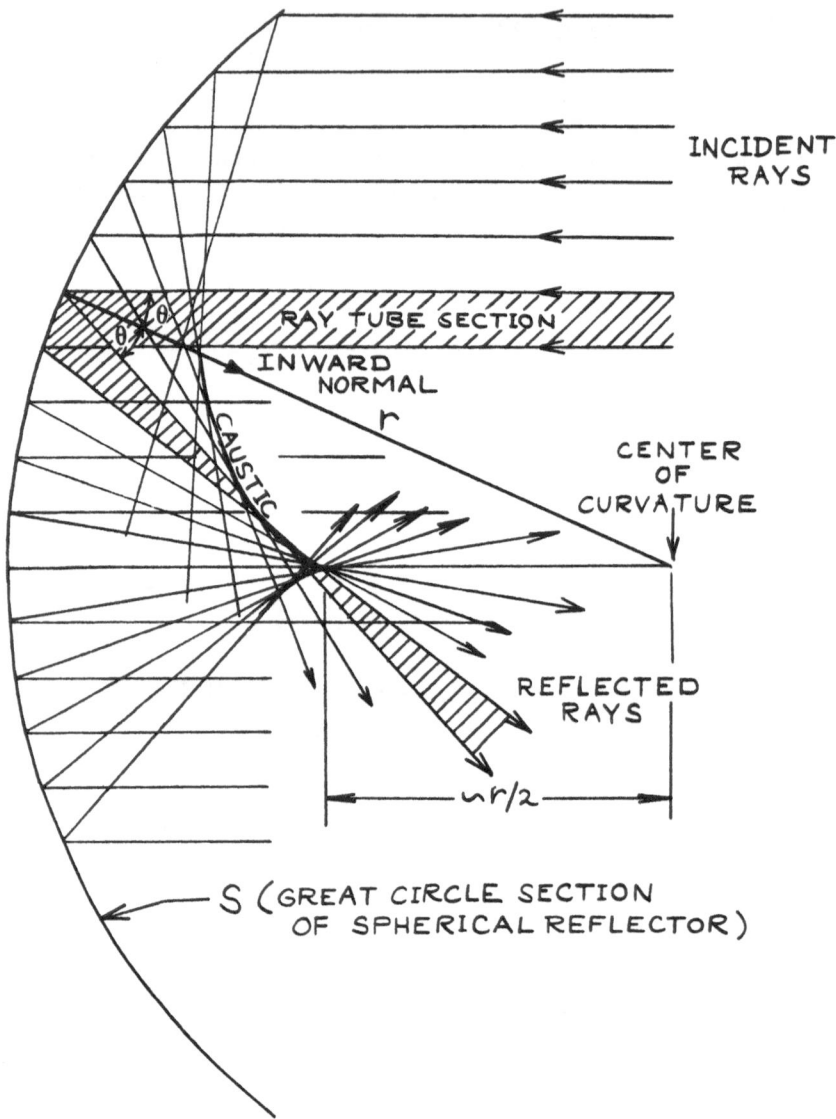

Figure 5-8. Reflection of plane wave by a spherical reflector and the associated caustic.

of the radiation being considered. An example of a smooth, gently curved, reflecting surface is shown in Figure 5-8. The reflecting surface, in the form of a concave spherical cap, is illuminated along its axis of rotational symmetry. Because the surface is smooth, an inward normal exists at each point, and a family of reflected rays can be drawn by use of the law of reflection. Previously, it was pointed out that a conducting plane, like a short circuit at the end of a transmission line, has a reflection coefficient of unity magnitude. Similarly, it is assumed that total reflection occurs at each point of the spherical reflector. If a tube of rays (shaded area) is selected, the energy of the plane wave is scattered backwards, and the rays that bound the tube first converge and then diverge.

Examination of Figure 5-8 also shows that there is a curve, or locus, at which rays cross; this is called a *caustic*. Because the cross section of any tube of reflected rays vanishes at some point on the caustic, one might conclude that the corresponding local power density becomes infinite; however, this is not true. Geometrical optics, which is only an approximation, breaks down at the caustic. Qualitatively speaking, however, the field is relatively large in the vicinity of the caustic. The relative intensity of the far field can be calculated by considering the divergence of a small tube of rays reflected from the curved surface. The procedure requires elementary (three-dimensional) geometry only, but nevertheless may involve tedious calculations. The details are not discussed here.

Another configuration of great practical importance, the parabolic reflector, is shown in cross section in Figure 5-9. The surface of the reflector is a paraboloid of revolution. It is mathematically expressed by

$$r^2 = 4fz, \tag{5-18}$$

in which f is the focal distance.* The inward unit normal vector for this surface is easily calculated, and it is not difficult to demonstrate that any ray that passes through the focus is related by reflection to a ray parallel to the z axis, and any ray that is incident parallel to the z axis is reflected through the focus. A source of spherical waves at the focus will, after reflection from the paraboloid, produce a family of parallel rays. This family resembles a plane wave directed out along the z direction. Similarly, a plane wave coming in from the z direction will converge at the focus before diverging in spherical wave fashion.

Figures 5-8 and 5-9 bear some similarity that is not accidental. A shallow spherical section or cup is a close approximation to a paraboloid of focal

---

*This f, which is a length, should not be confused with the frequency, such as in Eq. (3-22).

length equal to half the spherical radius—a result that is evident from examination of the figures. This spherical reflector property is of practical importance, because some optical systems for telescopes use spherical primary mirrors that are far easier to fabricate than parabolic mirrors. The small aberration introduced by the spherical mirror is then corrected by another optical element in the system.

As noted earlier, geometrical optics cannot describe the EM field where the rays converge, because geometrical optics suggests that the power density become infinite at such points (e.g., a caustic or focus). Other limitations of geometrical optics are illustrated in Figure 5-9. First, a ray that strikes the edge of the reflector cannot be accounted for, because the direction of the normal is undefined at the edge, and energy is scattered in all directions. Furthermore, geometrical optics cannot account for fields that exist in the shadow region behind the reflector, which is clearly inaccessible to rays. The

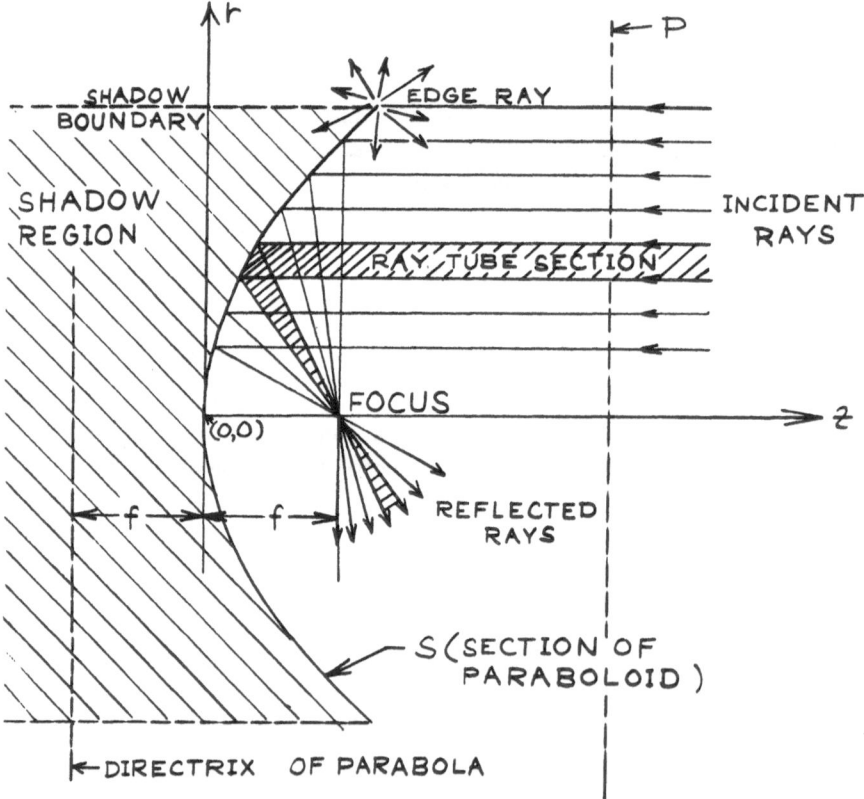

Figure 5-9.   Reflection of a plane wave by a parabolic reflector.

ray-optic method predicts sharp shadows at the reflector edges—a result that disagrees with the observed propagation of waves around obstacles by diffraction. A third inadequacy of geometrical optics is more subtle. Consider a family of radial rays emanating from the focus of Figure 5-9. These rays are reflected to form a parallel set of rays that illuminate a circular patch on the reference plane, *P*, which is normal to the *z* axis. Geometrical optics can be used to compute the amplitude and phase of the field at the illuminated patch, as long as the plane, *P*, is not too far from the reflector. As the plane moves farther away (e.g., into the far field region), the field calculated by geometrical optics diverges radically from the field obtained by actual measurements. Therefore, the far field of a *collimated beam* (i.e., a family of parallel rays of finite extent) cannot be predicted by geometrical optics, because parallel rays are tantamount to a focus at infinity.

If the reflectors of Figures 5-8 and 5-9 are reversed (i.e., if the illuminating rays fall on the convex side and diverge without forming caustics or foci) the relative intensity of the scattered field can be obtained from geometrical optics for any direction that lies in the illuminated regions, as long as the reflecting objects are smooth and gently curved. In later chapters it will be shown how geometrical optics can be used to design reflector antennas and to calculate the radar cross section under certain circumstances.

PROBLEMS

PROBLEM 5-1. A plane electromagnetic wave may be expressed as in Eq. (5-2) and (5-3). Show the following.

a. $\mathbf{E}_0$, $\mathbf{H}_0$, and $\mathbf{k}$ are mutually perpendicular.

b. $|\mathbf{E}_0| = \eta|\mathbf{H}_0|$.

c. $\dfrac{\partial^2 \mathbf{E}}{\partial x^2} + \dfrac{\partial^2 \mathbf{E}}{\partial y^2} + \dfrac{\partial^2 \mathbf{E}}{\partial z^2} + k^2 n^2 \mathbf{E} = 0.$

d. $\dfrac{\partial^2 \mathbf{H}}{\partial x^2} + \dfrac{\partial^2 \mathbf{H}}{\partial y^2} + \dfrac{\partial^2 \mathbf{H}}{\partial z^2} + k^2 n^2 \mathbf{H} = 0$

e. $\mathbf{P} = \dfrac{1}{2} \mathbf{E} \times \mathbf{H}^* = \dfrac{|\mathbf{E}_0|^2}{2n} \mathbf{k}_0.$

PROBLEM 5-2. Figure 5-10 shows the trajectory of a particle that moves from point *A* to point *B*. The particle speed is v_1, in the $y < 0$ half space and v_2 in the $y \geqslant 0$ *half space*.

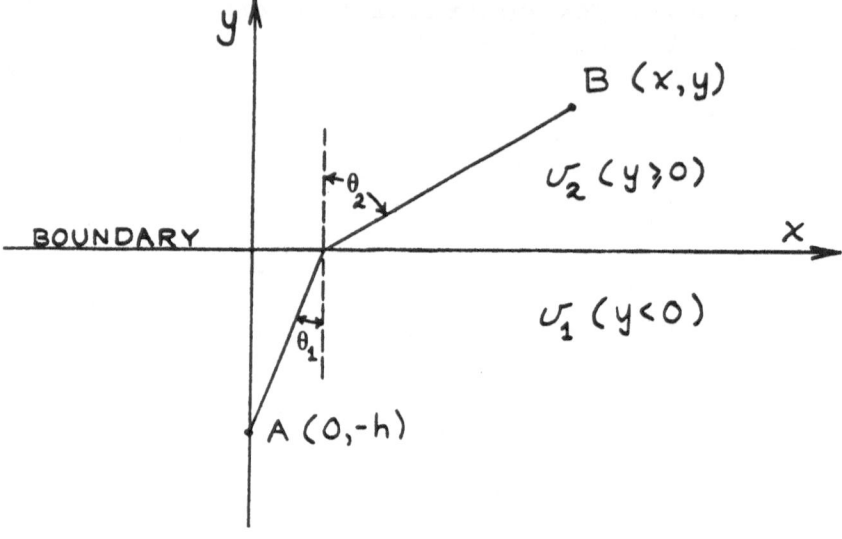

Figure 5-10.   The trajectory of a particle that moves from point A to point B.

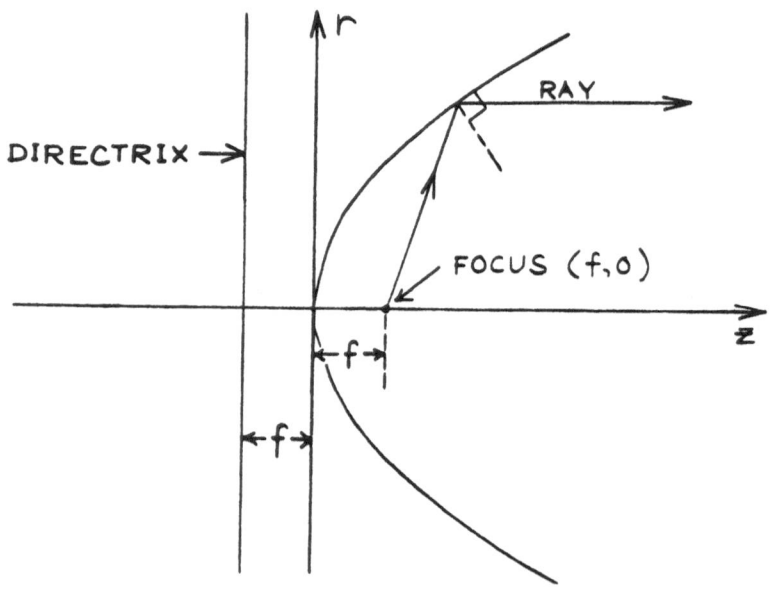

Figure 5-11.   Illustration of a parabola for Problem 5.4.

a. Show that the time required for the particle to pass from A to B (or from B to A) is minimized when

$$\frac{1}{v_1} \sin \theta_1, = \frac{1}{v_2} \sin \theta_2,$$

and relate this result to Snell's law of refraction.

b. Show that if points A and B are both in the upper half space, and the trajectory must contact the x axis, the principle of least time leads to the law of specular reflection.

PROBLEM 5-3. Refer to Figure 5-6 and show that when the Brewster angle condition, $n_1 \cos \theta_2 = n_2 \cos \theta_1$, is satisfied, $\theta_1 + \theta_2 = 90°$.

PROBLEM 5-4. The parabola in Figure 5-11 can be expressed by the formula $r^2 = 4 fz$. A parabola is defined as the locus of points that are equidistant from a point called the focus, and a straight line, called the directrix.

a. Show that the formula above for the parabola is correct.

b. Show that a ray that starts at the focus is reflected by the parabola in a direction parallel to the z axis.

PROBLEM 5-5. Show that a shallow spherical cup is a good approximation for a paraboloid of revolution that has a focal length equal to half of the radius of the cup.

PROBLEM 5-6. Show that the linearly polarized field,

$$\mathbf{E} = \mathbf{x}_0 E_0 \cos (\omega t - ky - \phi),$$

can be decomposed into two circularly polarized fields of equal amplitude and opposite sense.

PROBLEM 5-7. A diver is 10 m below the calm surface of a pond. When looking up at the air-water interface, the diver sees a distorted view of the surroundings above the water surface, because of the refraction of light. Indeed, all light incident from above the pond is confined to a finite circular disk centered on that point on the surface directly above the diver's head.

a. Explain the origin of this effect.

b. Calculate the diameter of the circular disk (use $n = 1.33$ for water).

CHAPTER 6

Waveguides

Waveguides are commonly used to direct the flow of EM energy and in this sense are a type of transmission line. Because waveguides can handle high power with low losses and operate well into the millimeter wave region, they are frequently used to feed antennas and are often integral components in antenna systems. Therefore, being familiar with the basic characteristics of waveguides is necessary to understand the operation of microwave and millimeter wave antennas.

6.1 PARALLEL PLANE WAVEGUIDE

The simplest conceivable waveguide is formed by two parallel perfectly conducting plates. A plane wave can propagate in the space between the plates if its electric field is normal to the plates. For mathematical convenience, it is assumed that the plates are parallel to the x-z plane and that propagation occurs along the z axis. Then by use of Eqs. (5-1) through (5-3),

$$\mathbf{E} = \mathbf{y}_0 \, E_0 \, e^{-jkz}, \quad k = 2\pi/\lambda$$

$$\mathbf{H} = -\mathbf{x}_0 \, \frac{E_0}{\eta} \, e^{-jkz}, \quad \eta = \sqrt{\mu_0/\varepsilon_0\varepsilon}$$

$$\mathbf{k} = \mathbf{z}_0 \, k. \tag{6-1}$$

The energy flow is given by Eq. (5-5).

$$\mathbf{P} = \mathbf{z}_0 \, E_0^2/2\eta \qquad (\text{W/m}^2)$$

and is limited only by arcing between the plates, which may occur when the electric field in the gap becomes too intense.

The parallel plate waveguide described above cannot be realized in practice, because the plates are unbounded. For a practical rectangular waveguide, the solution represented by Eqs. (6-1) may be replaced by the sum of two plane waves with wave vectors given by

$$\mathbf{k}_1 = k \, (\mathbf{x}_0 \sin \alpha + \mathbf{z}_0 \cos \alpha)$$

and

$$\mathbf{k}_2 = k \, (-\mathbf{x}_0 \sin \alpha + \mathbf{z}_0 \cos \alpha), \qquad (6\text{-}2)$$

in which α, which can take on any value between $0°$ and $90°$, is the angle between $\mathbf{z}_0$ and $\mathbf{k}_1$. (Fig. 1-5 provides an illustration of this geometry, if one replaces α by $\alpha/2$.) In accordance with Eq. (5-3),

$$\mathbf{E}_1 = \mathbf{y}_0 \, E_0 \, e^{-jk(x \sin \alpha \, + \, z \cos \alpha)},$$

$$\mathbf{E}_2 = \mathbf{y}_0 \, E_0 \, e^{-jk(-x \sin \alpha \, + \, z \cos \alpha)},$$

$$\mathbf{H}_1 = (-\mathbf{x}_0 \cos \alpha + \mathbf{z}_0 \sin \alpha) \frac{E_0}{\eta} e^{-jk(x \sin \alpha \, + \, z \cos \alpha)},$$

$$\mathbf{H}_2 = (-\mathbf{x}_0 \cos \alpha - \mathbf{z}_0 \sin \alpha) \frac{E_0}{\eta} e^{-jk(-x \sin \alpha \, + \, z \cos \alpha)}.$$

The combined wave is the vector sum of these two plane waves, which is

$$\mathbf{E} = \mathbf{E}_1 + \mathbf{E}_2 = 2\mathbf{y}_0 \, E_0 \cos \, (kx \sin \alpha) \, e^{-jkz \cos \alpha}$$

$$\mathbf{H} = \mathbf{H}_1 + \mathbf{H}_2 = -\frac{2E_0}{\eta} \, [\mathbf{x}_0 \cos \alpha \cos \, (kx \sin \alpha),$$

$$+ \; \mathbf{z}_0 \, j \sin \alpha \sin \, (kx \sin \alpha)] \, e^{-jkz \cos \alpha}. \qquad (6\text{-}3)$$

Examination of the above result reveals that the interference, or addition, of the two plane waves produces a standing wave pattern along the $\mathbf{x}_0$ direction. Standing waves, with their regularly spaced nodes and peaks, were discussed in Chapter 1 (see Fig. 1-5). In the present example, the E field has maxima at the positions,

$$x = \frac{m\lambda}{2 \sin \alpha}; \quad m = 0, \pm 1, \pm 2 \dots$$

and nodes (i.e., zeros or nulls) at the positions,

$$x = \frac{(2m + 1)\lambda}{4 \sin \alpha}; \quad m = 0, \pm 1, \pm 2 \dots \tag{6-4}$$

6.2 RECTANGULAR WAVEGUIDE

Because the electric field lies along the y_0 direction, it is tangent to any wall that is normal to the x axis. If conducting walls are erected perpendicular to the x axis at the node points specified by Eq. (6-4), the field is not perturbed. If conducting walls are placed at any two of these nodes, a metallic pipe of rectangular cross section is formed. For example, if the walls are specified at the adjacent nodes given by $m = 1,2$, then

$$x_1 = \frac{3\lambda}{4 \sin \alpha}, \quad x_2 = \frac{5\lambda}{4 \sin \alpha},$$

$$x_2 - x_1 = a = \lambda/2 \sin \alpha, \tag{6-5}$$

and the configuration illustrated in Figure 6-1 is obtained. The width of the waveguide, a, is clearly related to the wavelength (or frequency) of the field and must always exceed $\lambda/2$, because $0 < \sin \alpha < 1$. Theoretically, the height, b, is arbitrary; however, practicality requires that b be usually less than half the width, a. The fields inside the waveguide propagate along the z_0 direction, and in analogy with Eq. (1-6),

$$k \cos \alpha = k_g = \frac{2\pi}{\lambda_g} = \frac{\omega}{c_p}, \tag{6-6}$$

where λ_g is the wavelength inside the waveguide, and c_p is the phase velocity. Because $\omega/k = c_0$, the phase velocity in the waveguide is given by

$$\boxed{c_p = c_0/\cos \alpha}$$

and clearly exceeds the speed of light in free space. As noted in Chapter 4 however, this result does not violate the laws of special relativity, because phase velocity is simply the velocity of a point of constant phase, which does not transport matter or energy.

Also from Eq. (6-6),

$$\lambda_g = \lambda/\cos \alpha \Rightarrow \boxed{\cos \alpha = \lambda/\lambda_g} \qquad (6\text{-}7a)$$

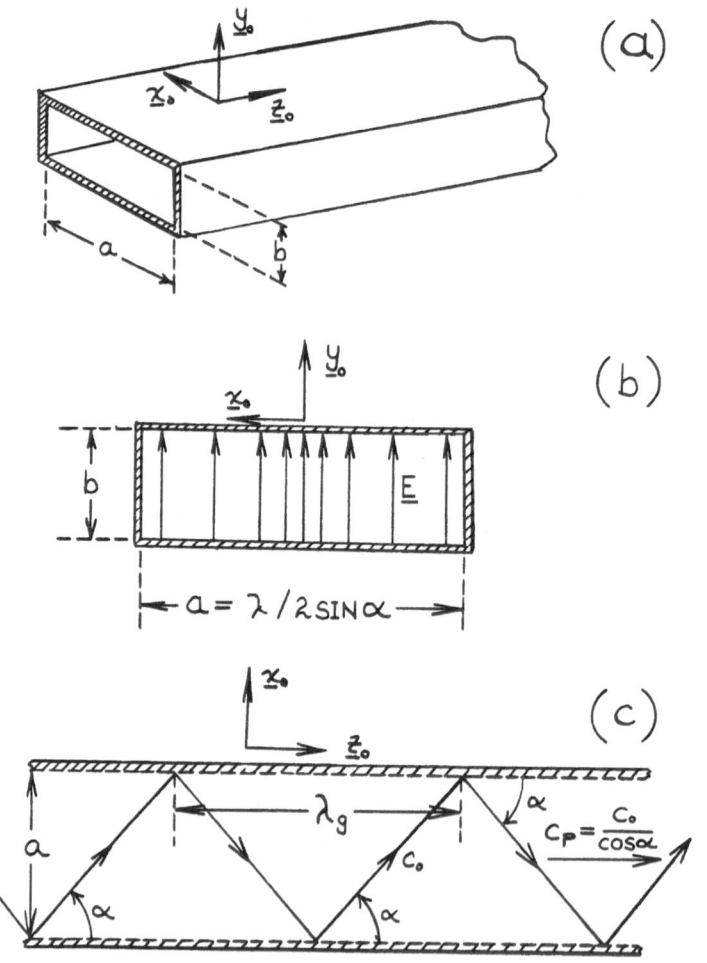

Figure 6-1. Rectangular wave guide (a). TE_{10} field distribution in the cross sectional plane (b). Path of plane wave component of TE_{10} (c).

and from Eq. (6-5),

$$a = \lambda/(2 \sin \alpha) \Rightarrow \boxed{\sin \alpha = \lambda/2a} \qquad \textbf{(6-7b)}$$

Therefore, by a common trigonometric identity,

$$\frac{1}{(2a)^2} + \frac{1}{\lambda_g^2} = \frac{1}{\lambda^2} \qquad \textbf{(6-7c)}$$

This expression relates the free space wavelength, λ, and the waveguide wavelength, λ_g, to the waveguide width. Eq. (6-7c) can be reexpressed

$$\boxed{\lambda_g = \lambda/\sqrt{1 - (\lambda/2a)^2}} \qquad \textbf{(6-8)}$$

This equation reveals that as λ is increased, a point is reached for which λ_g becomes imaginary. Clearly, when $\lambda > 2a$,

$$\lambda/2a = \sin \alpha > 1,$$

$$k_g = \frac{2\pi}{\lambda_g} = -j\frac{2\pi}{\lambda}\sqrt{\left(\frac{\lambda}{2a}\right)^2 - 1}. \qquad \textbf{(6-9)}$$

Substitution of Eq. (6-9) into Eq. (6-3) shows that the fields decay exponentially with z:

$$\mathbf{E} \propto \mathbf{H} \propto \exp\left[-\frac{2\pi z}{\lambda}\sqrt{\left(\frac{\lambda}{2a}\right)^2 - 1} \right].$$

Therefore, energy is not transported for any significant distance through the waveguide if $\lambda > 2a$. The frequency limit, below which propagation cannot occur, is given by

$$\boxed{f_c = \frac{c_0}{2a} = 14.99/a \qquad \text{(GHz)}} \qquad \textbf{(6-10)}$$

in which a is in centimeters. For example, if $a = 0.9$ in. = 2.286 cm, the cutoff frequency is about 6.56 GHz.

There is no upper frequency limit for this propagation mode. However, if the wavelength is such that $a > \lambda$, another mode of propagation, which places a node at the center of the waveguide, along the $\mathbf{x}_0$ direction, can occur. This new mode corresponds to wall positions specified, for example, by $m = 1$ and $m = 3$ in Eq. (6-4). Thus,

$$x_1 = \frac{3\lambda}{4 \sin \alpha}, \qquad x_3 = \frac{7\lambda}{4 \sin \alpha},$$

and

$$a = x_3 - x_1 = \lambda/\sin \alpha \qquad \sin \alpha = \lambda/a. \tag{6-11}$$

In analogy with Eq. (6-8), it is easily shown that the wavelength for the new mode is given by

$$\lambda_g' = \lambda/\sqrt{1 - (\lambda/a)^2}, \tag{6-12}$$

and the cutoff frequency is twice that of the mode that was considered previously. Thus, if $a > \lambda$, the waveguide can support two modes simultaneously—one with λ_g as in Eq. (6-8), and one with λ_g' as in Eq. (6-12). In general, as the frequency is increased, more propagation modes are possible, and the standard notations TE_{mn} and TM_{mn}, which describe the field distributions in the waveguide, are used.

The mode with the lowest cutoff frequency was discussed first and is designated TE_{10}, which indicates that the field is *transverse electric*, (i.e., has no E component along the direction of propagation). The subscripts indicate that the field has one standing wave half-cycle of variation in the $\mathbf{x}_0$ direction and no variation in the $\mathbf{y}_0$ direction. The second mode is labeled TE_{20} in the standard notation. In Figure 6-2a, the $\mathbf{E}$ and $\mathbf{H}$ fields for the TE_{10} and TE_{20} modes are illustrated. In order to predict and control the field distribution in a waveguide, the working frequency should be confined to a range within which only the fundamental mode, TE_{10}, can exist. The most common rectangular waveguide for x-band, for example, is the WR-90, which has inner dimensions of 0.9 in. by 0.4 in. Theoretically, propagation is limited to the

Figure 6-2.   Field configurations TE_{10}, TE_{20} (a) and current flow in walls for TE_{10} in a rectangular waveguide (b). (From S. Ramo, J. R. Whinnery, and T. VanDuzer, *Fields and Waves in Communication Electronics*, 2nd ed., New York: Wiley, 1984, pp. 414, 421; copyright © 1984 by John Wiley & Sons, reprinted by permission.)

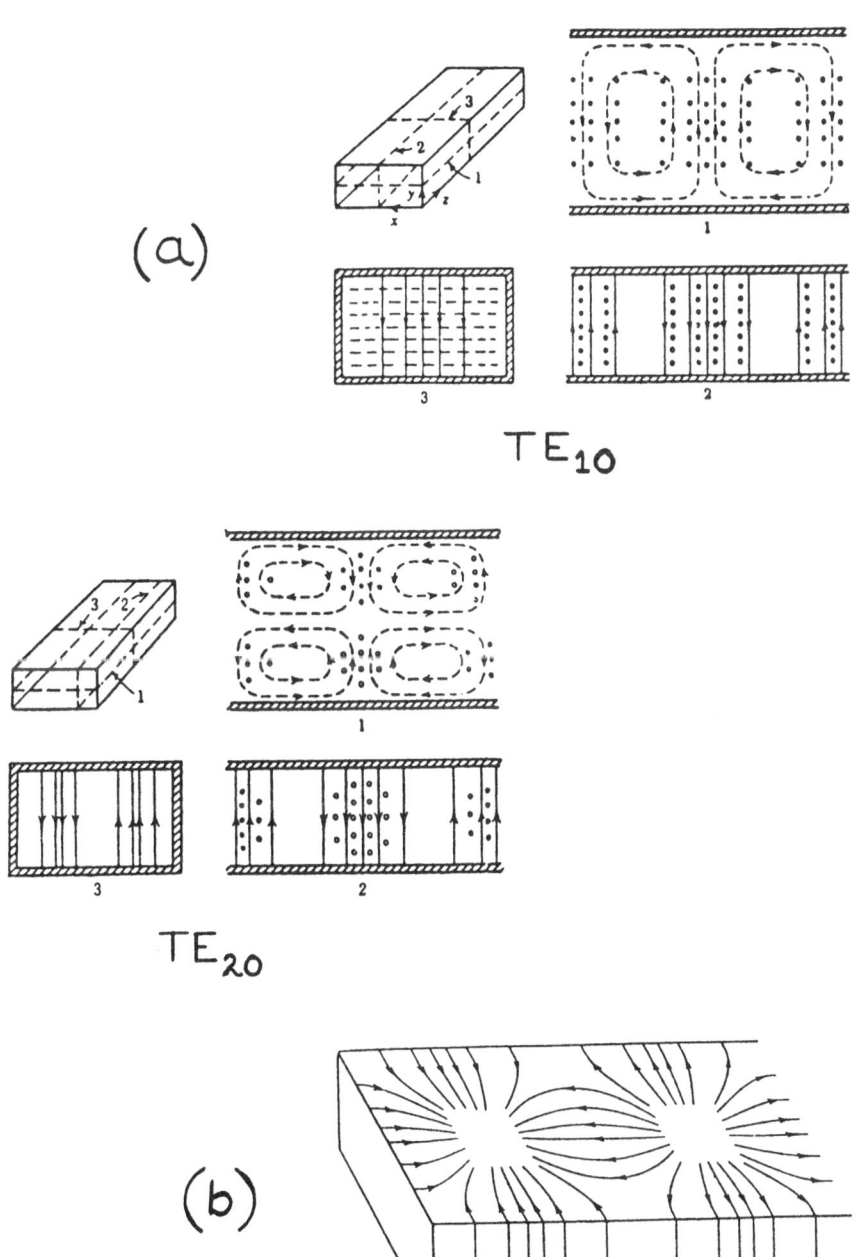

(a)

TE_{10}

TE_{20}

(b)

TE_{10}

fundamental mode between frequencies of 6.56 GHz and 13.12 GHz for this waveguide; however, to avoid cutoff frequency proximity, a working frequency range between 8.2 GHz and 12.4 GHz is recommended.

6.3 GENERAL WAVEGUIDES

The rigid, air-filled coaxial line (coax) can be visualized as a parallel plate waveguide that has been formed into concentric cylinders. In contrast to the rectangular waveguide, propagation modes exist for which both **E** and **H** are transverse, as was the case for a single plane wave in the parallel plane waveguide. Consequently, the lowest propagation mode in the coax is the transverse electric and magnetic (TEM), which has the same phase velocity and wavelength as a wave in free space and no cutoff. Rigid, air-filled coax lines can handle lower frequencies and wider bandwidths than can hollow waveguides. On the other hand, because they consist of two conductors, rigid coax lines are more complicated than hollow waveguides and, in addition, cannot accommodate as high a power level.

The rectangular waveguide, as well as other hollow waveguides, can be regarded as a transmission line in which current and voltage are represented by H_x and E_y, respectively, for the TE_{10} mode. The wave number in the propagation ($\mathbf{z}_0$) direction is $k_g = k \cos \alpha$, and the characteristic impedance (the ratio E_y/H_x) is

$$z_0 = \frac{\eta}{\cos \alpha} = \frac{\eta}{\sqrt{1 - (\lambda/2a)^2}}; \qquad (TE_{10}). \qquad \textbf{(6-13)}$$

The main difference between this transmission line and those discussed in Chapter 2 involves the frequency behavior: Characteristic impedance, wave number, and propagation velocity are all frequency dependent for a rectangular waveguide, which has a high-pass characteristic. When the propagation velocity is frequency dependent, the associated medium or transmission line is said to be dispersive.

From Eq. (3-22), the EM field boundary conditions at the surface of a conductor require that

1. The tangential component of **E** vanish and that
2. An electric surface current, $\mathbf{J} = \mathbf{n} \times \mathbf{H}$, exist on the surface, where **n** is the unit normal to the waveguide wall, and **H** is the total magnetic field at the wall.

The first condition is satisfied by the interference of the two plane waves, as in Eq. (6-2), which produces nulls at those waveguide walls that are parallel

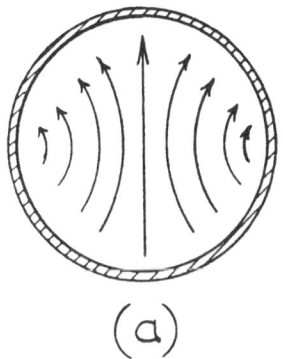

(a)

Figure 6-3.   Waveguides: (a) circular (TE_{11}), (b) ridged, (c) double ridged.

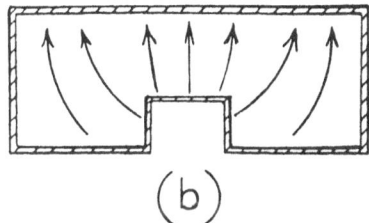

(b)

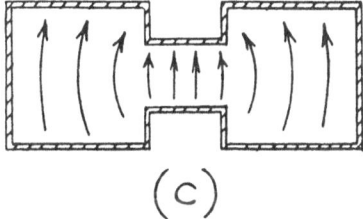

(c)

to the **E** field. The second requirement accounts for currents that travel along with the fields at the inner waveguide surfaces. The instantaneous current distribution for the TE_{10} mode is shown in Figure 6-2b as an example. Because waveguide walls are not perfect conductors, these currents cause ohmic losses, and the propagating waves suffer attenuation. High-quality waveguides are sometimes silver plated on the inside to minimize such losses. The current distribution along the waveguide must be considered in order to understand radiation from slots or holes, which may be cut in the waveguide walls for various purposes.

Rather than a detailed discussion of the various waveguide network elements, a short description of some other types of waveguides is presented here. Many good books describe waveguide elements in considerable detail.*

The hollow circular waveguide can support various *TE* and *TM* modes (modes for which either **E** or **H** is perpendicular to the axial direction of propagation). The lowest mode is TE_{11}, whose **E** field distribution is shown schematically in Figure 6-3a. Although a circular waveguide is the easiest to fabricate, the **E** field direction (polarization) is unstable because of the sym-

---

*See Annotated Bibliography.

metry and can change as a result of imperfections. Thus, if circular waveguides are used over long distances, the relation between output and input polarization is unpredictable. On the other hand, rectangular waveguides (as well as others of lower symmetry as shown in Fig. 6-3) can maintain their polarization over long distances despite bends and twists. Therefore, circular waveguides are employed only over short distances or where their use is necessary for mechanical reasons, such as in rotary joints.

The ridged and double-ridged waveguides, whose cross sections are shown in Figures 6-3b and 6-3c, respectively, cannot handle as much power as the other waveguides discussed, but they have the advantage of lower cutoff frequencies and wider useful bandwidths.

Propagating modes in the rectangular, metallic waveguide can be constructed by the addition of two plane waves that propagate at angles α and $-\alpha$, respectively, with respect to the waveguide axis. Wave interference provides nulls along those walls for which **E** must vanish to satisfy boundary conditions. Such waveguide modes exhibit a cutoff frequency below which propagating solutions do not exist. A waveguide that operates by a different physical principle is illustrated in Figure 6-4. Previously total internal reflection was described as a phenomenon that occurs at the interface between two dielectrics under certain conditions. If a plane wave is sent into a dielectric slab at an angle, α, such that

$$\alpha < \alpha_c = \cos^{-1}(1/\sqrt{\varepsilon}), \qquad \textbf{(6-14)}$$

in which α_c is the critical angle as in Eq. (5-16), it will be reflected in zigzag fashion with low loss as shown in Figure 6-4a. In contrast to the metallic guide, which is closed, the dielectric waveguide is open (i.e., the fields in the regions immediately beyond the boundaries [$x > a/2, x < -a/2$] are not zero, but instead decay exponentially, as shown in Fig. 6-4b). Furthermore, in the dielectric waveguide, there are modes that have no low frequency cutoff. These modes may exist in slabs that are much thinner than a wavelength. The phase velocity of these modes (in the $\mathbf{z}_0$ direction) is smaller than the free space velocity of light, but larger than the speed of light in the dielectric: $c_0/\sqrt{\varepsilon} < c_p < c_0$. This result follows from the restriction on α:

$$1/\sqrt{\varepsilon} < \cos \alpha < 1$$

and

$$c_p = \frac{\omega}{k \cos \alpha} = \frac{c_0}{\cos \alpha \, \sqrt{\varepsilon}}. \qquad \textbf{(6-15)}$$

As in the case of the metallic waveguides, the dielectric waveguide must have limited extension along the $\mathbf{y}_0$ direction. Most practical dielectric waveguides may have either a circular or rectangular cross section. In the optical frequency range, circular dielectric waveguides are called optical fibers, which permit the transmission of optical signals over long and arbitrarily curved paths. In the microwave regime, a propagating mode can be launched along a dielectric rod, as shown in Figure 6-5a. If the dielectric rod is tapered, as illustrated in Figure 6-5b, the angle of incidence at the air-dielectric interface increases, and the trapped mode gradually leaks or radiates its electromagnetic energy. This leakage can start as soon as the condition of Eq. (6-14) is no longer satisfied. (That condition specifies the range of angles for which waves undergo total internal reflection.)

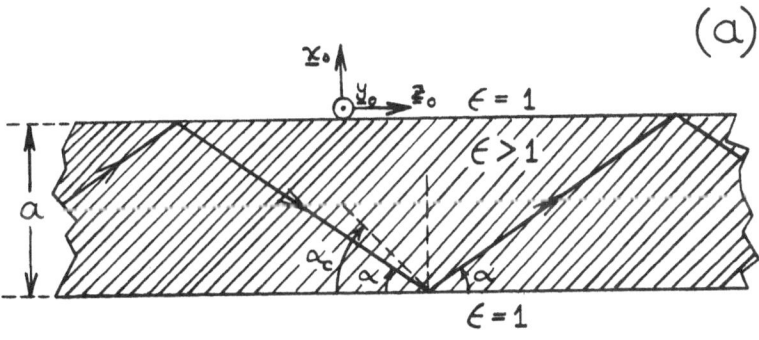

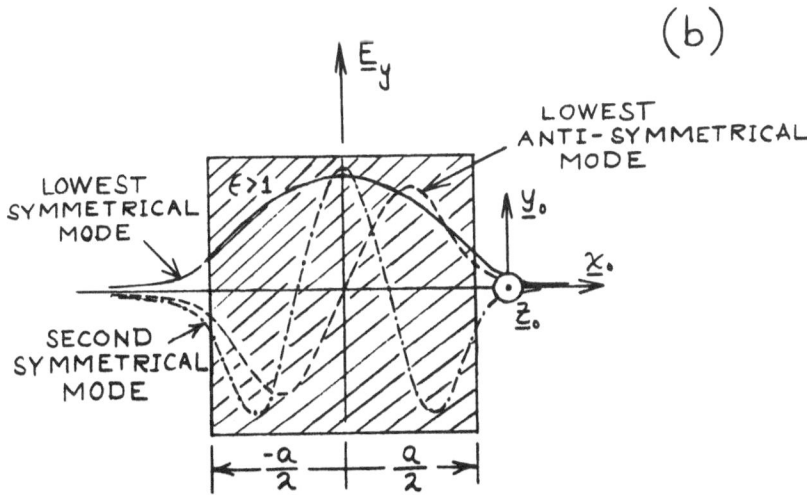

Figure 6-4. Plane wave in dielectric slab (a). Field distributions (b).

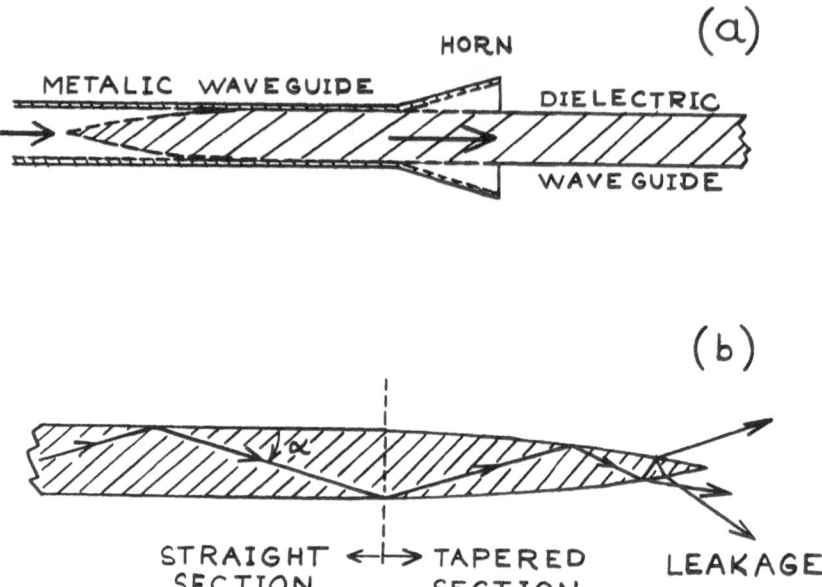

Figure 6-5. Modes in a dielectric waveguide: (a) launching and (b) leakage.

This last example illustrates the integration of a waveguide and antenna, in which the waveguide functions as both feeding line and radiating elements. Several classes of microwave antenna provide examples of the transmission line integration referred to above.

PROBLEMS

PROBLEM 6-1.   A hollow metallic rectangular waveguide has dimensions $a = 12$ mm and $b = 3$ mm.

  a. Calculate the band of operation for the TE_{10} mode.

  b. Calculate λ_g at a frequency of 15 GHz.

  c. Derive a general formula for the power transported by the waveguide in terms of a, b, E_0, and λ.

  d. If the maximum allowable field is 10^4 V/cm, calculate the power that the waveguide with the above dimensions can transport at 15 GHz.

PROBLEM 6-2.* Derive a general expression for the waveguide wavelength (λ_g) and cutoff frequency for the TE_{no} mode in hollow rectangular waveguides.

PROBLEM 6-3.* The phase velocity in a hollow metallic waveguide exceeds the speed of light and is given by

$$c_p = c_0/\cos \alpha = c_0 \lambda_g/\lambda.$$

Obtain a similar expression for the speed of energy flow or group velocity in the waveguide (Hint: review Fig. 6.1c).

PROBLEM 6-4. Under certain conditions, the atmosphere can behave like a giant dielectric waveguide. This phenomenon, called ducting, may occur when a cool dense atmospheric stratum lies between layers with lower refractive indices. If the index within the ducting stratum is 1.003, versus 1.0004 for the air outside the stratum, calculate the critical grazing angle at the layer boundaries for total internal reflection. (Just as in the case of dielectric waveguides, energy will leak out of the duct if this angle is exceeded.)

---

*Advanced problem.

CHAPTER 7

Horn, Slot, and Microstrip Patch Antennas

AC circuit and transmission line theory and techniques form the basis for understanding wire antennas, which represent a natural extension of those concepts. Many antennas, however, do not employ wire type radiating structures and consequently are not easy to explain in terms of filamentary electric currents. Such antennas belong to the large class of *aperture antennas,* which includes horns, slots, and microstrip patches. An example from acoustics sufficiently demonstrates the phenomenon of aperture radiation.

Sounds that originate in a room may be heard quite clearly through a window by an outside listener who does not lie along a line of sight from the source. The window aperture becomes a secondary source for sounds generated within the room by the (primary) source. The physical principle that accounts for aperture radiation was formulated in the seventeenth century by the Dutch physicist Christian Huyghens, who was one of the early proponents of the wave theory of light. Essentially, the Huyghens' principle states that every point on a wavefront may be regarded as a new point source, and a new wavefront can be computed by adding the contributions from all point sources on the original wavefront.

After light had been identified as electromagnetic (EM) radiation, Huyghens' principle was rigorously incorporated into electromagnetic theory. In the language of EM theory, the principle asserts that the field vectors at an arbitrary point of observation can be determined, if the tangential components of **E** and **H** at every point of the source, which is the aperture area, are given.

7.1 APERTURE RADIATION

Figure 7-1 shows an infinitesimal area, $ds = dx'\, dy'$, which is part of an aperture that lies in the x-y plane. The aperture is illuminated by an EM field whose tangential components $E_x(a)$ and $H_y(a)$ are given. The area ds represents an elementary source, sometimes referred to as a Huyghens' source.

The **E** field components at a point (r,θ,ϕ) in the far field, which arise from the tangential components $E_x(a)$ and $H_y(a)$ in the elementary aperture ds, are given (without proof) by

$$dE_\theta = -jk/4\pi[E_x(a) + \eta H_y(a)\cos\theta]\cos\phi\, u(r)\, ds$$

$$dE_\phi = jk/4\pi[E_x(a)\cos\theta + \eta H_y(a)]\sin\phi\, u(r)\, ds \qquad \textbf{(7-1)}$$

in which $\eta = 377\Omega$, and $u(r) = e^{-jkr}/r$.

If the aperture is illuminated by a plane wave propagating along the z axis, then $E_x(a) = \eta H_y(a)$. In general, however, the fields in the aper-

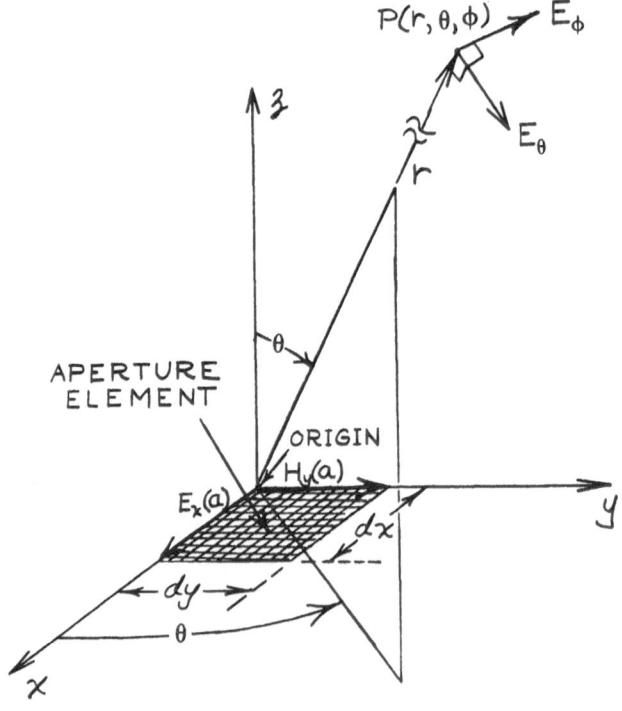

Figure 7-1. Coordinate system and electric field for an aperture element.

ture do not have to conform to this relation, as is the case, for example, when the aperture is illuminated by a plane wave that is incident at an oblique angle. Eqs. (7-1) imply that if the tangential components of the EM field in a plane aperture are known, the far field may be obtained by integration over the aperture in analogy with the approach used in Chapter 4 to calculate the fields due to a known current distribution.

The open ended waveguide, which is treated next, is both a practical device and a good example of this technique. For the aperture formed by an open rectangular waveguide, as shown in Figure 7-2, the aperture plane is defined by $z = 0$, and the waveguide a dimension lies along the y axis. If the reflection coefficient at the open end ($z = 0$) is denoted Γ, the TE_{10} field components in the aperture are

$$E_x(a) = E_0 (1 + \Gamma) \cos(\pi y'/a)$$

$$H_y(a) = -E_0\, \eta\frac{\lambda}{\lambda_g} (1 - \Gamma) \cos(\pi y'/a)$$

$$= -\eta\frac{\lambda}{\lambda_g} \left[\frac{1 - \Gamma}{1 + \Gamma}\right] E_x(a). \tag{7-4}$$

The above relations can be derived from the discussion leading to Eq. (6-3). The ratio of the free space and waveguide wavelengths is a geometrical factor, because $\lambda/\lambda_g = \cos \alpha$. It was shown in Chapter 6 that α is the angle between the waveguide axis and the free space wave vectors, which combine to form the waveguide mode. The relation between the fields and the reflection coefficient is analogous to that derived for voltage and current in Chapter 2. If the open end were a true open circuit, then Γ would be unity, and the magnetic field would vanish at the aperture. In fact, $\Gamma < 1$, but because the mismatch between the waveguide and free space is considerable, Γ is close to unity, and $H_y(a)$ can be neglected without introducing too serious an error. Therefore, Eqs. (7-4) can be replaced by the approximations

$$E_x(a) = E_0 \cos(\pi y'/a)$$

and

$$H_y(a) = 0, \tag{7-5}$$

in which E_0 is the E field amplitude at the open end.

The calculation required to obtain the far field produced by the aperture distribution of Eq. (7-5) is identical, in principle, to that employed for wire antennas. In the present case, however, integration is performed over area elements rather than over line (current) elements. The point of observation is in the far field ($a/r \ll 1$, $b/r \ll 1$), so that

$$R = \sqrt{(x - x')^2 + (y - y')^2 + z^2}$$
$$\cong r - x' \sin \theta \cos \phi - y' \sin \theta \sin \phi,$$

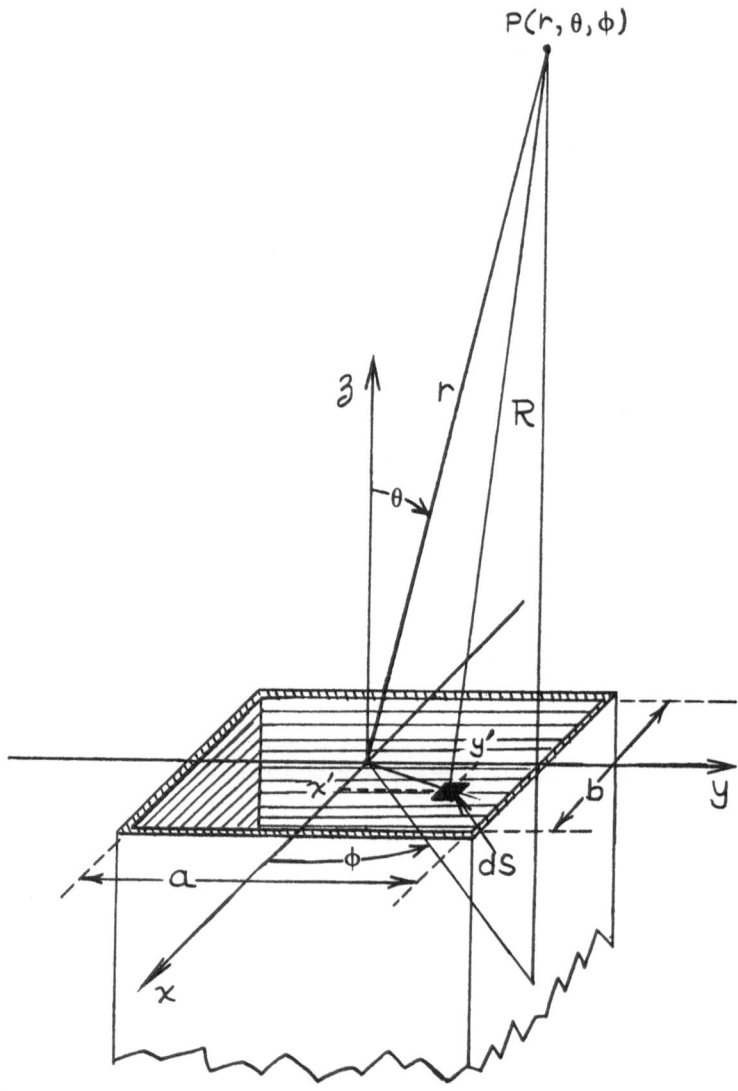

Figure 7-2.   Open-ended waveguide aperture.

in which x', y' are the position coordinates of the element of area $ds = dx'$ dy'. As discussed in Chapter 1, this approximation is tantamount to regarding **R** and **r** as parallel for points in the far field. By use of the above approximation,

$$\frac{e^{-jkR}}{R} \cong \frac{e^{-jkr}}{r} e^{jk(x' \sin \theta \cos \phi + y' \sin \theta \sin \phi)} = u(r)e^{\psi(x',y')}, \qquad \textbf{(7-6)}$$

in which

$$\psi(x', y') = jk(x' \sin \theta \cos \phi + y' \sin \theta \sin \phi).$$

From Eqs. (7-1), (7-5), and (7-6),

$$E_\theta \cong \frac{-jkE_0}{4\pi}u(r) \cos \phi \int_{-b/2}^{b/2} dx' \int_{-a/2}^{a/2} dy' \cos(\pi y'/a)e^{\psi(x',y')},$$

$$E_\phi \cong \frac{jkE_0}{4\pi}u(r) \cos \theta \sin \phi \int_{-b/2}^{b/2} dx' \int_{-a/2}^{a/2} dy' \cos(\pi y'/a)e^{\psi(x',y')}. \qquad \textbf{(7-7)}$$

The above integrals are not difficult to evaluate, and the following are the computations for the fields in the principal planes, that is, the x-z and y-z planes. The x-z plane is called the E plane, because the electric field in the aperture is parallel to this plane, which may also be specified by $\phi = 0$ in the polar coordinate system of Figure 7-2. Thus, the electric field in the E plane is, from Eqs. (7-7):

$$E_\phi = 0,$$

$$E_\theta = \frac{-jkE_0}{4\pi} u(r) \int_{-b/2}^{b/2} dx'e^{jkx' \sin \theta} \int_{-a/2}^{a/2} dy' \cos(\pi y'/a)$$

$$= \frac{jkE_0}{2\pi^2} u(r)ab \left[\frac{\sin(\pi b \sin\theta/\lambda)}{\pi b \sin\theta/\lambda} \right]. \qquad \textbf{(7-8)}$$

In the y-z plane, which is also called the H plane (specified by $\phi = \pi/2$),

$$E_\theta = 0,$$

$$E_\phi = \frac{jkE_0}{4\pi} u(r)\cos\theta \int_{-b/2}^{b/2} dx' \int_{-a/2}^{a/2} dy'\, \cos(\pi y'/a)e^{jky'\sin\theta}$$

$$= \frac{jkE_0}{2} u(r)\cos\theta\, ab \left[\frac{\cos(\pi a\sin\theta/\lambda)}{\pi^2 - 4(\pi a\sin\theta/\lambda)^2} \right]. \qquad (7\text{-}9)$$

Therefore, the *normalized* **E** field can be expressed:

$$\frac{|\mathbf{E}|}{|\mathbf{E}_{max}|} = \begin{cases} \left| \dfrac{\sin(\pi b\sin\theta/\lambda)}{\pi b\sin\theta/\lambda} \right| & \text{(in the E-plane)} \\[4mm] \left| \dfrac{\pi^2\cos\theta\cos(\pi a\sin\theta/\lambda)}{\pi^2 - 4(\pi a\sin\theta/\lambda)^2} \right| & \text{(in the H-plane).} \end{cases} \qquad (7\text{-}10)$$

From this discussion and Eqs. (7-7) to (7-10), it is seen that the field has the form of a spherical wave with the **E** vector parallel to the x-z plane. The **H** vector, which also can be calculated, is perpendicular to **E** (in the far field) that is, is parallel to the y-z or H plane. If $\lambda > a > \lambda/2$ and $b < \lambda/2$, as is typically true for rectangular waveguides, the radiation pattern will be broad in the E plane and narrower in the H plane, as illustrated in Figure 7-3.

The calculation of the radiation pattern in Eq. (7-10) is somewhat crude, because the aperture illumination plane was specified by the **E** vector of the TE_{10} mode, and all other effects were neglected. Despite this considerable approximation, the results represented by Eqs. (7-10) have been found to be in good agreement with measurements. On the other hand, the above calculations do not permit estimation of the reflection coefficient, Γ, at the open end of the waveguide or the corresponding radiation resistance. Determination of these quantities requires a more detailed analysis.

7.2 HORNS

An open waveguide end is rarely used as an antenna, because it has low gain and, due to the poor impedance match with free space, causes a high-standing wave ratio in the waveguide. If the aperture dimensions (a, b, or both) were larger, Eq. (7-10) would predict that the radiation pattern would become more peaked in the forward (z_0) direction (i.e., higher gain could be

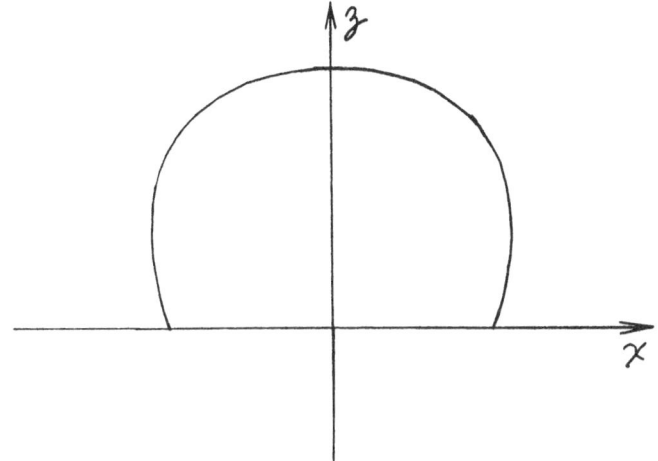

E-PLANE POWER PATTERN (φ = 0)

(a)

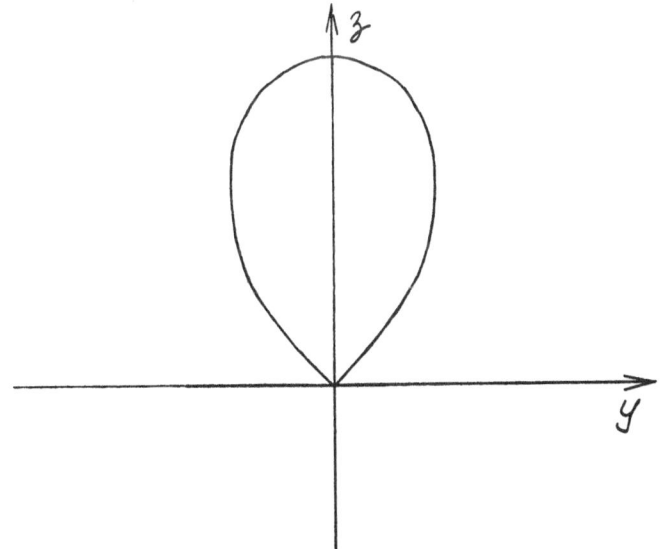

H-PLANE POWER PATTERN (φ = π/2)

(b)

Figure 7-3.   Power patterns for open-ended waveguide.

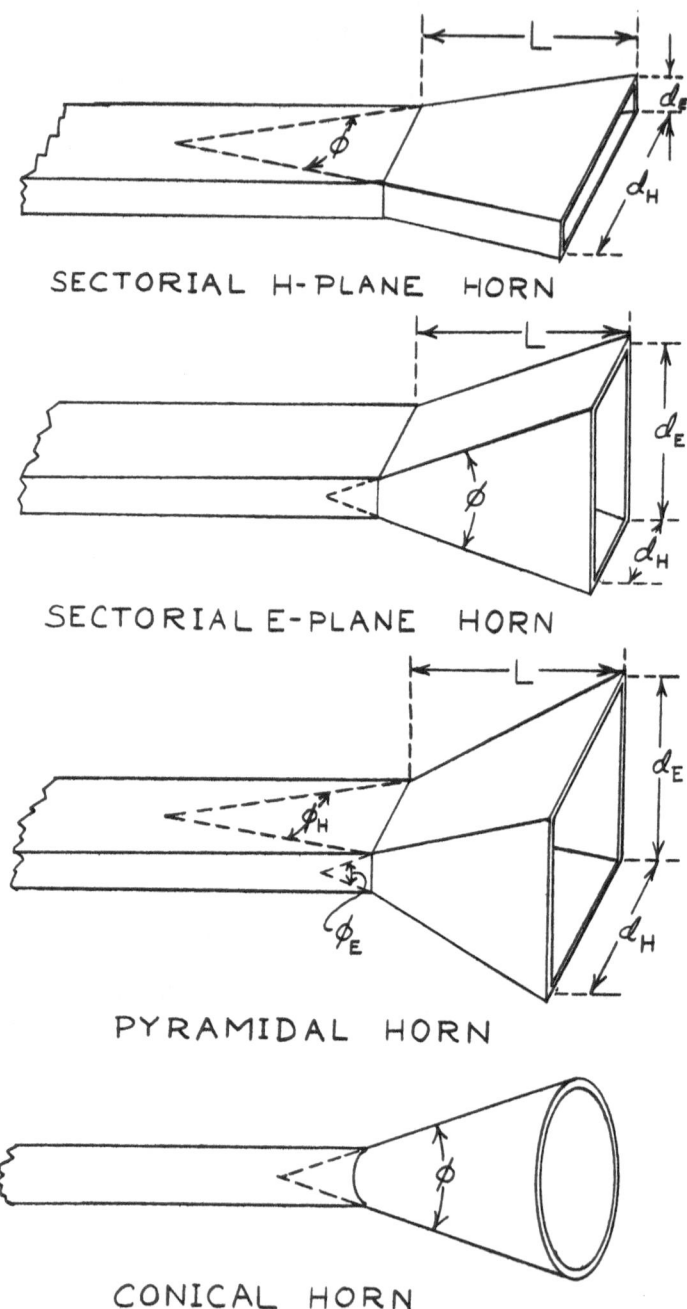

Figure 7-4.   Some waveguide horn types.

realized). Because the waveguide cross section must be limited to provide suppression of higher modes, not much gain increase may be achieved. However, if the end of the waveguide is flared out at the open end to form a horn, a very common and useful antenna results. The horn shape affords a gradual transition to free space for the TE_{10} mode, thereby reducing mismatch (and therefore reflection) at the open end.

Horns may also be attached to other types of waveguides (e.g., circular), and their shapes may be selected to produce a variety of radiation patterns. A few examples of the many types of horns in use are shown in Figure 7-4.

Among the most important antenna properties are gain and beam width, as discussed in connection with wire antennas in Chapter 4: high gain corresponds to narrow beam width. In general, the computation of gain and beam width requires a detailed knowledge of the field. A rough approximation for the gain of a horn can be obtained from

$$g \approx 7.5 \frac{d_E d_H}{\lambda^2} = 7.5 \frac{A}{\lambda^2}, \tag{7-11}$$

in which A is the aperture area. For all large aperture antennas, the directivity is related to the normalized aperture area by

$$g = c A/\lambda^2 \tag{7-12}$$

in which c is a constant. For a large aperture of a given area, the highest directivity is achieved when the aperture is uniformly illuminated by a plane wave, in which case $c = 4\,\pi$, and

$$g = g_{max} = \frac{4\pi A}{\lambda^2} \tag{7-13}$$

This result lends itself to the concept of *aperture efficiency*, which is defined as the ratio of real antenna gain to the maximum theoretical directivity that could be achieved by an ideal antenna of the same aperture. For the horn antennas, whose gain is represented by Eq. (7-11), the aperture efficiency, e_a, is approximately, $e_a = g/g_{max} \approx 7.5/4\pi = 0.6$.

The gain of an aperture of a given area increases with the square of the frequency (6 dB per octave), as long as other factors remain unchanged. The aperture size of a horn depends upon its length, L, and flare angle, ϕ. A small flare angle produces a good aperture distribution, but for a given length, the aperture area remains relatively small. On the other hand, a large flare angle causes the wavefront at the aperture to curve, introducing phase errors, which reduce the gain and broaden the beam. The antenna designer must determine the best L and ϕ for a particular application. Appropriate design formulas and graphs are found in antenna engineering handbooks (see annotated bibliography).

Horn antennas have many applications because of their ruggedness and simplicity. Although a horn antenna used alone does not provide high gain, it is often used as a primary antenna to illuminate high-gain lens or reflector antennas. The radiation pattern described by Eq. (7-10) has fairly high side-lobes, especially in the E plane. Reduction of sidelobe levels is very desirable in some applications, and various means have been devised to achieve this goal. The corrugated horn has become popular because of its low sidelobe pattern, which is produced by a conical horn with a grooved inner wall. (This subject is again discussed with reflector antennas in Chapter 10.)

7.3 WAVEGUIDE SLOTS

Another common type of aperture antenna is a slot that is cut into a waveguide wall. In general, cutting an opening in a conducting wall with an EM field on one side will allow EM energy to radiate to the other side. If the opening is large compared to a wavelength, it is assumed that the field components in the aperture are the same as those of the incident field. Occasionally, this assumption is justified even if the aperture is relatively small, as in the case of the open waveguide end discussed above. More often than not, however, edge effects will dominate the fields in small apertures.

To acquire a qualitative understanding for the fields in a slot requires picturing the fields in the waveguide and the currents induced in its walls. Figures 6-2a and 6-2b show the fields and currents for the TE_{10} mode in a rectangular waveguide. Note that along the narrow wall the induced current flows straight up (or down), because the adjacent H field is parallel to the z axis, and has no y variation. At the broad walls, the current and field configurations are not as simple, but along the centerline the current flows along the z (or $-z$) direction, because the adjacent H field is normal to the z axis. Thus, if a very narrow slot were cut straight along the broad wall centerline or up along the narrow wall, the induced current pattern would remain nearly unperturbed, and therefore, the slot would not radiate. Slots having any other orientation, as shown in Figure 7-5b, disturb the wall current pattern, which causes radiation to occur.

Slots that ordinarily do not radiate, such as those of Figure 7-5a, will radiate if the symmetry of the TE_{10} field is deliberately upset. A screw penetrating the waveguide, or any other asymmetrical obstacle in the vicinity of the slot, can provide the necessary perturbation. Of the radiating slots shown in Figure

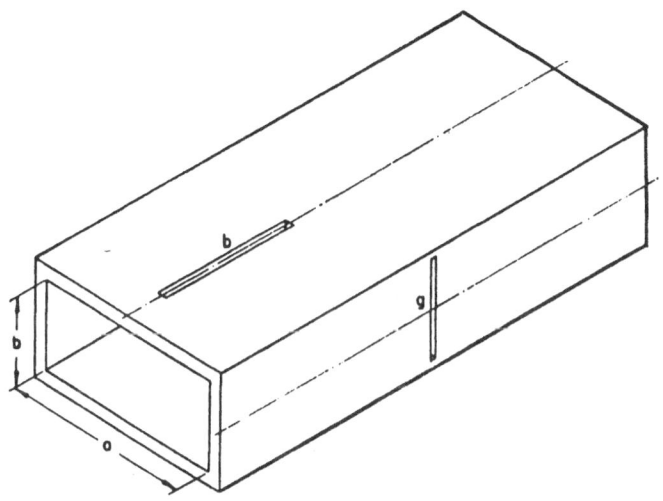

(a) NONRADIATING SLOTS CUT INTO THE WALLS OF A
RECTANGULAR WAVEGUIDE

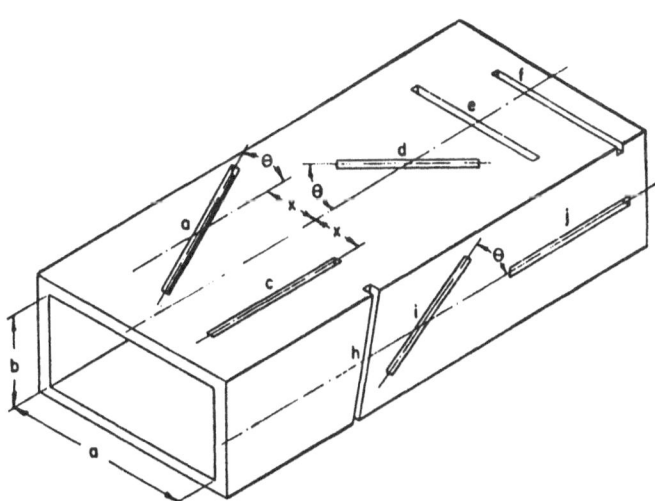

(b) RADIATING SLOTS CUT INTO THE WALLS OF A
RECTANGULAR WAVEGUIDE

Figure 7-5. Waveguide slot configurations. (From R. C. Johnson and H. Jasik, eds.,
Antenna Engineering Handbook, 2nd ed., New York: McGraw-Hill, 1984, pp. 9-3,
9-4; copyright © 1984 by McGraw-Hill Book Company, reprinted by permission.)

7-5b, the most frequently used are represented by c and h, the axial offset broad-wall slot and the angled narrow-wall slot, respectively. The field within the slot is predominantly electric and is directed across the narrow dimension of the slot. It is not obvious how to guess the E field distribution along the slot. In general, for short resonant slots, which are about half a wavelength long, it can be assumed that the field has a distribution of half a sinusoidal cycle. For the case illustrated in Figure 7-6a, the aperture field may be expressed,

$$\mathbf{E}_a = \mathbf{x}_0 \, E_0(x_0) \cos(\pi z/l), \qquad (7\text{-}14)$$

in which E_0 is the amplitude, which is a function of the slot-centerline offset distance. Because a slot positioned on the centerline does not radiate, $E_0(0) = 0$. The field on the outside of the waveguide can be calculated using a modified version of Eq. (7-1). For the inside of the waveguide the slot can be modeled

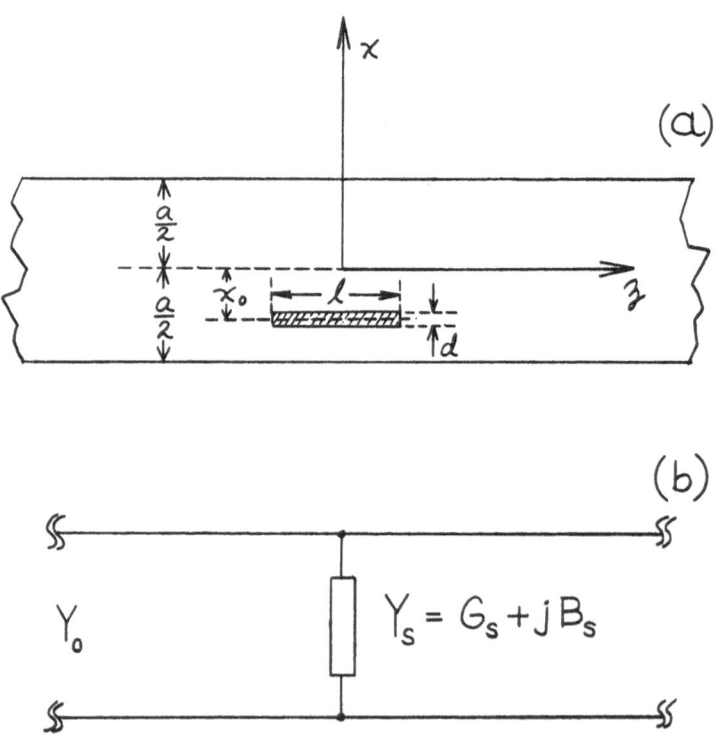

Figure 7-6.   Axial offset broadwall slot and equivalent circuit.

as a lumped shunt admittance across a transmission line, as shown in Figure 7-6b. The real part of this admittance represents the radiated energy. The imaginary part can be positive or negative and vanishes at resonance, which occurs when the slot length is approximately (but not exactly) equal to $\lambda/2$. Values for G and B, which are functions of the various parameters (λ, a, b, d, and l in this case) are provided in antenna design handbooks.* Once these are known, the interaction of the slots with the TE_{10} mode can be calculated by using transmission line formulas. From the outside, a resonant slot resembles the half-wave dipole discussed in Chapter 4, however, the polarization is parallel to the **E** field in the slot, not the slot axis. Antenna engineers sometimes say that the slot behaves like a half-wave dipole of *magnetic current*.

Waveguide slot radiators, which are rarely used alone, are usually used as elements in waveguide slot arrays, which form high-gain antennas. (This is further discussed in Chapter 9.)

Slot radiators are also used in other configurations. For example, holes or slots may be cut at appropriate positions along the conducting shield of coaxial line or stripline or the walls of a resonant cavity. Furthermore, slots and other types of holes are used to couple energy from one waveguide to another. Detailed descriptions of these applications can be found in books on microwave technology.*

7.4 MICROSTRIP PATCHES

A class of radiators that are related to slots, and which have become very popular in recent years, are microstrip patches. A section of microstripline is shown in Figure 2-4c, and a cross section of its field distribution is illustrated in the edge-on view of Figure 7-7. For a straight, infinitely long strip (and its image in the ground plane), virtually no radiation will occur, as long as the separation, h, is small compared to a wavelength. In the presence of discontinuities, however, the field in the gap between the strip and the ground plane becomes unbalanced, and the gap will radiate. Any patch of microstrip, such as shown in Figure 7-8a, has a radiating aperture around its rim. If fields and currents are excited, for example, by a stripline or feed-through coaxial line connection, as shown in Figure 7-8, the patch will radiate. (For a given patch, only one excitation method or the other is used, not both simultaneously.)

The shape of the patch and the method and location of its feed determine the field distribution around the rim and, therefore, its radiation characteristics. A general analysis of the fields and currents around a microstrip patch is not simple and requires sophisticated mathematical methods. Often it is much

---

*See Annotated Bibliography.

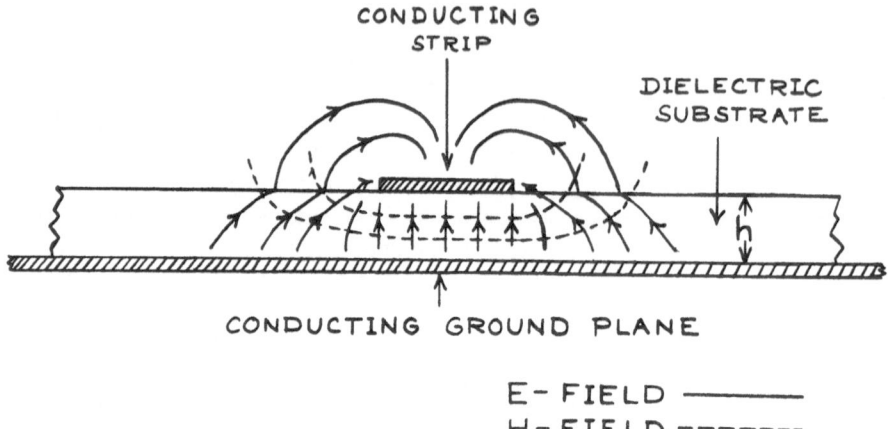

Figure 7-7. The transverse field components in a microstrip.

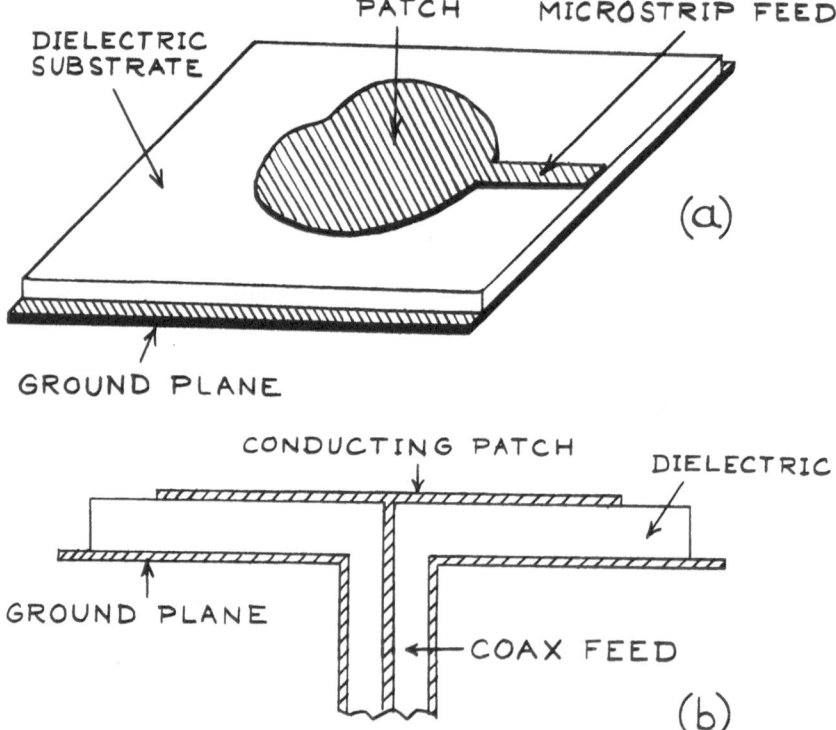

Figure 7-8. Exciting methods for microstrip patch.

easier for the antenna designer to fabricate various configurations and ascertain their properties experimentally. Indeed, analytical work often follows empirical research.

Because the dimensions of an individual patch are usually comparable to a wavelength, the radiation pattern of the patch is rather wide. The most common patch radiators are either square or circular and produce a single lobe of radiation in the direction normal to their surfaces. Many other shapes have been reported, and both linear and circular polarizations are common. Most patch radiators have a narrow useful frequency range (bandwidth) of about 1–3%. Patches are most frequently used in light-weight planar or conformal arrays; however, they do see some application as individual antennas as well.

Microstrip antennas and arrays may assume a variety of shapes, such as circular rings, rectangular frames, meandering lines, fishbones, and comb-like patterns, which have been reported in the literature. The number of possible shapes is infinite and new configurations are constantly under investigation. A few are shown in Chapter 9.

The attractive features of a microstrip antenna include ease of fabrication, light weight, and flat profile. The latter two properties make microstrip a popular choice for aircraft, spacecraft, and missile antennas. The main drawbacks of microstrip antennas include narrow bandwidth, low efficiency, and low power handling capability.

PROBLEMS

PROBLEM 7-1. Consider a horn antenna that has a far field pattern given by Eq. (7-10). The main lobe nulls occur at $\pm 60°$ in both the E and H planes.

    a. Determine the dimensions of the horn.

    b. Calculate the beam width between the -10-dB points.

PROBLEM 7-2. Figure 7-6a shows a narrow slot that is cut into a conducting plane $(d \ll \lambda, \lambda/2 < l < \lambda)$. The aperture field is given by

$$E_x = E_0 \cos\left(\frac{\pi z}{l}\right), \qquad H_z = 0.$$

    a. Calculate the far field.

    b. Compare the result of part (a) with the far field of a small current element and a small current ring source.

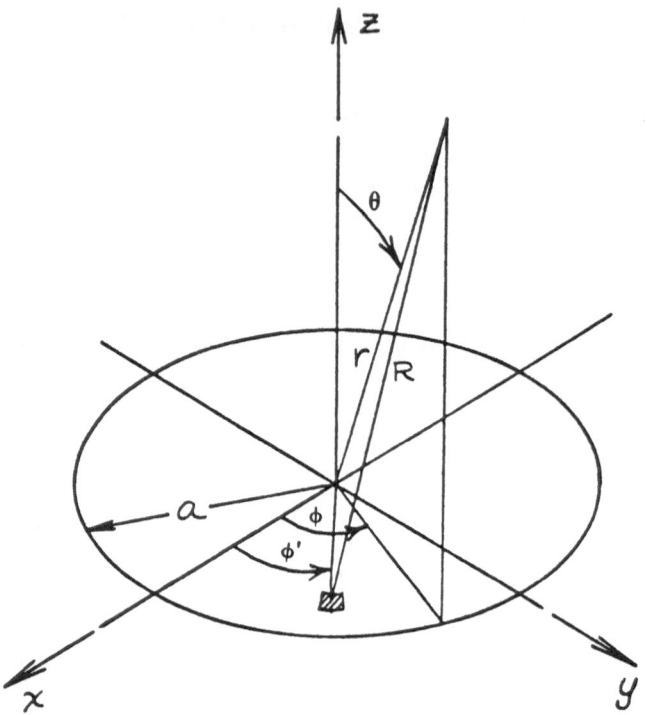

Figure 7-9. Illustration of a round aperture situated in a system of coordinates for Problem 7-3.

PROBLEM 7-3.* Consider a round aperture situated in a system of coordinates as shown in Figure 7-9.

a. If the aperture field depends only on the radial distance ρ

$$E_x (a) = \eta \, H_y (a) = E (\rho),$$

show that the far field components are proportional to $(1 + \cos \theta) u(r) \int_0^a E(\rho) J_0(k\rho \sin \theta)\rho d\rho$ in which $u(r) = e^{jkr}/r$, and

$$J_0(z) = \frac{1}{\pi} \int_0^{2\pi} e^{jz \cos a} \, da$$

is the zero-order Bessel function of the variable.

---

*Advanced problem.

b. Determine the exact expressions for E_ϕ and E_θ in the E plane $(x\text{-}z)$ and the H plane $(y\text{-}z)$.

c. Use the identity

$$\int J_0(z)\, dz = z\, J_1(z)$$

to obtain a closed formula (in terms of J_1) for E_ϕ and E_θ in the case of uniform aperture illumination $(E(\rho) = \text{constant})$.

d. Determine the level of the first sidelobe (in decibels relative to the main beam) for the case of part (c).

CHAPTER **8**

Antenna Arrays

Many applications require antennas that produce narrow concentrated beams and little radiation in other (nonbeam) directions. In Figure 8-1, a simple keyhole radiation pattern for such applications is shown as an approximation of a more realistic sidelobe pattern. The keyhole beam is assumed to be a narrow cone with a rectangular cross section, with sides of angular width, θ_1, and θ_2 (radians). Directional gain (directivity) is then approximately $g_d = 4\pi/\theta_1\theta_2$. As seen in Figure 8-1, this is simply the ratio of the area of a unit sphere to the beam area, as projected on the unit sphere. For more realistic antenna patterns, such as the low side-lobe pattern shown in Figure 8-1, beam width is usually defined as the angular separation of points of the pattern that are 3 dB down from the peak (3 dB beam width). For beams that occupy only a few square degrees, directivities on the order of 10,000 (40 dB) can be realized. Very large directional gains can be achieved by a single antenna (e.g., large dish) or by the use of antenna arrays, which operate on the interference principle to obtain high gain.

Chapter 1 (section 4) described how the radiation from two individual sources of spherical waves, operating at the same frequency, creates an interference pattern with maxima and minima in directions of constructive and destructive interference. To produce high gain while minimizing stray radiation, the interference that results from an array of many individual radiators is exploited. An antenna array is a collection of two or more identical

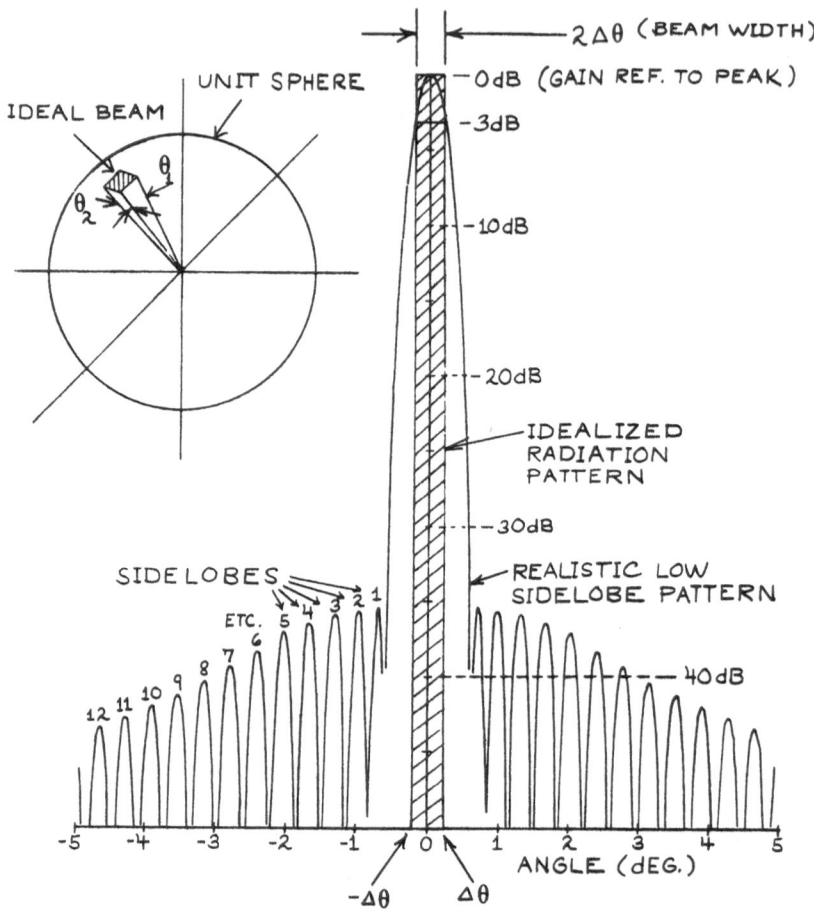

FIGURE 8-1.   Ideal versus real radiation pattern.

antennas that are arranged and excited to produce high-gain beams in certain directions.

8.1 GENERAL ARRAYS

Consider a number of identical antennas, $A_1, A_2 \ldots A_i, \ldots A_n$ situated parallel to each other within a volume of radius a, which is much smaller than the distance, r, from the origin, 0, to the point of observation,

$$a/r \ll 1, \tag{8-1}$$

as illustrated in Figure 8-2. A reference antenna at the origin (or phase center) radiates an electromagnetic field with far field components that are proportional to

$$F_0 = I_0 \frac{e^{-jkr}}{r} f(\theta,\phi), \tag{8-2}$$

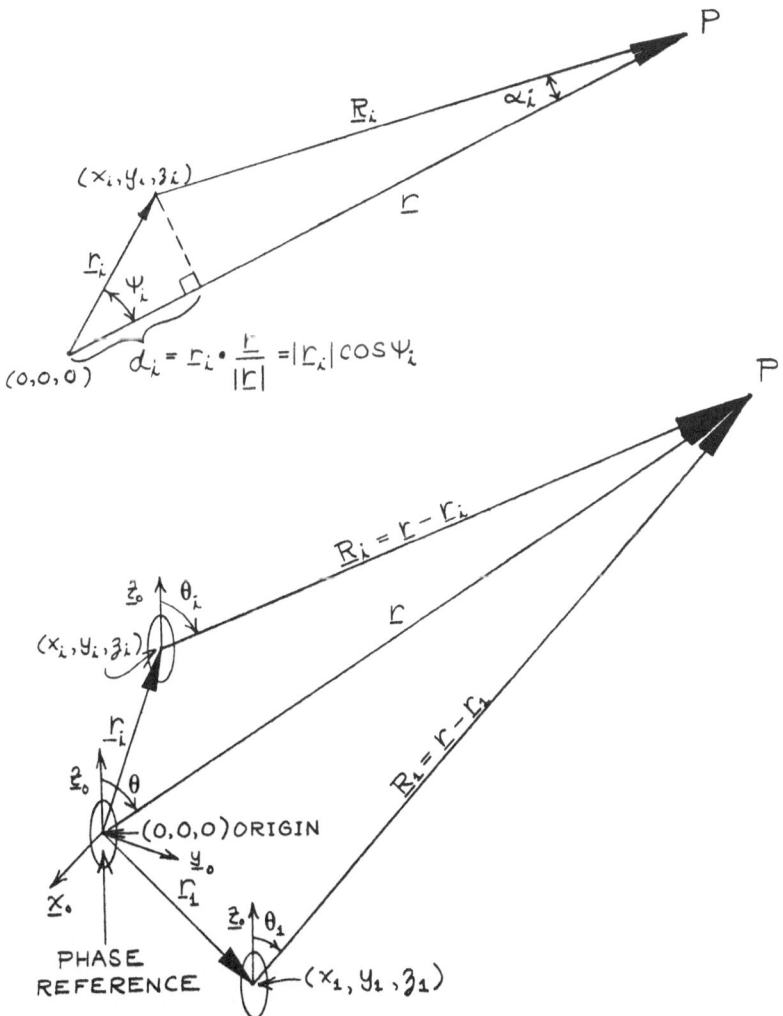

FIGURE 8-2.   Antenna array geometry.

in which I_0 is a complex amplitude, and $f(\theta,\phi)$ is the radiation pattern. Consider a typical antenna, A_i, whose position vector with respect to the origin is $\mathbf{r}_i$:

$$\mathbf{r}_i = \mathbf{x}_0\, x_i + \mathbf{y}_0\, y_i + \mathbf{z}_0\, z_i.$$

The far field components that arise from this antenna are proportional to

$$F_i = I_i \frac{e^{-jkR_i}}{R_i} f(\theta_i,\phi_i), \tag{8-3}$$

in which

$$R_i = |\mathbf{r} - \mathbf{r}_i| = \sqrt{(x - x_i)^2 + (y - y_i)^2 + (z - z_i)^2}$$
$$= r\sqrt{1 + (x_i^2 + y_i^2 + z_i^2)/r^2 - 2\,(xx_i + yy_i + zz_i)/r^2}.$$

In the spherical coordinate system of Figure 8-2, $x = r\sin\theta\cos\phi$, $y = r\sin\theta\sin\phi$, $z = r\cos\theta$, and

$$(x_i^2 + y_i^2 + z_i^2)/r^2 = |\mathbf{r}_i|^2/r^2 \le a^2/r^2 \ll 1.$$

By means of a derivation entirely analogous to that represented by Eqs. (1-33) and (1-34), it can be shown that, to a very good approximation,

$$e^{-jkR_i} = e^{-jkr}\, e^{jk\,(x_i\sin\theta\,\cos\phi\,+\,y_i\sin\theta\,\sin\phi\,+\,z_i\cos\theta)}. \tag{8-4}$$

The definition of a unit vector, $\mathbf{r}_0$, along the $\mathbf{r}$ direction is

$$\mathbf{r}_0 = \frac{\mathbf{r}}{|\mathbf{r}|} = \mathbf{x}_0\sin\theta\cos\phi + \mathbf{y}_0\sin\theta\sin\phi + \mathbf{z}_0\cos\theta.$$

The exponent of the second term in Eq. (8-4) may be expressed as

$$k(x_i\sin\theta\cos\phi + y_i\sin\theta\sin\phi + z_i\cos\theta)$$
$$= k(\mathbf{r}_i \cdot \mathbf{r}_0) = 2\pi d_i/\lambda = \psi_i(\theta,\phi), \tag{8-5}$$

in which d_i is the projection of $\mathbf{r}_i$ on $\mathbf{r}$. If it is assumed that P lies at a great distance (i.e., that α_i is very small), $\mathbf{r}$ and $\mathbf{R}_i$ are very nearly parallel. Under these circumstances, Eq. (8-5) simply expresses the phase difference between the origin and the ith antenna as measured along the path of a plane wave that propagates in the $\mathbf{r}_0$ direction. This phase difference clearly depends on the location of the observation point, P, and therefore is a function of θ and ϕ.

The arguments of Chapter 1 (section 4) showed that, to a very good approximation, $1/R_i \cong 1/r$ and $\theta_i \cong \theta$, $\phi_i \cong \phi$, for all antennas situated in accordance with the relationship (8-1). By combining all of these approximations we may write, in lieu of Eq. (8-3),

$$F_i = \frac{e^{-jkr}}{r} f(\theta,\phi) \, I_i e^{j\psi_i}. \tag{8-6}$$

By virtue of the superposition principle, the total field due to all of the antennas (or elements) is, therefore,

$$S(\theta,\phi) = \sum_{i=1}^{n} F_i = I_0 \frac{e^{-jkr}}{r} f(\theta,\phi) \sum_{i=1}^{n} \frac{I_i}{I_0} e^{j\psi_i(\theta,\phi)} \tag{8-7}$$

This important formula, which is a generalization of Eq. (1-39), indicates that the far field that arises from an array of n identical parallel antennas (or elements) appears as that of the reference antenna at the origin as given by Eq. (8-2), multiplied by a sum called the *array factor*. The reference element simply represents the phase reference point of the array and does not necessarily coincide with a physical element of the array. The array factor depends only on the normalized complex amplitudes I_i/I_0, and the phases, ψ_i, of the elements with respect to the phase reference. Linear and planar arrays, which are special cases of general array theory, are of great practical importance and will be studied in the following two sections.

8.2 LINEAR ARRAYS

A linear array, consists of n equally spaced elements along a straight line as shown in Figure 8-3. The reference antenna, A_1, is located at the origin of coordinates, and the remaining $n - 1$ elements lie along the z axis with the equal interelement separation, d. The phase at the ith element, for a far field observation point is, in accordance with Eq. (8-5),

$$\psi_i = k(\mathbf{r}_i \cdot \mathbf{r}_0) = (i - 1)kd \cos\theta. \tag{8-8}$$

For a special case in which the complex amplitudes associated with the elements are equal in magnitude and the relative phase between adjacent elements assumes a constant value, the phase slope, or gradient, may be expressed

$$I_i = I_0 e^{-j\beta_i}; \qquad \beta_i = (i - 1)\beta. \tag{8-9}$$

Substitution of Eqs. (8-8) and (8-9) into Eq. (8-7) yields, for the array factor,

$$S(\theta) = \sum_{i=1}^{n} \frac{I_i}{I_0} e^{j\psi_i} = \sum_{i=1}^{n} e^{j(i-1)(kd\cos\theta - \beta)}$$

$$= 1 + e^{j\xi} + e^{2j\xi} + \ldots e^{j(n-1)\xi} \tag{8-10}$$

in which

$$\xi = kd \cos\theta - \beta. \tag{8-11}$$

The right-hand side of Eq. (8-10) is a geometric series, whose sum may be expressed

$$1 + x + x^2 + x^3 + \ldots + x^{n-1} = \frac{1 - x^n}{1 - x}; \qquad x = e^{j\xi}.$$

Therefore,

$$S(\theta) = \frac{1 - e^{jn\xi}}{1 - e^{j\xi}} = \frac{1 - \cos n\xi - j\sin n\xi}{1 - \cos\xi - j\sin\xi}. \tag{8-12}$$

Usually only the power pattern is of interest, which is given by the squared modulus of $S(\theta)$:

$$|S(\theta)|^2 = S(\theta)S^*(\theta) = \frac{(1 - \cos n\xi)^2 + \sin^2 n\xi}{(1 - \cos\xi)^2 + \sin^2\xi}$$

$$= \frac{1 - \cos n\xi}{1 - \cos\xi} = \left[\frac{\sin(n\xi/2)}{\sin(\xi/2)} \right]^2. \tag{8-13}$$

For the last step of Eq. (8-13), the trigonometric identity $1 - \cos\gamma = 2\sin^2 (\gamma/2)$ was used. Although Eqs. (8-8) through (8-13) are valid only for this particular array, several general features of the array factor can be illustrated by further investigation of these results.

Note that the power pattern of the array factor is a periodic function of ξ, with period 2π (i.e., it has equal maxima at $\xi = 0, 2\pi, 4\pi \ldots$) These can be evaluated by L'Hospital's rule:

$$\lim_{\xi \to 0} \frac{\sin^2(n\xi/2)}{\sin^2(\xi/2)} = n^2.$$

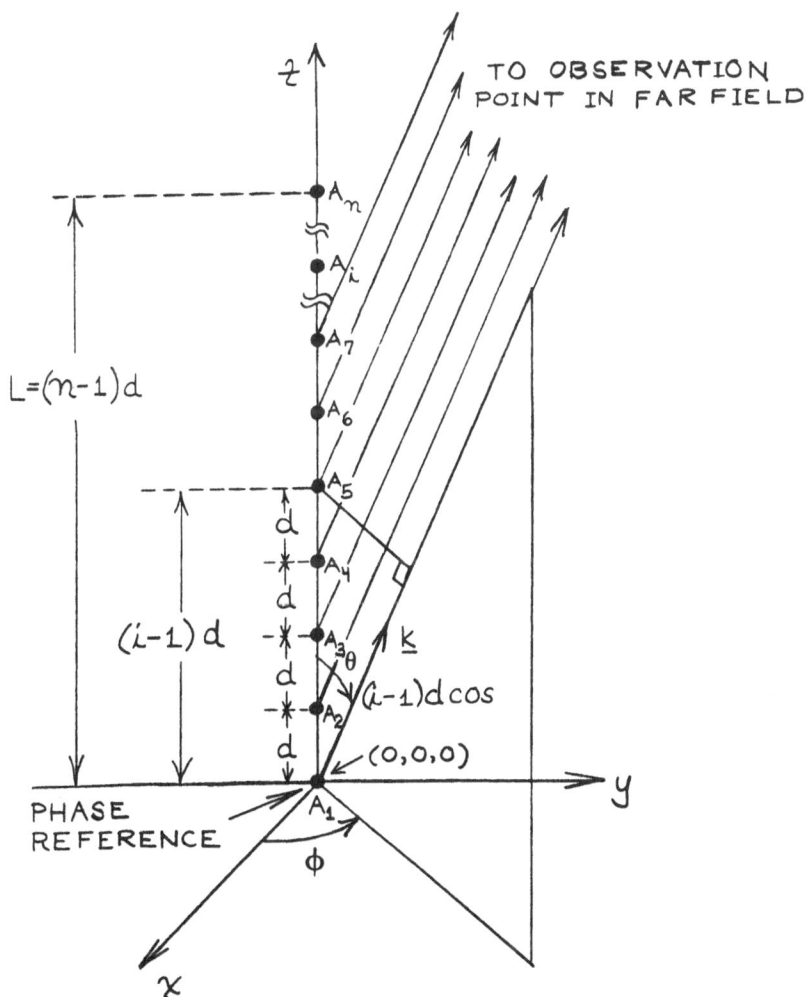

FIGURE 8-3. Linear array geometry.

These maxima correspond to directions for which the contributions from all of the antennas add up in phase (modulo 2π radians or 360°). The total field is therefore proportional to n, and the power, which goes as the square of the field, is proportional to n^2. The maximum that corresponds to $\xi = 0$ is designated as the peak of the main beam, and from Eq. (8-11), this peak occurs at an angle

$$\theta_{peak} = \theta_0 = \cos^{-1}[\beta/kd]. \qquad (8\text{-}14)$$

Thus, the main beam direction depends on the phase gradient across the array. In other words, if β (the phase difference between adjacent elements) is changed, the beam will move to a new position in accordance with Eq. (8-14). In particular, if $\beta = 0$ (all elements in phase) then $\theta_0 = \pi/2$, and the array is said to be broadside (i.e., its peak lies in the x-y [equatorial] plane). On the other hand, if $\beta = kd = 2\pi d/\lambda$, then $\theta_0 = 0$, and the array is said to be operating in endfire (i.e., the main beam points along the array or the z axis). The array factor of the linear array has cylindrical symmetry about the z axis, that is, it is independent of the azimuth angle ϕ.

Other pattern maxima occur when $\xi = 2m\pi$, where m is an integer. These maxima are equal in height to the main beam and are called grating lobes. The angles at which these maxima occur are, from Eq. (8-11), $kd\cos\theta_m - \beta = 2m\pi$, and by use of Eq. (8-14),

$$\cos\theta_m = \cos\theta_0 + m\lambda/d. \qquad (8\text{-}15)$$

If the ratio λ/d is sufficiently large, $\cos\theta_m > 1$ for all nonzero values of m, and the additional maxima do not occur at real angles (i.e., θ values between 0 and π radians). This condition is guaranteed when

$$\boxed{d < \frac{\lambda}{1 + |\cos\theta_0|}} \qquad (8\text{-}16)$$

For a broadside array, $\theta_0 = \pi/2$, and condition (8-16) requires that $d < \lambda$ to suppress the unwanted grating lobes. For an endfire array, $\theta_0 = 0$, and condition (8-16) stipulates that $d < \lambda/2$. For any other main beam position, the limit lies between λ and $\lambda/2$. If an equality sign is used in condition (8-16), the grating lobe lies at endfire (i.e., $\theta_m = 0$ or $\theta_m = \pi$). In an actual array design, the element spacing would usually be made somewhat smaller, to ensure that no part of the grating lobe beam appears at a real angle from the array axis. The grating lobe beam is thus excluded from real space.

The above discussion shows that the elements of an array must be quite close to completely suppress grating lobes. As a consequence of this proximity, the elements are electromagnetically coupled to each other. Strictly speaking, the elements cannot be seen as entirely independent of each other. Mutual coupling between elements is always present to some extent and is usually difficult to quantify. The pattern decomposition into array and element factors represented by Eq. (8-7) is the classical result, which is strictly valid only for uncoupled elements. The array environment, through mutual coupling, causes modification of the isolated element pattern, $f(\theta, \phi)$.

A straight wire antenna is an example of a continuous linear array, in which the elements are small current dipoles, of length Δz, which compose the wire. From Eqs. (4-15) and (4-17), the Fourier integral is analogous to the array factor discussed above. Indeed, the array factor is a discrete Fourier transform. The analogy to analog and digital filter theory is quite obvious. For the idealized case of a uniform current distribution,

$$|S(u)|^2 = c \left[\frac{L \sin u}{u} \right]^2,$$

where $u = (kL/2)\cos\theta$, and c is a constant. This expression can be obtained from the formula for the discrete array factor, Eq. (8-13), in the limit as $d \to 0$, $n \to \infty$ and $nd \to L$. For the sake of mathematical analysis, a discrete array may sometimes be represented by a continuous array, because Fourier integrals are often easier to evaluate than discrete sums of the type

$$S = \sum_{l=1}^{n} (I_l/I_0) e^{j\psi_l},$$

which usually have no closed form.

The angular width of the main beam, as distinguished by the nulls that flank it, can be obtained from Eq. (8-13). At broadside ($\theta_0 = \pi/2$), the nulls are symmetrically located about the main beam and are given by,

$$\frac{n\xi}{2} = m\pi; \quad m = \pm 1, \pm 2, \ldots$$

from which, by use of Eq. (8-11), with $\beta = 0$ (broadside),

$$\theta_m = \cos^{-1}\left(\frac{m\lambda}{nd}\right) = \cos^{-1}\left(\frac{m\lambda}{L}\right). \tag{8-17}$$

The angular separation of the nulls that lie closest to the main beam ($m = \pm 1$) is, therefore,

$$\Delta\theta \cong 2\,\frac{\lambda}{L}.$$

The approximation is good as long as $\theta_1 \cong \pi/2$, which is true when $L \gg \lambda$. If the main beam width is defined at the half power (3 dB down) points, then

$$\boxed{\text{3-dB broadside beam width} = 0.886 \;\lambda/L \cong \lambda/L \text{ (radians)}} \qquad \textbf{(8-18)}$$

As the main beam is scanned toward endfire, it broadens considerably, and at endfire ($\theta_0 = 0$)

$$\Delta\theta = \sqrt{2\lambda/L}. \qquad \textbf{(8-19)}$$

If the elements of the linear array described above are isotropic radiators, the directive gain, according to Eqs. (4-8) and (8-13), is given by

$$g_{max} = \frac{4\pi|S(\theta)|^2_{max}}{2\pi \int_0^{\pi} |S(\theta)|^2 \sin\theta\, d\theta}. \qquad \textbf{(8-20)}$$

Although the integral cannot be evaluated in closed form, for $L \gg \lambda$ the approximate result is

$$\boxed{g_{max} \approx 2L/\lambda} \qquad \textbf{(8-21)}$$

which is independent of θ_0. When considering an idealized linear array, it must be remembered that the beam is cylindrically symmetrical about the array axis. In qualitative terms, the angular volume (gain) of the beam remains constant despite the noted broadening in the θ dimension, because the beam occupies a smaller azimuthal cone as it is steered off broadside (recall that $dS = r^2 \sin\theta\, d\phi\, d\theta$). It is emphasized, however, that for arrays of real elements (nonisotropic), directivity is in most cases scan angle dependent.

Eqs. (8-18), (8-19), and (8-21) are applicable to equispaced linear arrays with uniform illumination (i.e., arrays for which elements have the same amplitude: $|I_i/I_0| = 1$, $i = 1, 2, \ldots n$). Despite this stipulation, the results presented are fairly general. Gain is proportional to an array's normalized length, L/λ, and, although the beam width is not constant as the scan angle changes, at any given scan angle it is proportional to λ/L, or given by some simple function of λ/L.

Side lobe level is another important array parameter. If an array's normalized length is large, $L/\lambda \gg 1$ or $n \gg 1$, the numerator of Eq. (8-13) oscillates, producing the close-in sidelobe peaks before the denominator changes appreciably. The first side lobe peak occurs when the argument of the sine in the numerator is $3\pi/2$. Therefore, the first side lobe gain, relative to peak gain, is given by

$$\frac{|S(\theta)|^2_{SL}}{|S(\theta)|^2_{max}} = \left[\frac{\sin(3\pi/2)}{\sin(3\pi/2n)} \right]^2 \bigg/ n^2 \cong \left(\frac{2}{3\pi} \right)^2 \cong 0.045$$

or about -13.5 dB. By a similar approximation, the second side lobe's relative gain is $\sim(2/5\pi)^2$ or about -17.9 dB. Side lobe level is dependent on the amplitude distribution of the array or, in the language of antenna designers, the illumination. For arrays that are long compared to a wavelength, side lobe levels, in contrast to gain, are not significantly influenced by array length. Gain, on the other hand, is proportional to array length, but not as strongly affected by the illumination.

An interesting example of an equispaced, linear array with nonuniform illumination is the binomial array. This array is based, in principle, on the binomial theorem, namely

$$(1 + x)^n = \sum_{i=0}^{n} b_i x^i, \tag{8-22a}$$

in which

$$b_i = \binom{n}{i} \equiv \frac{n!}{(n - i)!\, i\,!} \tag{8-22b}$$

is the *i*th binomial coefficient. By making the substitution $x = e^{j\xi}$, $\xi = kd\cos\theta - \beta$, and through Eq. (8-22)

$$S(\theta) = b_0 + b_1 e^{j\xi} + b_2 e^{2j\xi} + \ldots + b_n e^{nj\xi} = (1 + e^{j\xi})^n$$

or

$$|S(\theta)|^2 = [2(1 + \cos \xi)]^n = 4^n \cos^{2n}(\xi/2) \qquad (8\text{-}23)$$

is obtained.

The normalized pattern may be expressed, therefore,

$$|S(\theta)/S(\theta)_{max}| = \cos^n[\pi d/\lambda(\cos\theta - \cos\theta_0)]. \qquad (8\text{-}24)$$

This result shows that the pattern of the binomial array is given by the nth power of the pattern of a two-element array, which was derived in Chapter 1, Eq. (1-41). If the element spacing is chosen to provide grating lobe suppression, as in Eq. (8-16), the pattern is very smooth and has no more than two nulls in the visible region, between $\theta = 0$ and $\theta = \pi$ radians. As the number of elements in the binomial array is increased, the beam becomes more narrow and the gain increases. In contrast to the uniformly illuminated array (or most arrays for that matter), an increase in the number of elements does not increase the number of pattern lobes. Despite its fascinating properties, the binomial array is seldom used in practice, because the required amplitude variation or *taper* becomes extreme as the number of elements becomes large. For example, if $n = 10$, the binomial coefficients, given by Eq. (8-22b), are 1, 10, 45, 120, 210, 252, 210, 120, 45, 10, 1. Thus, the central element must be fed by a current that exceeds that of an edge element by a factor of 252, which corresponds to a power difference of 48 dB (20 log 252). Such a strong amplitude taper greatly reduces the aperture efficiency (i.e., lowers the gain with respect to that produced by uniform illumination). Indeed, optimum aperture efficiency, or utilization, is provided by uniform illumination; however, the resultant side lobe level (e.g., -13.5 dB for the first side lobe) is too high for many antenna applications.

Most arrays use some amplitude taper in order to reduce side lobe levels. However, as noted, the price is loss of aperture efficiency. Tapers that lower side lobes weight the central elements more heavily than the edge elements. The ratio of the power of the central elements to that of the edge elements is called the edge taper, which typically lies in a range of 5 dB to 20 dB. Clearly, the antenna designer is faced with a trade-off between aperture efficiency and side lobe level. If gain is not critical and very low side lobes are desired, a strong taper would be advantageous. Mathematical array pattern synthesis methods have been developed to obtain the amplitude distribution that is consistent with a desired side lobe structure. Best known among these methods are Dolph-Chebyshev's (for discrete arrays) and Taylor's (for continuous arrays). More advanced books describe these and other synthesis methods in detail.*

8.3 PLANAR ARRAYS

The planar array, which is shown schematically in Figure 8-4a, is another important antenna configuration. The radiating elements, which lie in a common plane (the x-y plane in Fig. 8-4a), are arranged in a doubly periodic fashion. The rectangular and hexagonal lattices shown in Figures 8-4b and 8-4c, respectively, are frequently used. The hexagonal lattice is more compatible with round or elliptical aperture shapes than is the rectangular lattice. The hexagonal lattice also can provide significant advantages with respect to grating lobe suppression. More specifically, a hexagonal lattice requires fewer elements per unit area than a rectangular lattice for grating lobe suppression in a given scan volume. The rectangular arrangement is obviously more suitable for rectangular arrays and also is easier to analyze, because its pattern is, under certain circumstances, the product of two linear array patterns.

Although Eq. (8-7) is quite general and may be used to compute antenna patterns for arbitrary element configurations, the rectangular lattice can be easily visualized as a column array of linear row arrays, provided that each row has the same taper. By this conceptual shortcut, the array factor of the rectangular lattice can be regarded as the product of two linear array factors. The array factor separability, as noted, does not always apply. However, rectangular array performance parameters can often be estimated if separability is assumed. In practice, nonseparable tapers are often advantageous, because they provide greater aperture efficiency.

For the mathematical treatment of the planar array Eq. (8-7) is used, and the angle conventions of Figure 8-4a are also necessary to understand. The angle θ is measured from the z axis, which is perpendicular to the array face. With this consideration and Eqs. (8-5) and (8-7), the array factor for the rectangular lattice can be expressed

$$S(\theta,\phi) = \sum_{m} \sum_{n} \frac{I_{mn}}{I_{00}} e^{jk \, \sin\theta(md_x\cos\phi \, + \, nd_y\sin\phi)}, \qquad (8-25)$$

in which the indices run over the row, m, and column, n, positions as shown in Figure 8-4b. The amplitude taper is separable if the amplitude of each

---

*See Annotated Bibliography.

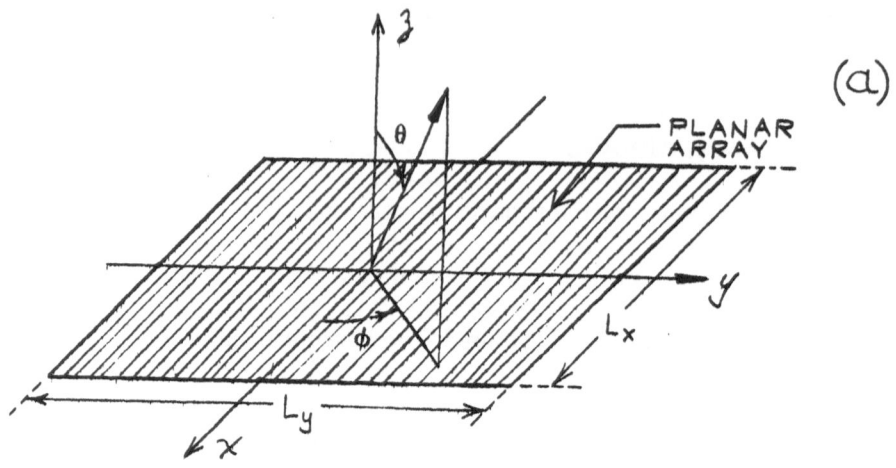

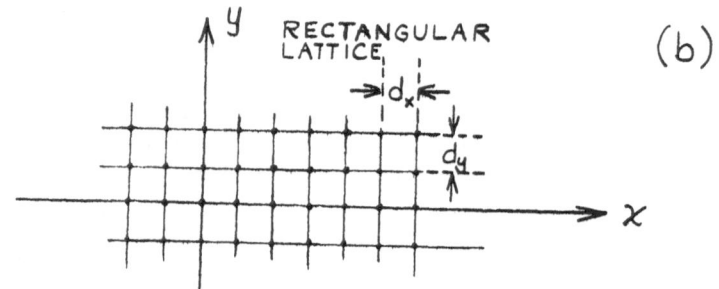

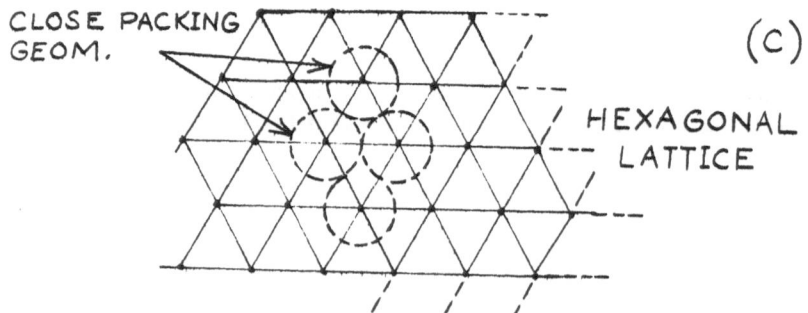

FIGURE 8-4. Planar array geometry.

element can be expressed as the product of two factors, one that depends only on n (column) and one which depends only on m (row). This stipulation is expressed by

$$I_{mn} = I_{m0} I_{0n}/I_{00} \text{ (for all } m \text{ and } n),$$

so that

$$S(\theta,\phi) = S_x(\theta,\phi) S_y(\theta,\phi), \tag{8-26}$$

where

$$S_x = \sum_m \frac{I_{m0}}{I_{00}} e^{jmkd_x \sin\theta \cos\phi},$$

$$S_y = \sum_n \frac{I_{0n}}{I_{00}} e^{jnkd_y \sin\theta \sin\phi}.$$

The amplitude factors I_{m0} are the row reference amplitudes, and the factors I_{0n} are the column reference amplitudes. The common amplitude of the central element (or elements, if n and m are not both odd numbers) provides the overall reference amplitude, I_{00}.

As in the example of the linear array, a uniform phase progression may be applied across the planar array. The elemental amplitudes may be expressed

$$I_{m0}/I_{00} = I_m e^{-jm\beta_x},$$

$$I_{0n}/I_{00} = I_n e^{-jn\beta_y},$$

in which I_m and I_n are real numbers. Pattern maxima occur if

$$k\sin\theta\cos\phi \, d_x - \beta_x = 0, \ \pm2\pi, \ \pm4\pi, \ldots$$

and

$$k\sin\theta\sin\phi \, d_y - \beta_y = 0, \ \pm2\pi, \ \pm4\pi, \ldots \tag{8-27}$$

The main beam corresponds to the zero value on the left-hand side of both these equations, and all other combinations represent grating lobes. As in the case of the linear array, lattice spacings can be chosen to exclude grating lobes from real space, but the mathematics is more complicated because of the two-dimensional array geometry. Simultaneous solution of Eqs. (8-27) provides expressions for the main beam angles:

$$
\left.
\begin{aligned}
\tan\phi_0 &= \frac{\beta_y d_x}{\beta_x d_y} \\
\sin\theta_0 &= \sqrt{(\beta_x/kd_x)^2 + (\beta_y/kd_y)^2} \\
&= \frac{\lambda}{2\pi} \sqrt{(\beta_x/d_x)^2 + (\beta_y/d_y)^2}
\end{aligned}
\right\}
\qquad \textbf{(8-28)}
$$

Because $\sin\theta_0 \leq 1$, not all values of β_x and β_y will produce a main beam. If λ, d_x and d_y are given, useful values of β_x and β_y are constrained by the relation

$$
\left(\frac{\beta_x}{d_x}\right)^2 + \left(\frac{\beta_y}{d_y}\right)^2 \leq \left(\frac{2\pi}{\lambda}\right)^2
\qquad \textbf{(8-29)}
$$

in accordance with Eqs. (8-28). If other values of β_x and β_y are used, a grating lobe, as defined in Eq. (8-27), may replace the main beam, or no main lobe may exist.

Although the beam direction can be controlled by adjustment of the phase progression parameters β_x and β_y, the beam shape and directional gain do not remain constant, in contrast to the beam of a mechanically steered antenna. When the beam is normal to the array (broadside) it is elliptical in cross section, with angular dimensions (subtended by its major and minor axes) inversely proportional to the physical dimensions of the array, L_x and L_y. However, when the beam is scanned off broadside, its cross section changes, as illustrated in exaggerated fashion in Figure 8-5. For example, the beam is elliptical at P_1 with the minor axis along the x direction, because the array is longer in that direction. When the beam is scanned toward the x axis, the array is effectively foreshortened along its x dimension, and the beam will eventually assume a circular shape, as at P_2. Generally, the area within the main beam contour is proportional to the reciprocal of $\cos\theta$, except for near-endfire scan angles.

A planar array's peak directional gain depends on the array dimensions, illumination, and scan angle. For a uniformly illuminated array and small-to-moderate scan angles,

$$
\boxed{g_{max}(\theta_0) = \pi\cos\theta_0 \, g_x \, g_y}
\qquad \textbf{(8-30)}
$$

in which θ_0 is the scan angle, and g_x and g_y are the gains of the rows and columns, respectively. From Eq. (8-21), $g_x = 2L_x/\lambda$ and $g_y = 2L_y/\lambda$. Because the array area is given by $A = L_x L_y$, Eq. (8-30) may be reexpressed

$$g_{max} = \cos \theta_0 \left(\frac{4\pi A}{\lambda^2}\right),$$

so that at broadside ($\theta_0 = 0$):

$$\boxed{g_{max} = 4\pi A/\lambda^2} \tag{8-31}$$

This result was also discussed in Chapter 7, Eq. (7-13). This represents the highest gain attainable and does not depend on the array shape, provided

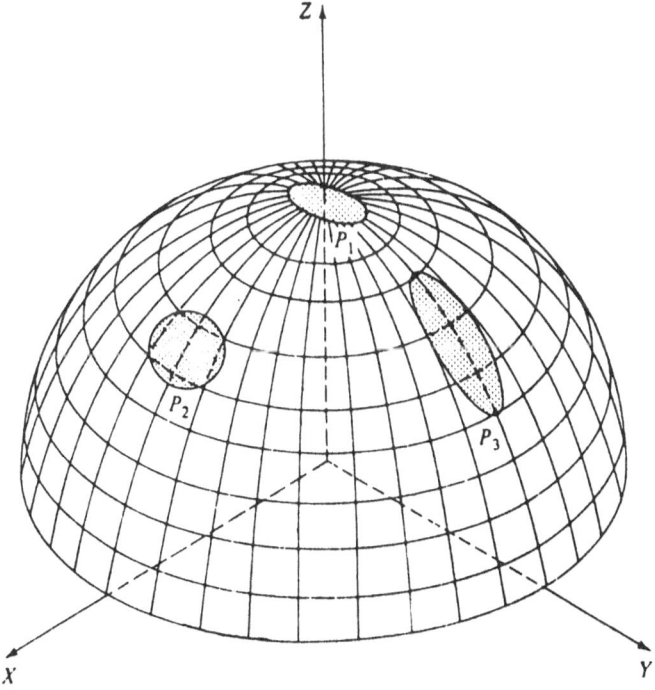

FIGURE 8-5. Beam shape versus scan position ofr a pencil beam. (*From R. C. Hansen, ed., Microwave Scanning Antennas, vol. 2, New York: Academic Press, 1966, p. 43; copyright © 1966 by Academic Press, reprinted by permission.*)

that the aperture dimensions are large compared to λ. Many practical arrays require low side lobes and, therefore, have tapered illuminations, so that the directivity is somewhat less than indicated by Eq. (8-31). To account for the relation between net gain and peak directivity in a convenient and conceptually simple way, the notion of area efficiency and effective area are used. The effective area is defined $A_{eff} = eA$, in which A is the physical array area and e, which is always less than unity, is the area efficiency. All array losses are included in e, including illumination loss, ohmic losses, mismatch losses, and random error losses. For all arrays, the peak gain can be expressed

$$g_{max} = \frac{4\pi eA}{\lambda^2} = 4\pi A_{eff}/\lambda^2. \qquad \textbf{(8-33)}$$

Furthermore, for small antennas such as slots, patches, or even dipoles, Eq. (8-31) or (8-33) are not valid, but may, nevertheless, be used to define an effective area using the gain:

$$A_{eff} = \frac{\lambda^2 g_{max}}{4\pi} \qquad \textbf{(8-34)}$$

For example, consider a half-wave dipole, which has a maximal gain of 1.64, as was shown in Chapter 4. If it is assumed that the dipole effective area is given by the product of its length and some value b (to be determined) the effective area is

$$A_{eff} = \frac{\lambda b}{2} = \frac{1.64\lambda^2}{4\pi}$$

$$b = 0.26\lambda.$$

This notion is meaningful for calculation of a receiving dipole's maximum capture area. It is attained when the dipole is perpendicular to the direction of the incoming radiation and is aligned with the polarization direction.

In the relationship between main beam width and gain for a linear array, gain is proportional to normalized length, L/λ; for a planar array, gain is proportional to normalized area, A/λ^2. Beam area, BA, is inversely proportional to gain, and a good approximate formula connecting these quantities is provided by

$$BA \cong 32{,}400/g \qquad \text{(8-35)}$$

in which g is the peak gain, and beam area is measured in square degrees.* If, for a planar array, the beam is circular in cross section, the beam width is then,

$$BW \cong \sqrt{BA/\pi} \cong 100/\sqrt{g} \qquad \text{(8-36)}$$

For a linear array at broadside, the beam is fan shaped so

$$BW \cong \frac{BA}{360} \cong 90/g \text{ (degrees)} \qquad \text{(8-37)}$$

The more exact result for a uniformly illuminated linear array is given by Eq. (8-18).

The relationship among gain, aperture area, beam area, and beam width can also be comprehended as follows. First, assume that the beam is sharply defined or idealized, as illustrated in Figure 8-1, so that it occupies a discrete solid angle, Ω, within which the power density is constant. If no power is radiated in directions outside Ω, the gain, which is the ratio of the peak to isotropic (or average) power, is simply

$$g = 4\pi/\Omega. \qquad \text{(8-38)}$$

As noted earlier, the beam width of a linear array is approximately λ/L radians. If the beam area for a planar array is the product of the beam widths of its row and column arrays,

$$BA = \Omega = \left(\frac{\lambda}{L_x}\right)\left(\frac{\lambda}{L_y}\right) = \lambda^2/A, \qquad \text{(8-39)}$$

and Eq. (8-31) is recovered by combination of Eqs. (8-38) and (8-39).

---

*There are about 41,253 square degrees in 4π steradians.

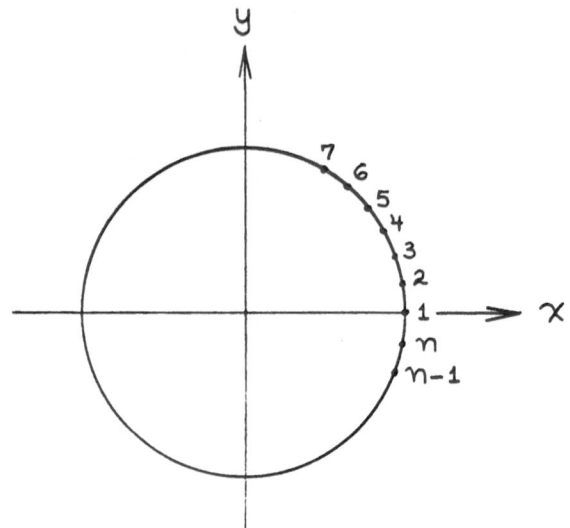

FIGURE 8-6.   Illustration for problem 1.

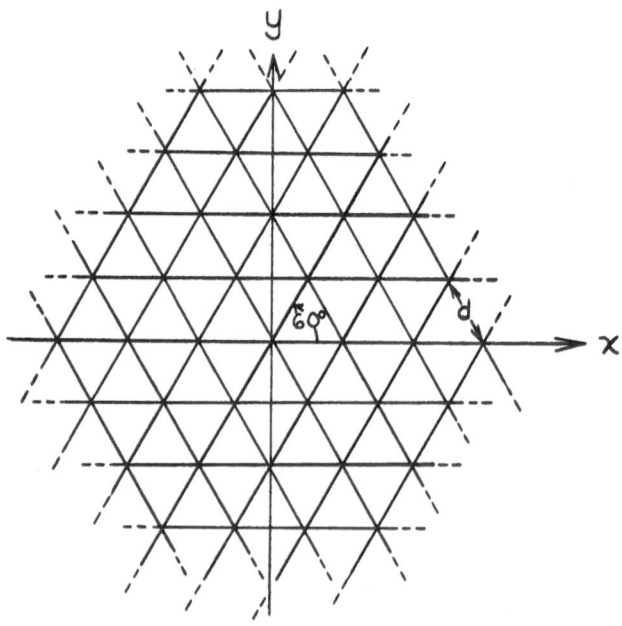

FIGURE 8-7.   Illustration for problem 3.

PROBLEMS

PROBLEM 8-1. A circular array in the x-y plane consists of n equispaced elements on the circumference of a circle of radius, a, as shown in Figure 8-6. Using Eq. 8-7, show that the array factor for this array is

$$\sum_{i=1}^{n} I_i \exp\left\{ j\frac{2\pi a}{\lambda} \sin\theta \cos\left[(i-1)\frac{2\pi}{n} - \phi \right] \right\}.$$

PROBLEM 8-2. Prove Eqs. (8-4), (8-16), (8-19), (8-28), (8-29), (8-36), and (8-37).

PROBLEM 8-3. A planar array in the x-y plane has a hexagonal arrangement as shown in Figure 8-7, with a distance, d, between any two neighboring elements.
   a. Show that for any element in the array, the phase ψ of Eq. (8-5) is given by

$$\psi_{mn} = 2\pi \frac{d}{\lambda} \sin\theta \left[\left(m + \frac{n}{2} \right)\cos\phi + n \frac{\sqrt{3}}{2} \sin\phi \right]$$

   where ϕ and θ are the usual spherical angles, and m and n are integers.
   b. Show that any point in the hexagonally shaped area can be reached by a combination $0 \le |m + n| \le p$, where p is a positive integer. How many elements are there in an area defined by $p = 5$?

PROBLEM 8-4. A linear array of colinear dipoles should operate in the 200- to 250-MHz band, provide a gain of 27 dB at broadside, and scan the beam over $\pm 60°$ from broadside.

Find the minimal necessary number of elements, the maximum element spacing, and calculate the beam width for the case of uniform excitation.

PROBLEM 8-5. Calculate and graph the normalized power pattern of a binomial array with 3, 4, 5, and 6 elements.

PROBLEM 8-6. Estimate the maximum gain of a rectangular array of 2 by 3.5 m at both 1.2 and 6 GHz. What shape does the beam have at $\theta = 0$ (zero scan).

PROBLEM 8-7. A square planar array with element spacing $d_x = d_y = d$ should scan a beam throughout a cone $0 \le \theta \le 60°$ about the z axis. Calculate the element spacing, d, required to suppress grating lobes and the range of phase shifts, β_x and β_y, required.

CHAPTER 9

Practical Arrays: Elements, Feed Methods, Scanning, and the Effects of Errors

When many antennas are combined in an array, constructive interference of their fields creates a main beam of highly concentrated radiation. Outside the main beam, the fields interfere destructively to produce side lobes, which are manifestations of stray radiation. In order to produce the desired patterns, the individual elements that constitute the array are characterized, and each element is energized with the proper phase and amplitude, relative to the other array elements. These issues are covered in the first six sections of this chapter. The last three sections are devoted to the important effects of imperfections in the array illumination. These imperfections may be due to random errors, systematic errors, or element failures. Understanding performance degradation by random errors requires some familiarity with probability theory concepts. Sections 9.7, 9.8 and 9.9 may be omitted without loss of continuity.

9.1 LINEAR ARRAY FEEDING METHODS

In the linear arrangement of array elements shown in Figure 8-3, the element separation, d, must be less than the wavelength, λ, to avoid grating lobes. Because the elements are closely spaced, it is preferred, if possible, to feed all elements with a single transmission line, as shown in Figure 9-1. Dipole elements, as in the illustration, are occasionally used in arrays, especially at the frequencies below the microwave region (high, very high, and ultra high frequencies—HF, VHF, and UHF). But many other element types also find application, as described in the following sections. The coupling networks can be simple sections of transmission line or can include other circuit elements. Furthermore, the element separation, d, is not necessarily equal to d'—the separation of the interconnections at the transmission line.

Each radiating element has an associated radiation impedance that can be viewed as a load impedance Z_i ($i = 1, 2, 3 \dots n$) at the point of connection to the transmission line. The power delivered to each element can be controlled by the coupling networks, which permit the shunt impedance to be varied.

A wave traveling down the array feed line can be described as

$$U = Ae^{-jkx}, \qquad I = \frac{1}{Z}Ae^{-jkx}, \qquad (9\text{-}1)$$

in which U and I are the voltage and current along the line, respectively. If the shunt impedances, Z_i, are large compared to the line impedance, Z, reflections from the interconnection points can be neglected, at least to a first

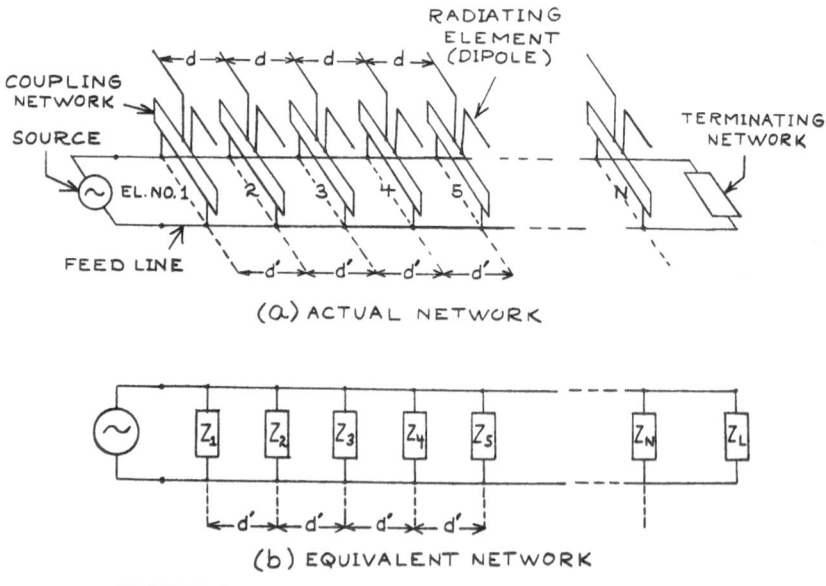

FIGURE 9-1.   Linear array fed by single transmission line.

approximation. The phase difference, β, between the points i and $i + 1$ is given by

$$\beta = kd' = 2\pi f d'/c \quad \text{(radians)}, \qquad (9\text{-}2)$$

in which f is the frequency and c is the phase velocity along the line. Thus, there is a constant phase increment or step between adjacent array elements, as was stipulated in Eq. (8-9). If the coupling networks do not change this relationship, Eq. (9-2) can be substituted into Eq. (8-11) to obtain

$$\xi = 2\pi f \left(\frac{d}{c_o} \cos\theta - \frac{d'}{c} \right), \qquad (9\text{-}3)$$

in which $k_o = 2\pi f/c_o \equiv$ free space wave number. The peak of the main beam corresponds to $\xi = 0$, so that the main beam location, θ_o, is given by

$$\cos\theta_o = d'c_o/(dc), \qquad (9\text{-}4)$$

which is not explicitly frequency dependent.

The frequency dependance of θ_o can enter, however, through c, the phase velocity along the feed (transmission) line. Many transmission media, such as waveguides, are dispersive (i.e., phase velocity in the media is a function of frequency). In such cases, a limited amount of beam scanning can be realized by changing the frequency, but the signal bandwidth cannot be too large or the main beam will be "smeared out."

Although it has been assumed that the individual element impedance, Z_i, is large, a certain amount of energy is coupled out of the transmission line and radiated by each element. Consequently, the wave amplitude, A, in Eq. (9-1) must gradually diminish. Ideally, no wave remains in the feed line beyond the last (nth) element. Because this is difficult to achieve in practice, the feed line is usually terminated with a matched load, $Z_L = Z_o$, which absorbs any residual energy and thereby prevents reflection from the end of the feed. Such a reflection would give rise to an undesirable secondary beam or lobe at $\pi - \theta_o$.

Another type of feed arrangement is shown in Figure 9-2. In this case the line is terminated by a short circuit, $Z_L = 0$, which gives rise to a standing wave distribution of voltage and current along the line, as illustrated. The radiating elements, which are shown as equivalent shunt impedances, are connected at the points of maximum voltage. Although a standing wave has maxima separated by $\lambda/2$, as discussed in Chapter 1, in-phase maxima are separated by λ. All of the elements are therefore in phase, and produce a broadside beam. If $c < c_o$ ($c_o \equiv$ free space speed of light), then $\lambda < \lambda_o$, and the spacing ensures that grating lobes are suppressed. In this standing wave

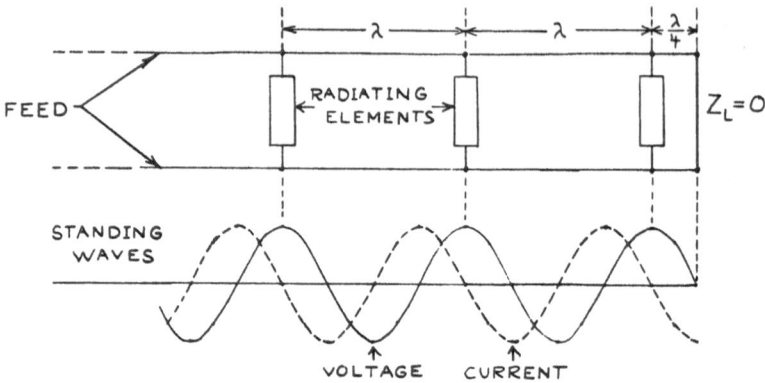

FIGURE 9-2.   Resonant feed arrangement.

arrangement, element spacing is clearly frequency dependent, so that the operation is constrained to narrow bandwidth.

An additional feed technique, illustrated in Figure 9-3, is a *corporate feed*. Elements are fed in parallel using power splitters and equal delay transmission lines, in contrast to the series arrangements discussed above. Phase shift between elements can be controlled by phase shifters. If phase shift can be varied, Figure 9-3 represents a phased array. (A later section describes the principles of operation of some common types of variable phase shifters.)

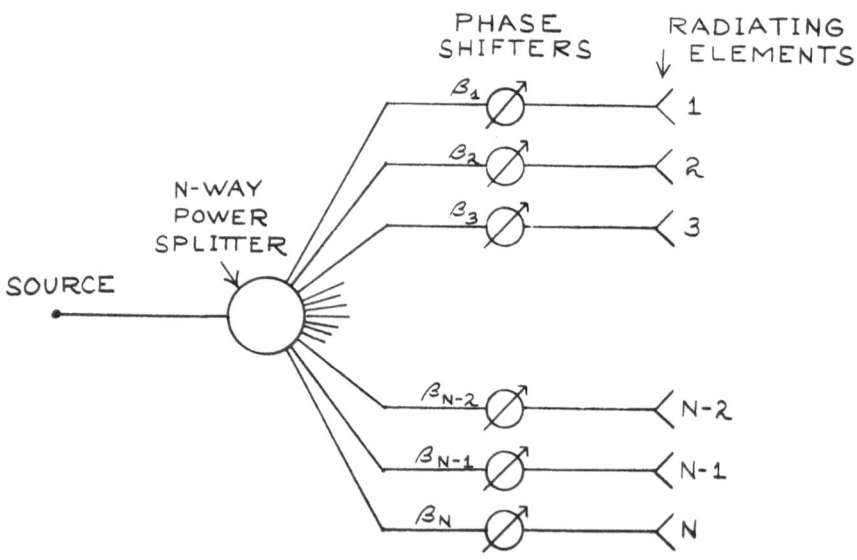

FIGURE 9-3.   Corporate-fed array.

In two-dimensional arrays, both series and parallel feed methods are often used. For example, each of the elements in Figure 9-3 may actually represent a series feed line of the type shown in Figure 9-1 or 9-2. For such arrays, two-dimensional beam scanning is performed by a combination of phase shifters in one plane and mechanical motion in the orthogonal plane.

9.2 WAVEGUIDE SLOT AND MICROSTRIP ARRAYS

The discussion in Chapter 7 includes descriptions of radiation from apertures, including waveguide slots and microstrip patches. Waveguide slots can be fed in series to form linear slot arrays that can handle the high RF power often required for radar. Both the traveling wave (Figure 9-1) and resonant or standing wave (Figure 9-2) feeds are in use. For a traveling wave feed, the beam direction in relation to the axis of the feed is controlled by the designer. The example in Figure 9-4 has small holes that act as elements in the narrow waveguide wall. The array pattern is given by

$$S(\theta) = A_1 + A_2 e^{j(kd'\cos\theta - k_g d')}$$

$$+ \quad A_3 e^{2j(kd'\cos\theta - k_g d')} + \dots$$

(9-5)

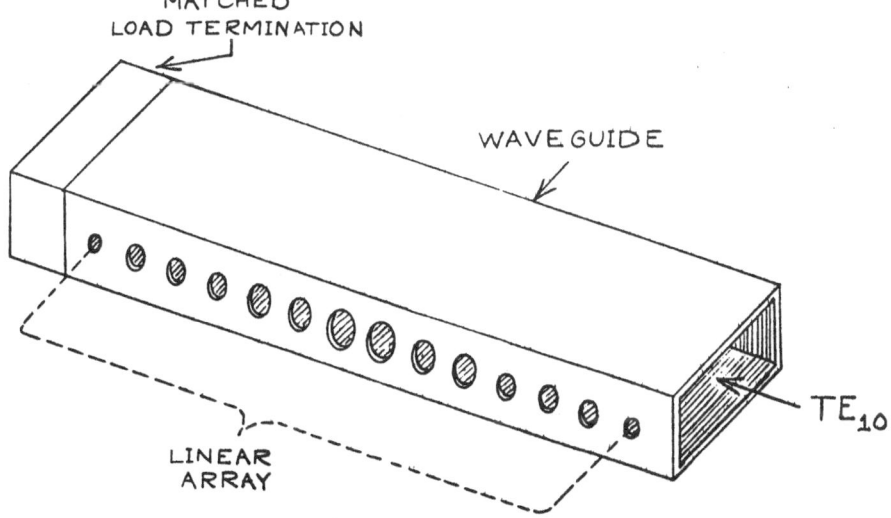

FIGURE 9-4. Traveling wave waveguide array.

in which $k_g = 2\pi/\lambda_g$ is the wave number inside the waveguide. The amplitudes, A_i, can be controlled by variation of the radii of the respective holes. The main beam peak lies in the direction θ_o (with respect to the waveguide axis), as determined by

$$\xi = kd' \cos \theta_o - k_g d' = 0,$$

or

$$\cos\theta_o = \lambda/\lambda_g. \qquad \textbf{(9-6)}$$

Note that the beam direction is apparently independent of the hole (element) spacing, d'; however, this is only approximately true, because the effects of reflections and mutual coupling have not been taken into account. Furthermore, Eq. (9-6) holds for the main beam. The location of grating lobes is definitely a function of d', thus, d' cannot be chosen arbitrarily. Furthermore, the waveguide wavelength, λ_g, is frequency dependent, as seen from Eq. (6-8). Thus, some beam scanning is feasible by adjustment of the frequency, as mentioned earlier.

A common resonant waveguide of broad wall slots is illustrated in Figure 9-5. As shown, the array is a waveguide "stick" of total length $2.5\lambda_g$, short circuited at both ends. The stick, which in effect is a resonant cavity, is fed by another waveguide, which can feed several additional sticks in series. Note

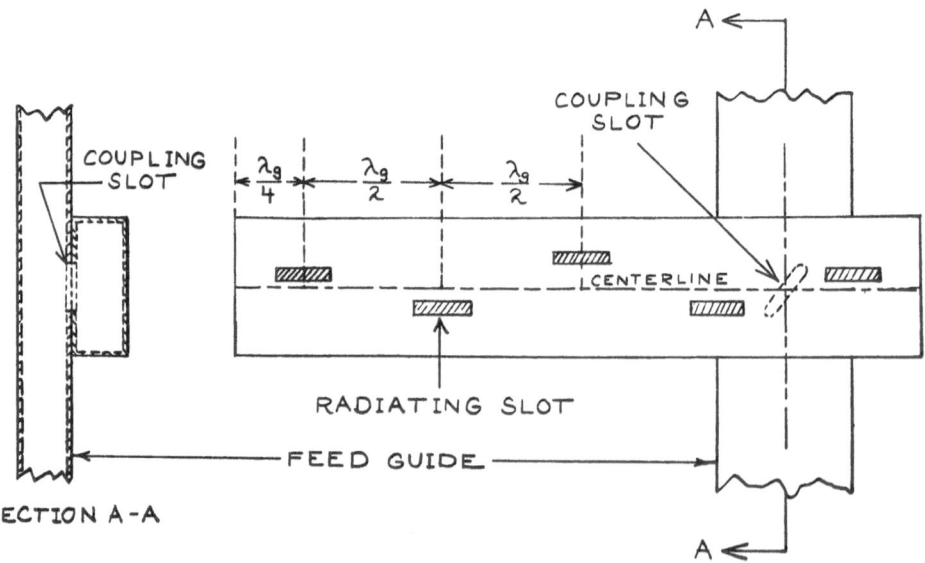

FIGURE 9-5. Resonant waveguide array with broadwall slots.

that the radiating slots are not configured as shown in Figure 9-2, but rather, are spaced at increments of $\lambda_g/2$. The half-wavelength spacing is necessary to suppress grating lobes, because in hollow metallic waveguides, $c > c_o$ (i.e., $\lambda_g > \lambda$). To compensate for the 180° phase difference between the adjacent crests of the standing wave, the slots are cut on alternating sides of the centerline. The desired illumination taper is obtained by variation of the displacement of the slots from the centerline. This resonant waveguide slot array is narrowband (typically less than 5%), and the beam is fixed at broadside (always lies perpendicular to the broad wall of the radiating sticks). Two-dimensional arrays can be built up by stacking one dimensional resonant waveguide arrays side by side. Such an arrangement can provide a light-weight, high-power array with a fixed beam. Scanning requires mechanical slewing of the radiating structure. This feature is used by the F-16 radar, shown in Figure 9-6. The 24 sticks of the radiating aperture conform to an elliptical outline that permits mechanical movement of the aperture within the aircraft's aerodynamic nose radome. A radiation pattern contour plot of this type of antenna is provided in Figure 9-7. The main beam is very narrow and the side lobes are strongly suppressed in state-of-the-art radar antennas of this type.

FIGURE 9-6. F-16 radar antenna. (*Photo courtesy of Westinghouse Electric Corporation.*)

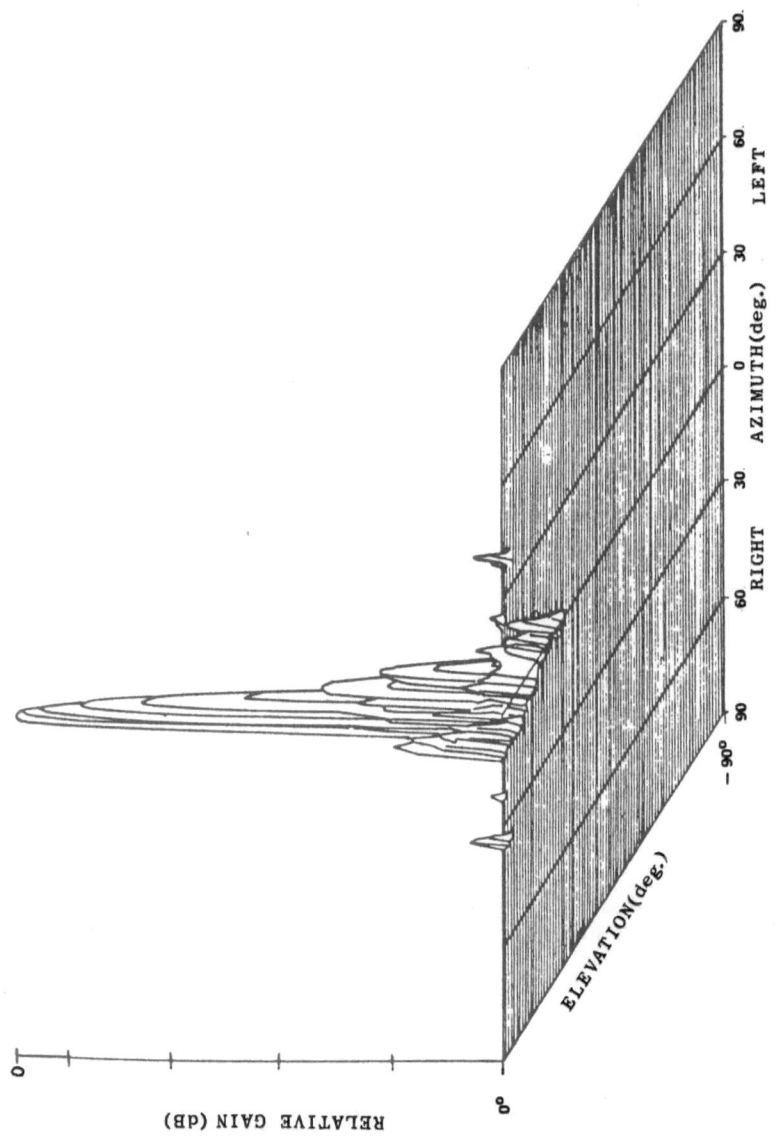

FIGURE 9-7. Contour plot of resonant waveguide stick array radiation pattern.

A resonant waveguide slot array can also use narrow wall slots as shown in Figure 9-8. The slots also have $\lambda_g/2$ spacing to avoid grating lobes. The slots are tilted alternately to the left and right to compensate for the standing wave phase reversal, and aperture illumination can be controlled by varying the tilt angle. Because the narrow dimension, b, of a waveguide usually is less than $\lambda/2$, the slots are "wrapped" around the edge to provide sufficient length for resonance. At resonance (slot length $\sim \lambda/2$) the slot impedance is purely real and the slot has the characteristics of a certain type of half-wave dipole antenna (a magnetic dipole, somewhat analogous to the electric dipole discussed in Chapter 4).

The microstrip patch element, discussed in section 7.4, is another common array antenna building block. It is convenient to feed such elements through a microstrip transmission line, which is printed on the same board. An example of such an array is shown schematically in Figure 9-9. The feed lines for the patch radiators can be varied in width to control their characteristic impedance. In this way, the desired array aperture taper may be realized. This array cannot be electronically scanned, and it is not suitable for high-power applications. However, because such arrays are light, inexpensive, and flat in profile, they are popular in many applications, including those involving aircraft, spacecraft, and missiles. Microstrip patch arrays take many forms that use various patch shapes and feeding schemes. This is an area of antenna design in which imagination and creativity continuously yield new and interesting products.

Microstrip patches are only one type of microstrip radiator in use today. Figure 9-10 shows some configurations in which segments of microstripline support filamentary currents similarly to wire antennas. Microstrip radiators (both patches and line segments) are mounted on dielectric substrate above a metallic ground plane (see Fig. 7-8), so that these types of arrays do not radiate into the half space below the ground plane. The array radiation impedance and other characteristics are affected by the substrate material properties (ε, $\tan\delta$) and thickness. In the high-microwave and millimeter wave regions, dielectric substrate losses significantly degrade the efficiency of microstrip

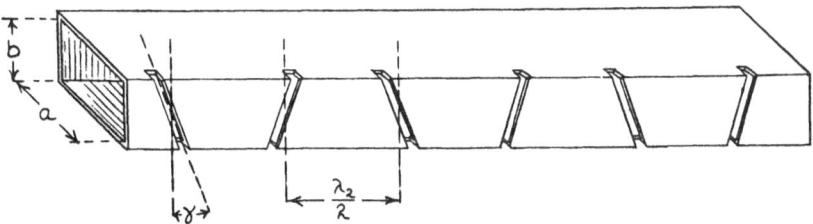

FIGURE 9-8. Section of edgewall slot array.

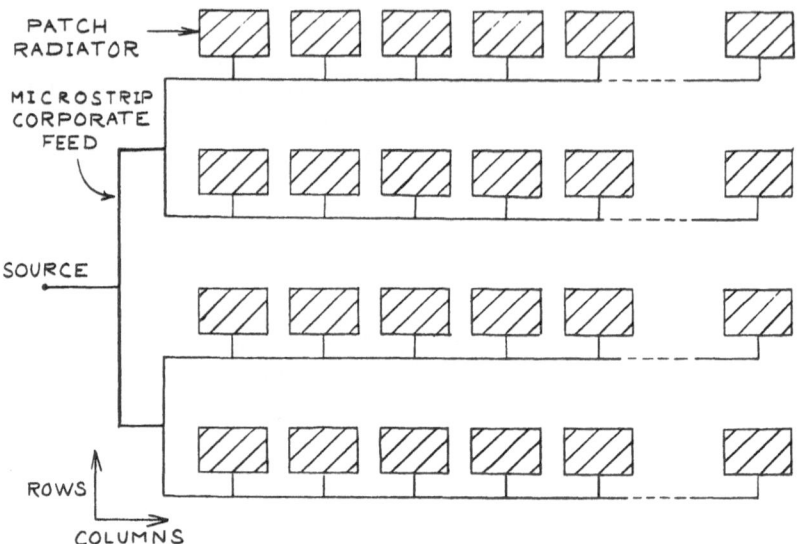

FIGURE 9-9. Rectangular planar microstrip patch array arrangement.

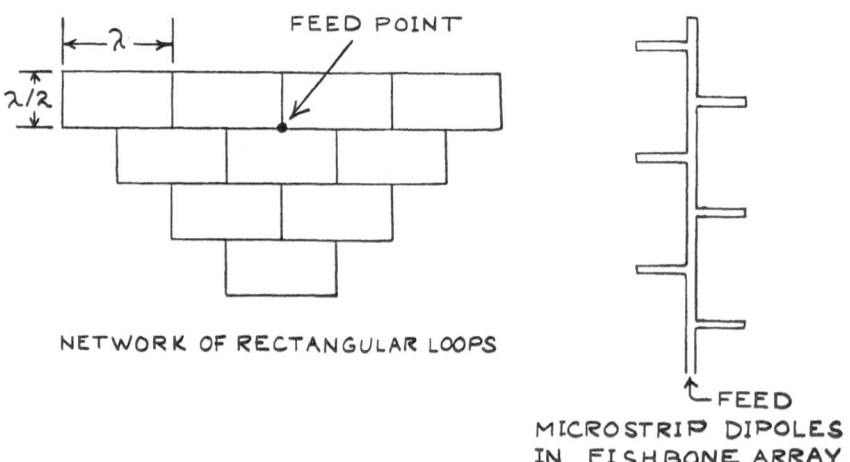

NETWORK OF RECTANGULAR LOOPS

MICROSTRIP DIPOLES
IN FISHBONE ARRAY

FIGURE 9-10. Examples of microstripline arrays.

antennas. However, in the 1- to 20-GHz regime, microstrip has been the medium employed in some of the most innovative antenna and array designs seen in the last decade.

9.3 PARASITIC ELEMENTS AND THINNED ARRAYS

Closely spaced array elements couple to one another electromagnetically. This mutual coupling is difficult to model mathematically and leads to complications in array design. Mutual coupling can sometimes be exploited, however, to feed array elements in lieu of a transmission line. The most common example is the Yagi-Uda array shown in Figure 9-11. This arrangement is widely used for home and commercial television and radio. The array uses

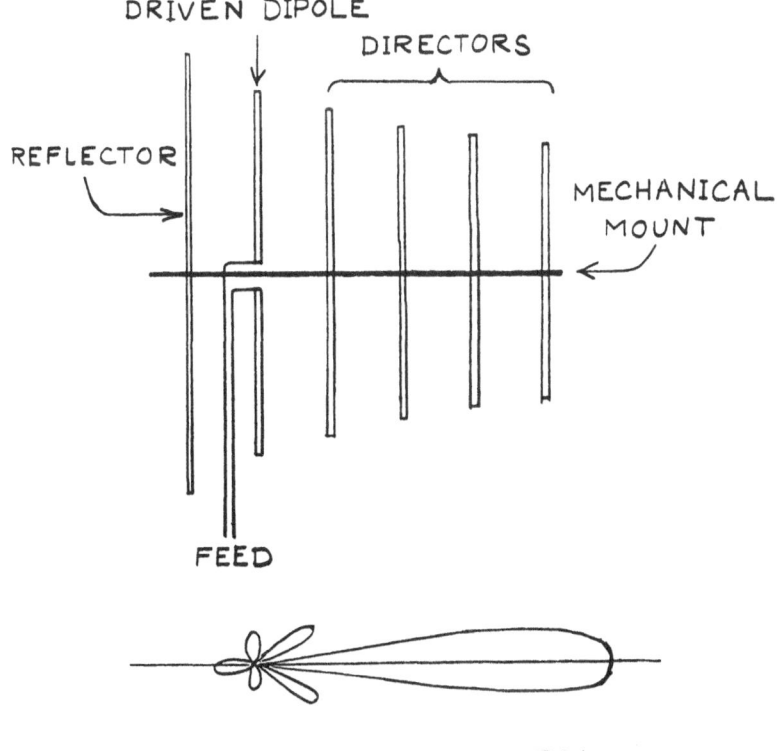

FIGURE 9-11. YAGI-UDA array.

one driven dipole and several (as many as 30) parasitic dipoles that are closely spaced and mutually parallel. One element, called the reflector, is slightly longer than the driven dipole. The other parasitic elements, called directors, are shorter than the driven dipole. The mechanical mount is electrically insulated from the dipoles. This type of antenna is widely used in UHF and VHF (30–1000 MHz). It can provide gains of up to about 20 dB and a radiation pattern with good front-to-back ratio. Because the theory of coupled parasitic elements is difficult, much knowledge about Yagi-Uda antennas is empirical. The reflector element acts as a simplified ground plane, and the directors, which are shorter than the resonance length, have a considerable reactive component, which affects the phases of the currents induced in them. The number, spacing, and length of the directors are all parameters that affect the antenna's radiation pattern, bandwidth, and input impedance. Typical Yagi antennas have from 3 to 12 elements with spacings of from 0.1 λ to 0.4 λ.

Another interesting application of inductively fed array elements is found in thinned arrays (Figure 9-12). If each driven element is given an equal

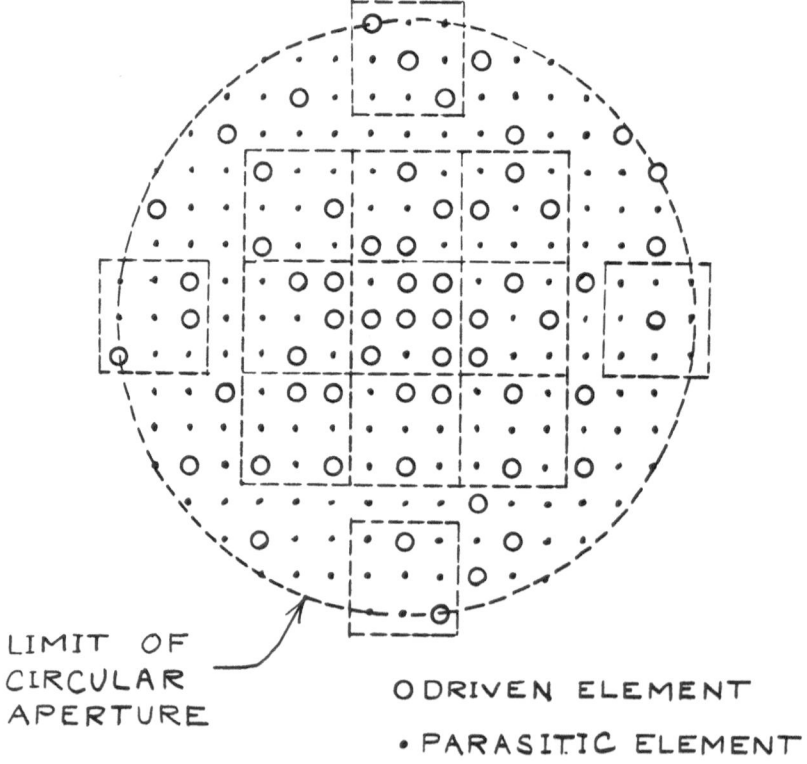

LIMIT OF
CIRCULAR
APERTURE

O DRIVEN ELEMENT

• PARASITIC ELEMENT

FIGURE 9-12.   Thinned planar array.

amount of power, array power distribution networks are greatly simplified. Illumination taper can be controlled by variation of the number of active elements per unit area. As seen in the simplified figure, the average number of driven elements in the center area is $7/9 \approx 0.78$. In the eight adjacent sections, the ratio falls to 0.36; in the four outlying areas, the ratio drops further to about 0.25. The average density of active elements for the whole array is approximately 0.28 in this example. If the array is phase steered, the element spacing should be about $\lambda/2$ to avoid grating lobes, and the driven elements should be thinned out randomly in order to suppress long-range periodicity in any direction. The parasitic elements occupy the unenergized lattice positions in order to make the environment of the driven elements appear uniform. In this case, the parasitic elements are not as close together as those of the Yagi-Uda antenna and are spaced evenly. The parasitic elements are terminated by matched loads, which makes all the element patterns identical, thus making the array environment appear uniform. This configuration increases losses through absorption in the matched loads, and thereby reduces array efficiency. Thinning provides a narrower main beam* for a given number of driven elements, but also incurs somewhat higher side lobes and losses.

FIGURE 9-13. PAVE PAWS phased array antenna. (*Photo courtesy of The MITRE Corporation.*)

---

*In sparse thinned arrays, peak directivity is proportional to the number of elements. Greater element dispersal narrows the main beam, but does not increase the directivity.

The PAVE PAWS long-range surveillance and tracking radar antenna shown in Figure 9-13 is a good example of a large, phase-steered, thinned array. This antenna has crossed dipole (dual polarization) elements and operates at about 450 MHz. The two large planar array apertures (size can be estimated by comparison with surrounding objects) are tilted with respect to each other to provide the desired angular coverage. In such arrays, considerable savings accrue from thinning, with only small sacrifices in performance.

9.4 MECHANICAL, ELECTRONIC, AND HYBRID SCANNING

If the beam of a planar array is fixed with respect to the aperture, as in the case of the resonant waveguide slot array (section 9.2), scanning is provided by mechanical movement of the structure. Such arrays must compete with reflector antennas, which are discussed in Chapter 10. Indeed, when choosing a reflector or an array, considerations include cost, available space, weight, and power handling. An array with full electronic beam steering has phase adjustibility at each active element, so that the beam may be scanned by application of the appropriate phase gradient. Because the physical aperture remains stationary, the beam can be scanned with great speed—much faster than is feasible with mechanical scanning. In addition to beam scanning with no inertia, phase steering permits the simultaneous formation of independent beams. Therefore, one aperture can be used to do the work of several fixed-beam antennas in applications such as radar tracking and satellite communications. This capability is costly, however, because thousands of phase shifters and their control networks must be built and maintained.

It is often feasible to combine both mechanical and electronic scans to cover the desired field of view. In the arrangement of Figure 9-14, a large elliptical antenna is formed by stacking many horizontal waveguides. Each waveguide has a series of edgewall slots and is fed by means of a variable phase shifter (phaser). The waveguides (usually fewer than 50) can be phased to provide elevation scan, but the azimuthal beam direction with respect to the array face is fixed. Azimuth scan is provided by rotating the entire antenna about a vertical axis. Understandably, the azimuth scan is much slower than the elevation scan. In this way the beam covers a portion of angle space defined by $0° \leqslant \phi < 360°$ and $\theta_{min} \leqslant \theta \leqslant \theta_{max}$ during each complete rotation. This hybrid scan technique represents a practical compromise between full electronic two-dimensional scan and two-axis mechanical scan: single-axis rotation is far simpler than two-axis rotation, and one-dimensional electronic scan is far simpler than a fully phased scan. The scan rate lies somewhere between that of the fully mechanical and fully electronic arrangements.

Another type of two-dimensional electronic scanning is provided by the

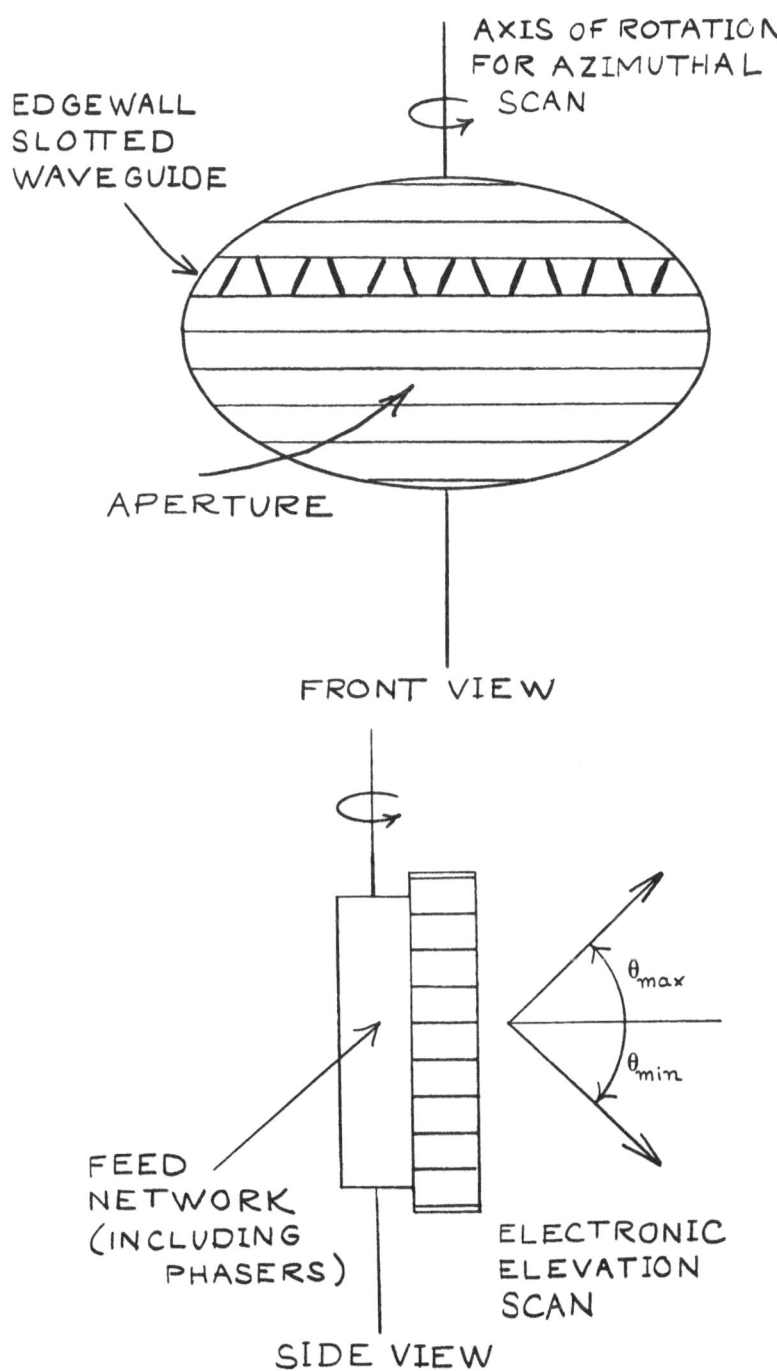

FIGURE 9-14. Arrangement with mechanical scan in azimuth and electronic scan in elevation.

173

Radant Lens®, a scanning lens that can be understood in array terms. The lens (Fig. 9-15) consists of an array of parallel plate waveguides, illuminated by a plane wave with the polarization that excites a transverse electromagnetic (TEM) mode in the waveguides. The phase shift introduced by each wave-guide is controlled by metallic obstacles that act as phase-switching reactances. On-off diode switches incorporated in these reactances provide control of the phase shift in the waveguide. Thus, the tilt of the emerging wavefront is electronically adjustable. Beam scanning in three-dimensions is achieved by two lenses combined in series. Beam pointing control is much simpler than in conventional fully phased arrays, which require N^2 phasers, because only $2N$ phase shifters are required for a square Radant array (for example). The Radant Lens® is a more economical alternative for some antenna applica-tions. It is, however, still an experimental system that has not yet found widespread use.

9.5 MONOPULSE AND OTHER SPECIAL BEAM-FORMING DESIGNS

The directive beam of an array is used to track a target either actively, as in radar, or passively, as for communications links (e.g., via satellite). This task may be performed by the popular monopulse technique as illustrated in

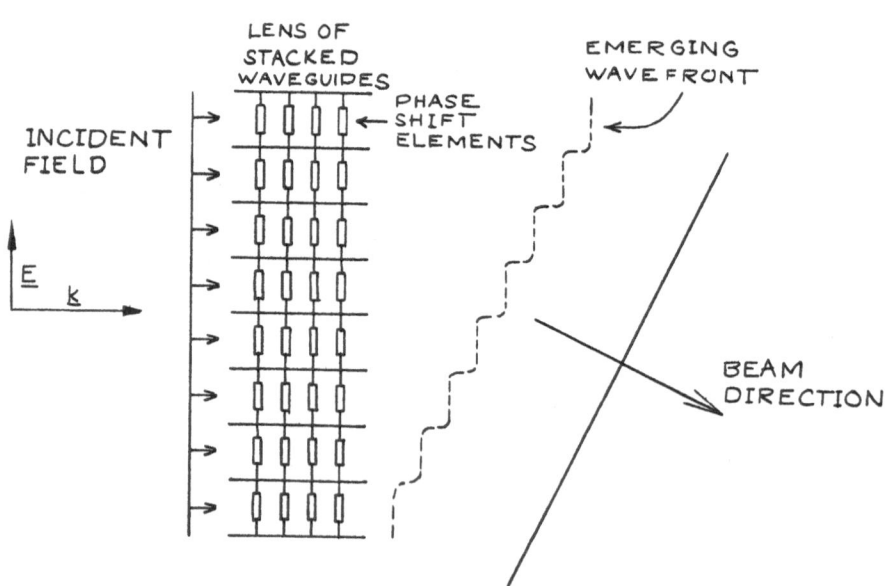

FIGURE 9-15.   Beam scanning by Radant Lens TM.

Figure 9-16. The two identical antennas, A and B, have main beams that point in slightly different directions, θ_A and θ_B. These antennas are linked by a special network, a hybrid, that has four ports. Signals from antennas A and B enter two of these ports. The hybrid delivers to the other two ports signals that are proportional to the sum, $\Sigma = A + B$, and the difference, $\Delta = A - B$, respectively. If a signal arrives from some direction, θ, between θ_A and θ_B, the ratio of the difference to the sum signal is a single valued and monotonic function of θ. Signals that arrive with angular deviations that are equal and opposite with respect to the center position, $(\theta_A + \theta_B)/2$, yield ratio or error functions of equal amplitude; however, their phases are separated by 180°. The difference signal can therefore be used as an error signal in a feedback loop that drives an antenna positioning servo. In this fashion, the antenna can be driven to follow the target using either the target's own signal (co-operating target) or energy reflected from the target (radar). Because mon-

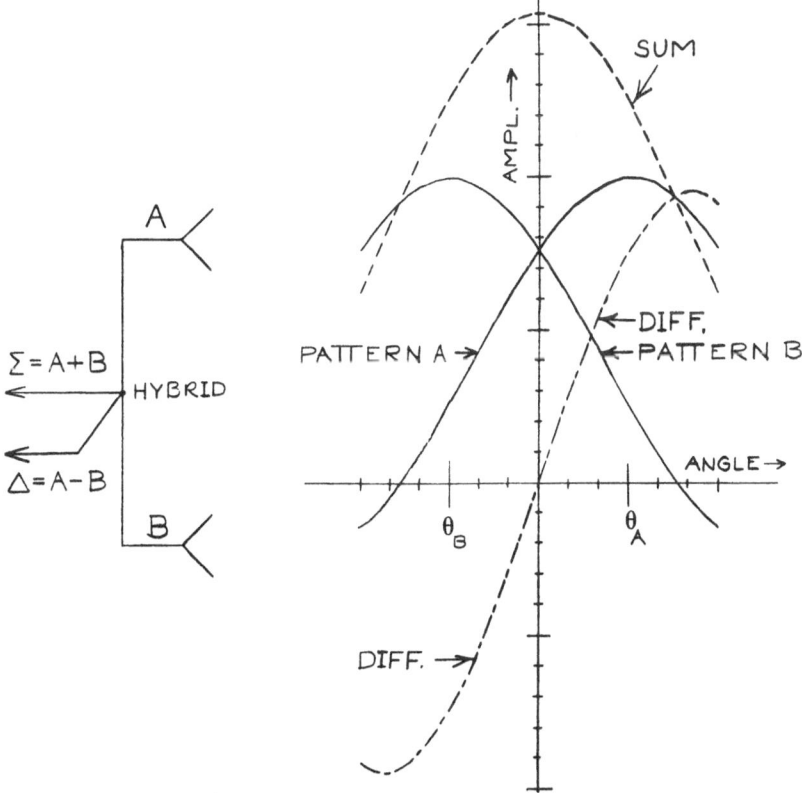

FIGURE 9-16. Monopulse network and patterns.

opulse only works in the narrow range between θ_A and θ_B, most systems include a separate search or acquisition mode that provides target information of sufficient accuracy to initialize a "lock" for the monopulse tracker.

With four antennas and four hybrids, the monopulse network can be generalized to perform two-angle (three-dimensional) tracking (illustrated in Figure 9-17). A planar array aperture is shown with four subarrays, A, B, C, and D. The network on the right of Figure 9-17 shows how the four hybrids are configured to produce three useful signals: Σ, Δ_{az}, and Δ_{el} (the difference output of the fourth port, $A + D - (B + C)$, is not needed). The difference signals can be used as error signals in a tracking loop that controls the position of the antenna beam, either electronically, as in a fully phased array, or mechanically. It is important to realize that monopulse can be implemented in a variety of ways: by using separate antennas, subdivided array apertures, or a single reflector antenna with multiple feeds.

In contrast to antennas whose single beam is mechanically or electronically scanned, multiple beam arrays provide a set of fixed beams that simultaneously cover the desired field of view. A beam-forming network causes the aperture to deliver, at each output port, the signal collected by a fixed, full-gain beam. Two beam-forming networks are illustrated in Figure 9-18. The equal time delay beam former (Fig. 9-18a) uses directional couplers in a network whose differential time delays provide beams with frequency-inde-

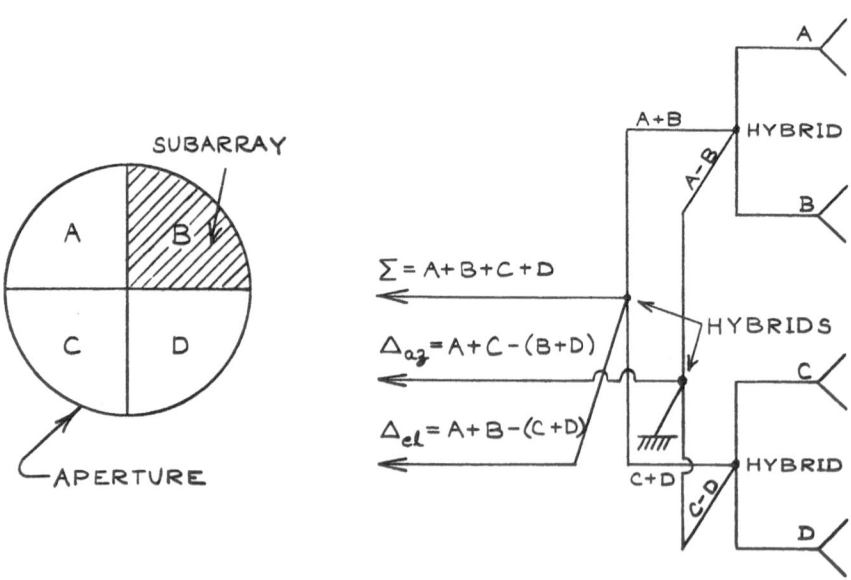

FIGURE 9-17.   Two-angle monopulse network for three-dimenional tracking.

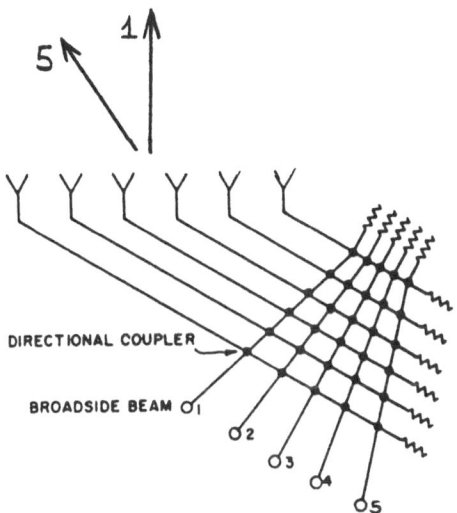

(a) EQUAL TIME DELAY BEAMFORMER

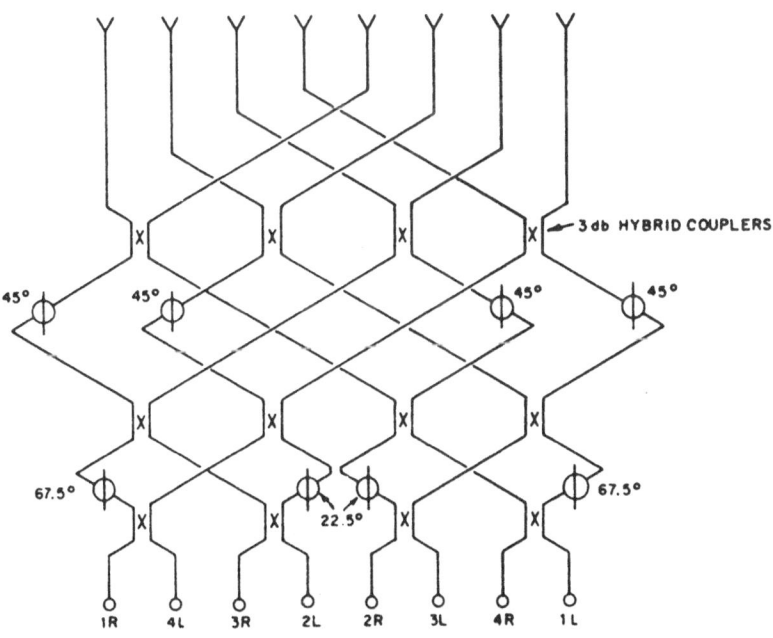

(b) EIGHT ELEMENT, EIGHT BEAM BUTLER BEAMFORMER

FIGURE 9-18. Multiple simultaneous beamforming networks. (*From M. I. Skolnik, ed., Radar Handbook, New York: McGrawp-Hill, 1970, p. 11–66; copyright © 1970 by McGraw-Hill Book Company, reprinted by permission.*)

pendent directions. Port 1 corresponds to a broadside beam, and higher numbered ports correspond to beams directed progressively farther to the left. The Butler matrix (Fig. 9-18b) uses hybrid couplers and fixed-phase shifters; it produces one beam per element. Problem 9-3 at the end of this chapter challenges the reader to explain the mode of operation of these two schemes in greater detail.

Multiple-beam arrays are most useful for receive-only applications, because transmission through the beam-forming network divides the available power among the beams, reducing the radiated power density. Unless a receiver is provided for each beam, multiple-beam antennas also require a beam-switching matrix to select the desired beams for reception. Many additional types of multiple-beam arrays have been invented. By and large, this class of antennas is being used only in special cases, where the advantages outweigh both cost and complexity.

9.6 PHASE SHIFTERS (PHASERS) FOR PHASED ARRAYS

Large phase-steered arrays would not be practical without reliable low-loss, reasonably priced variable phase shifters (often called phasers). These devices take many forms but are usually designed to provide up to 360° of differential phase shift in an RF signal over a specific frequency band. Phasers fall into two broad categories: digital and analog.

A digital phaser can exist in any one of 2^m discrete, equally spaced states of differential phase shift, in which m denotes the number of bits (binary digits). For example, a two-bit phaser has four phase states that may be designated as follows:

State	$\psi + 0°$	$\psi + 90°$	$\psi + 180°$	$\psi + 270°$
Binary Code	00	01	10	11

Note that ψ is a constant insertion phase shift that depends on details of the phaser design. Two conceptually simple examples of three-bit digital phasers are shown schematically in Figure 9-19. The parallel switched-line phaser (Fig. 9-19a) incorporates $N = 2^3 = 8$ lengths of line differing in length by units of

$$\Delta l = \frac{c}{2^3 f} = c/8f, \qquad (9\text{-}7)$$

in which c is the phase velocity in the line. The series switched-line phaser (Fig. 9-19b) requires only six lengths of switched line (i.e., two for each bit of phase control; $2 \times 3 = 6$). In both of these examples, diodes are used as on-off switches. Although series configuration uses fewer diodes, the parallel set-up has lower insertion loss because the signal traverses only two diodes for any phase state. In the popular series or cascaded arrangement, the individual sections are called phase bits and are designated by their differential phase shift.

Digital phasers require control circuitry to translate the phase commands into the appropriate switch control voltages. Weight- and space-efficient designs have the control circuitry incorporated in microwave integrated circuits (MICs). An especially efficient MIC, four-bit phaser noted in the literature

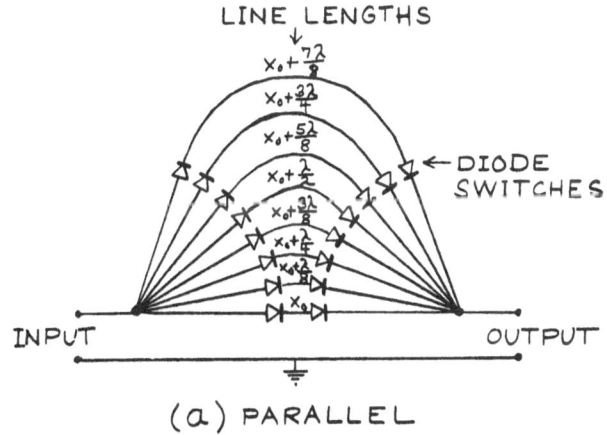

(a) PARALLEL

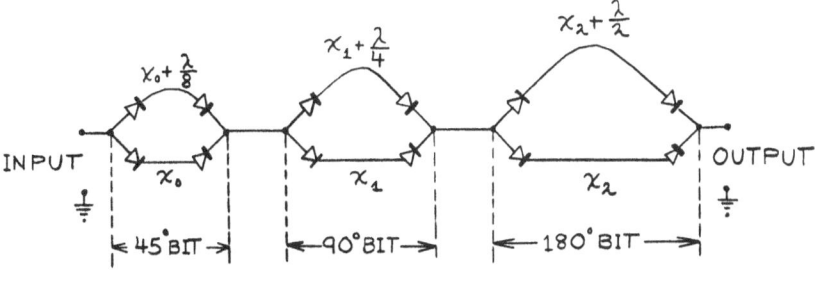

(b) SERIES

FIGURE 9-19. Switched line phase shifter configurations.

operates in the 11.7- to 12.2-GHz band, consumes only 35 mW on average, can switch phase states in 10^{-9} s, and has an insertion loss of only 1.6 dB. Phasers that are small, light, and use little power are desirable for aerospace applications. More typically, switching times are the order of microseconds and power consumption is on the order of 1 W for phase shifters of the above types. If higher power handling is required, switching time and power consumption increase.

In contrast to the digital types, which are quantized, analog phasers are continuously adjustable over a specified range. Their operation relies on a continuously variable physical parameter of some type that can be used to change the phase of an RF signal. For example, some designs are based on the electrical characteristics of the varactor diode, which has a capacitance that varies with an applied negative bias voltage. In modern phased arrays, analog phase shifters are less common than digital ones.

The operation of a ferrite phaser, which may be configured as analog or digital, depends on the properties of ferrites—metal-oxide insulator that can be magnetized. Propagation through the ferrite can be influenced by an applied direct current magnetic field. A digital toroidal latching phaser is a simple example, and one phase bit is illustrated in Figure 9-20. The ferrite toroid is secured in a section of waveguide and is threaded by a loop called the "drive" wire. A short duration pulse of current through the loop applies a magnetic field to the ferrite, which drives its magnetization to saturation. As a result of hysteresis in the ferrite, some magnetization remains after the pulse has subsided. This remanent magnetization influences the propagation through the ferrite-loaded waveguide. Differential phase shift is produced by reversal of the ferrite magnetization with a pulse of opposite polarity to that of the previous pulse. The size of the differential phase shift depends on the length of the toroid and the ferrite material. Power consumption is relatively low,

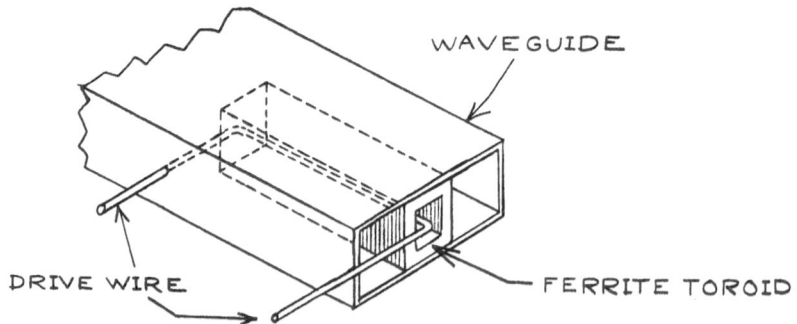

FIGURE 9-20.   One phase bit of toroidal latching phaser.

because continuous current is not required to maintain the magnetization of the ferrite.

The latching ferrite phaser is an example of a nonreciprocal phaser. (The phase shift produced does depend on the direction of propagation through the device.) The switched-line phasers discussed earlier are obviously reciprocal. Lack of reciprocity in a phaser is ordinarily not of great consequence, because there is usually sufficient time to reset the phaser for the receive mode, following transmission, in pulsed-radar systems, for example.

A ferrite phase shifter can, in principle, provide continuous (analog) phase shift as shown in Figure 9-21. A slab of ferrite is placed inside a section of circular waveguide. The electric field, **E**, which lies in the z direction, has parallel and perpendicular components $\mathbf{E}_\parallel$ and $\mathbf{E}_\perp$, with respect to the broad face of the ferrite slab. The ratio of the components depends on the slab angle, ϕ. Because the phase shifts of $\mathbf{E}_\parallel$ and $\mathbf{E}_\perp$ differ as the field propagates down the waveguide, the ferrite angle, ϕ, may be used to continuously vary the total phase shift. Of course, mechanical rotation of the slab would be slow, so in practice the ferrite is a specially shaped piece, whose anisotropic phase shift is controlled electrically from the outside by currents in coils (similar to the armature of an electric motor). Even though it has no moving parts, such a phaser is relatively slow because of the large inductance of the coils. These types of analog phasers are used in some radar systems that require high-power handling and highly accurate phase shift.

As phase shifter design and manufacturing technology advances and phased array costs decline, these versatile antennas will see wider application. One modern technology that holds great promise in bringing down the size and cost of phasers for array applications is Monolothic Microwave Integrated Circuitry (MMIC). At present, fully phase-steered arrays are still quite rare and have been in use only in a few highly sophisticated radar systems.

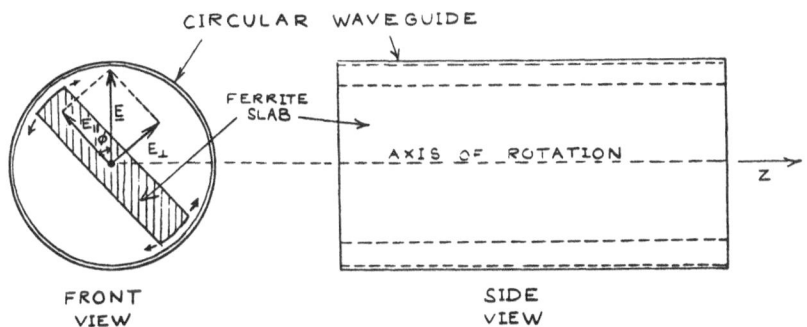

FIGURE 9-21. Analog ferrite phase shifter.

9.7 RANDOM ERRORS IN PHASED ARRAYS

For the "ideal" phased arrays, the illumination of the elements is exactly equal to that required by the beam-steering phase gradient and aperture taper. In practice, exact phase and amplitude control is unrealizable, and some degradation of the antenna pattern occurs as a result of phase and amplitude errors. For example, when digital phasers are used, phase accuracy is limited by the inherent spread of the least significant bit, which is given by

$$\delta = 2\pi/2^m \quad \text{(radians)}, \quad \text{(9-8)}$$

in which m denotes the number of bits. In well-designed arrays, errors due to phaser quantization are made to occur with nearly equal probability within the interval specified by Eq. (9.8) (random errors). (Section 9.9, discusses quantization lobes.) Other random errors include:

Errors in element phase center (translational) position
Errors in element rotational position
Feed or beamformer errors
Variations in element pattern
Radome inhomogeneites

To clarify the concept of random errors, an example of element positional errors in a waveguide slot array can be considered. As in any physical process, the machining of the slots is subject to mechanical tolerances (perhaps as small as a few thousandths of an inch). Any given array selected from a collection (or ensemble) of arrays produced using the same machine may be subjected to precise measurements that reveal the actual slot positions. If such measurements are performed for the whole ensemble, a statistical description of the slot position error can be found: the slot position becomes a random variable. Theoretically, the antenna pattern degradation caused by such errors can be described statistically, if a statistical characterization of the errors themselves is found. The error distribution parameters can be estimated experimentally, derived from experience, or based on proven theory. Slot machining errors can give rise to both phase and amplitude errors in the antenna illumination.

If the element contribution to a far field radiation pattern is considered as a phasor or complex number,

$$A_m \exp(j\phi_m) \quad \text{(mth element)},$$

in which A_m is the correct weight and ϕ_m is the correct phase, then the random error modified contribution is expressed

$$a_m\, A_m\, \exp[j(\phi_m + \phi_{em})],$$

in which a_m is an amplitude error factor and ϕ_{em} is the phase error. The phase error rotates the phasor, and the amplitude factor changes its length. The squared length of the mth error phasor, found by subtracting the ideal contribution from the modified contribution of the mth element, is approximately (for $a_m \simeq 1$, $\phi_{em} \ll \pi$)

$$A_m^2\, [(a_m - 1)^2 + \phi_{em}^2] = A_m^2 (\gamma_m^2 + \phi_{em}^2), \qquad \textbf{(9-9)}$$

in which γ_m is the amplitude ratio error. It can be shown that the average antenna pattern gain, or expected gain for a large ensemble of antennas, referenced to the ideal pattern peak power, is

$$*P(\theta) = P_i(\theta) + \frac{(*\gamma^2 + *\phi_e^2)\, \sum\limits_{m=1}^{n} A_m^2}{\left(\sum\limits_{m=1}^{n} A_m\right)^2} = P_i(\theta) + \frac{*\gamma^2 + *\phi_e^2}{D}, \qquad \textbf{(9-10)}$$

in which D is the array directivity if the element spacing is $\lambda/2$, the asterisks denote ensemble averages, and P_i is the ideal power pattern. In addition, for Eqs. (9-10) to be valid, $*\gamma_m = *\phi_{em} = 0$ must hold. There are two observations for this result:

The random errors cause an angle independent, average power contribution to the pattern.

The normalized random error power contribution is proportional to $1/n$, because D is proportional to n, the number of elements.

Thus, random errors are more easily tolerated in large arrays than in small ones. If the random error statistics remain fixed, every doubling of array element count leads to a 3-dB reduction in the relative random error power. Because the peak gain is proportional to n, the random error power is fixed with respect to the isotropic level for given error statistics.

Reduction of random errors becomes increasingly important as one attempts to produce antennas with lower side lobe levels. The potential impact of random errors on side lobe levels is illustrated by Figure 9-22. The ideal broadside pattern (dashed curve) is based on a 50-dB Taylor taper applied to a 100-element linear array with $d/\lambda = 0.513$. The solid pattern results when a random phase error characterized by a uniform probability density of $22.5°$ width ($*\phi_e^2 = 0.01285$ radian^2, least significant bit for a four bit phaser) is impressed on the illumination. A random number generator is used to simulate the errors. The error modified pattern has a chaotic side lobe structure that completely masks the 50-dB design taper side lobes. Eq. (9-10) predicts a mean error power of -37.4 dB, which is consistent with the simulated pattern. The 50-dB taper represents an overdesign of about 12.6 dB, in view of the hypothetical random error distribution. If random error power levels can be estimated in advance, tapers that provide overdesigns of 4–6 dB are typical in practice.

Random errors increase the relative gain in the side lobe region. Because a larger percentage of power is leaving through the side lobes, the gain of the main beam must be reduced. The average peak gain loss, in decibels, due to random errors is approximately

$$G_r = 10 \log[1 - (*\gamma^2 + *\phi_e^2)] \text{ (dB)}, \qquad \textbf{(9-11)}$$

which is independent of n. For the example of Figure 9-22, this peak loss only amounts to about -0.06 dB.

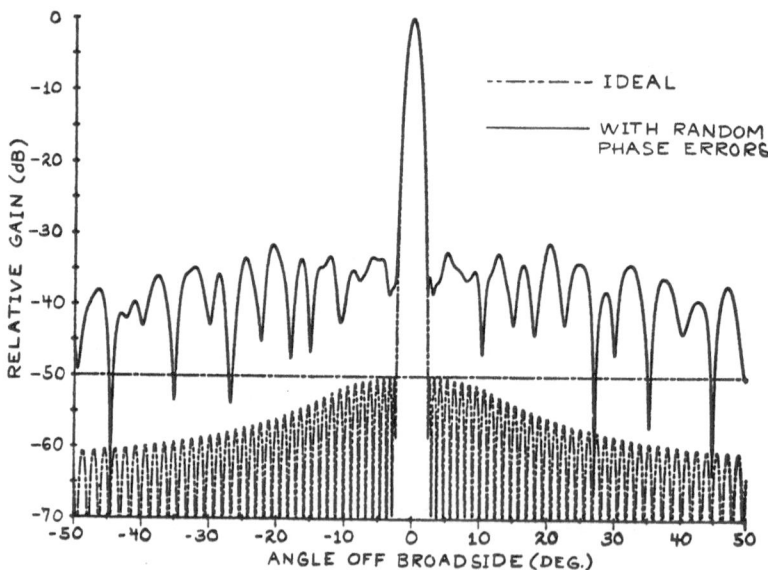

FIGURE 9-22.   Ideal and random error modified patterns for a 100-element array with 50 dB Taylor illumination.

9.8 ELEMENT FAILURES IN PHASED ARRAYS

A phased array usually has one phase shifter dedicated to each radiating element. Commands from the beam-steering computer are interpreted by phase shifter drivers, which apply the signals required to set the phasers to the desired states. Experience indicates that these driver units are the weakest link in the chain of array phase control. A driver failure can freeze the phase shifter in one of its discrete states, so that subsequent commands from the beam-steering computer are ignored. For some beam-pointing angles, the failure state is fortuitously correct; however, usually the failure degrades the pattern.

Element failure degradation can be very gradual and tolerable in arrays with many elements. Indeed, "graceful" degradation is a frequently touted advantage of array architecture. An array can be designed with sufficient margin to perform satisfactorily with 5% of its elements malfunctioning. However, when a strong aperture taper is applied to control side lobes, failure of the heavily weighted center elements is far more detrimental to antenna performance than is edge element failure. Indeed, for an ultralow side lobe array with a modest number of elements, the loss of a single element in the central region of the aperture can lead to a catastrophic increase in the side lobe level.

In order to treat element failures mathematically, first a model is specified. A reasonable and simple model includes the following assumptions:

The probability of element failure per unit time is a constant and is independent of element position.
A failure has the effect of randomizing the phase of the affected element, and failure phase states are uniformly distributed in 360° (phase quantization is neglected).
The amplitude (power) of the failed element is unaffected by the failure.
Any mutual coupling effects are neglected for simplicity.

Consider a linear array of n isotropic elements in which the mth element has failed and has a phase error of ϕ_e (referenced to its ideal phase). The amplitude pattern, referenced to the ideal pattern peak, can be expressed

$$S_f(\xi, m, \phi_e) = S_i(\xi) - A_m \left[\frac{e^{jm\xi} - e^{j(m\xi + \phi_e)}}{\sum\limits_{i=1}^{n} A_i} \right], \qquad (9\text{-}12)$$

in which $S_i(\xi)$ is the ideal relative amplitude pattern, and A_m is the weight of the failed (mth) element. It is assumed that the aperture taper is very strong, so that $S_i(\xi)$ can be neglected in the side lobe region of the pattern. The phasor diagram of Figure 9-23 shows that the power pattern can now be expressed (in the side lobe region) by

$$P_f(m,\phi_e) = |S_f(m,\phi_e)|^2 \cong \frac{(2A_m \sin (\phi_e/2))^2}{\left(\sum\limits_{i=1}^{n} A_i\right)^2}, \qquad \textbf{(9-13)}$$

which is a constant. This result pertains to the failure state of the mth element with phase error, ϕ_e. If all phase states are uniformly averaged (the second assumption), the result is

$$*P_f(m) \cong \frac{1}{\left(\sum\limits_{i=1}^{n} A_i\right)^2} \frac{(2A_m)^2 \int_0^{\pi} \sin^2 (\phi_e/2)\, d\phi_e}{\int_0^{\pi} d\phi_e} = \frac{2A_m^2}{\left(\sum\limits_{i=1}^{n} A_i\right)^2}. \qquad \textbf{(9-14)}$$

Thus, the average side lobe power associated with a single failed element is proportional to the square of that element's illumination weight, A_m, which is a reasonable result. The significance of the failed element position in the

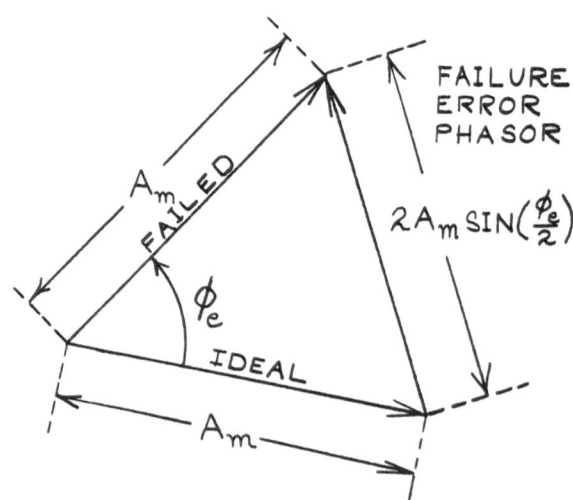

FIGURE 9-23.   Construction for element failure phasor.

aperture is obvious, because A_m can vary by one or two orders of magnitude, depending upon the specific taper employed.

By averaging over all element positions, m,

$$*P_f = \frac{2}{n} \frac{\sum\limits_{m=1}^{n} A_m^2}{\left(\sum\limits_{i=1}^{n} A_i\right)^2} = 2/nD, \qquad (9\text{-}15)$$

in which D is the linear array directivity, as encountered in section 9.7. Because the directivity and element count are related by $D = n \times$ taper loss factor for an array of isotropic radiators, Eq. (9-15) can be re-expressed

$$\boxed{*P_f = 2/[(n^2)\ (\text{taper loss factor})]} \qquad (9\text{-}16)$$

which explicitly shows the dependence of the average single element failure power on n^2. However, the above result tends to obscure the major role of element position. For a 50-dB Taylor taper, the ratio of edge-to-center element amplitude is about -13.5 dB, so that the failure powers can vary by 27 dB if the phase errors (ϕ_e) are equal.

Figure 9-24 displays simulated broadside radiation patterns based on a 100-element linear array with a 50-dB Taylor taper. The lowest curve represents the near-ideal pattern, which corresponds to the failure of an edge element. The upper curves represent the effects of single-element failures in intermediate and near-central positions. The importance of element position for this strongly tapered 100-element array is readily apparent. Side lobe levels are consistent with Eq. (9-13), especially at large pattern angles, for which the ideal side lobes have rolled off considerably.

Because failure power, like random error power, adds incoherently in the ensemble average, Eq. (9-16) can be generalized to account for p failures. If $p \ll n$,

$$\boxed{*P_f(p) = p\ *P_f = 2p/[(n^2)\ (\text{taper loss factor})].} \qquad (9\text{-}17)$$

The simple addition used to obtain the approximate result of Eq. (9-17) implies that the same element can have multiple simultaneous failures. A more rigorous analysis would avoid this problem.

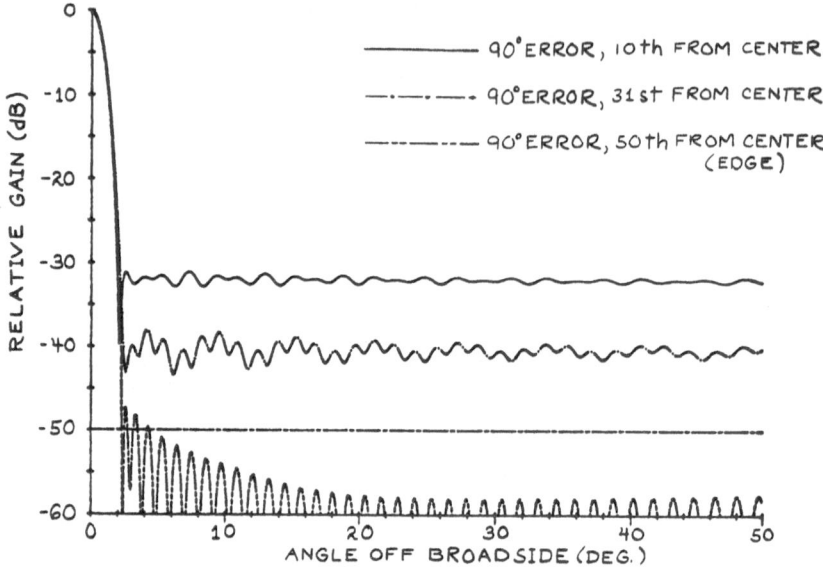

FIGURE 9-24. Single element failure sidelobe degradation for a 100-element linear array with 50 dB Taylor illumination.

9.9 SYSTEMATIC ERRORS IN PHASED ARRAYS

As seen in section 9.7, random errors remove power from the main beam and distribute it evenly, on the average, across the side lobe region. The even distribution of the error power is a consequence of its incoherence. In contrast to random errors, systematic errors remove power from the main beam and concentrate it into discrete grating-lobe-like peaks, which are usually most prominent in the immediate vicinity of the main beam. Often, the only manifestation of systematic errors is a broadening of the main beam itself, coupled with an elevation of the close-in side lobes. If the systematic errors are small, the pattern features they incur are usually masked by the chaotic random error side lobe structure, making pattern interpretation difficult.

Systematic errors are errors that are coherent over a sizable portion of the antenna aperture. In simple terms, this means that if the error at the mth element is known, the value of the error can be predicted at adjacent and more distant elements with confidence. Sources of systematic errors include:

Phase quantization errors, as can occur when digital phasers are employed
Aperture warpage due to mechanical or thermal stresses
Cumulative errors in traveling wave feeds
Cross polarization effects

Radome effects, such as boresight shift (see Chap. 12)
Amplitude or phase errors in subarray output combiners

Phase quantization leads to systematic errors when the round-off phase error
function obtained by subtracting the actual phase shifter setting from the ideal
linear phase slope function is periodic. If the amplitude of the periodic round-
off error is large enough, distinct quantization lobes appear in the antenna
pattern. This type of systematic error can be alleviated by reducing the size
of the least significant phase bit or by including, at each element, a known
randomly selected phase shift (e.g., line length) to break up the periodicity
of the round-off error. The basic character and pattern features of systematic
errors are illustrated by the following example.

A linear array of n isotropic elements has interelement spacing, d, and
element weights, A_m. The array illumination is perturbed by a sinusoidal phase
error with a wavelength, λ_e. The phase error at the mth element is

$$\phi_e(m) = \phi_o \sin(m\zeta). \qquad (9\text{-}18)$$

in which

$$\zeta = 2\pi d/\lambda_e. \qquad (9\text{-}19)$$

Such a phase error might represent a crude approximation for a quantization
phase error or a displacement error perpendicular to the array axis. The
amplitude of the radiation pattern is then

$$S' = \sum_{m=1}^{n} A_m e^{j(m\xi + \phi_e(m))} = \sum_{m=1}^{n} A_m e^{jm\xi} e^{j\phi_e(m)} \qquad (9\text{-}20)$$

in which

$$\xi = (2\pi d/\lambda)(\cos\theta - \cos\theta_o). \qquad (9\text{-}21)$$

If it is assumed that $\phi_o \ll \pi$, then the approximation $e^{j\phi_e(m)} \cong 1 + j\phi_e(m)$
can be used to obtain

$$S' \cong \sum_{m=1}^{n} A_m(1 + j\phi_o \sin m\zeta)e^{jm\xi}, \qquad (9\text{-}22)$$

and the phase modulation is seen to be nearly equivalent to a complex
amplitude modulation. A simple manipulation of Eq. (9-22) leads to

$$S' = \sum_{m=1}^{n} A_m \left[e^{jm\xi} + \frac{\phi_o}{2} e^{jm(\xi+\zeta)} - \frac{\phi_o}{2} e^{jm(\xi-\zeta)} \right], \qquad \textbf{(9-23)}$$

which is essentially the sum of three amplitude patterns: the unperturbed pattern, S, and two "satellite" patterns. For $\theta_o = 0$, the patterns have peaks at angles $\pm \theta_s$:

$$\xi \pm \zeta = 0 \rightarrow \theta_s = \cos^{-1}\left(\frac{\lambda}{\lambda_e}\right). \qquad \textbf{(9-24)}$$

If the array has a low side lobe taper, and the satellite amplitude peaks are in the side lobes of the main beam pattern, the power pattern may be approximated by

$$P' = |S'|^2 \cong |S(\xi)|^2 + \left(\frac{\phi_o}{2}\right)^2 |S(\xi+\zeta)|^2 + \left(\frac{\phi_o}{2}\right)^2 |S(\xi-\zeta)|^2. \qquad \textbf{(9-25)}$$

The sinusoidal phase modulation of the array illumination gives rise to a pair of lobes that are scaled replicas of the main beam of the unperturbed pattern. If the perturbation were known to be caused by quantization error, these

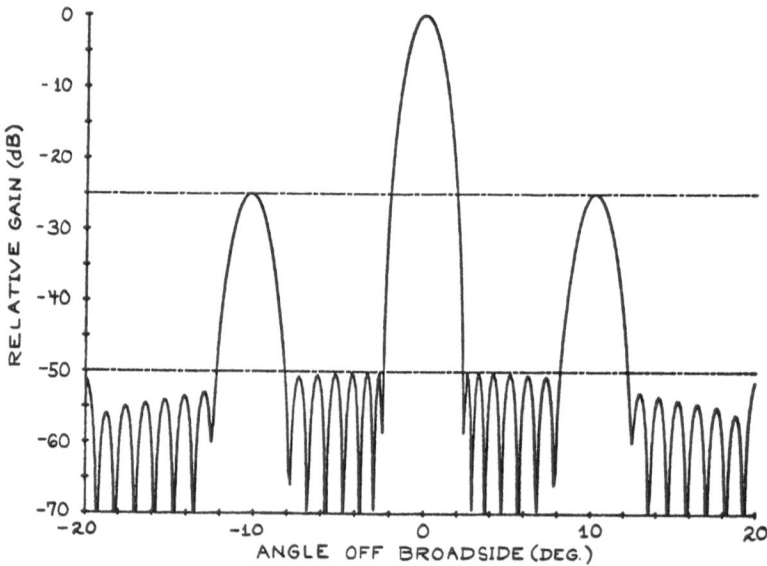

FIGURE 9-25. Systematic phase error modified pattern for a 100-element array with 50 dB Taylor illumination.

would be called quantization lobes. Also, the positions of the lobes are given by Eq. (9-24). They are not grating lobes in the strict sense; however, these new peaks arise from a periodicity in the array illumiination.

The pattern of Figure 9-25 is a pattern perturbed by a sinusoidal phase error for a 100-element array with a 50-dB Taylor taper. Estimating λ_e/λ and ϕ_o for this case proves a good exercise for students of antennas.

The example illustrates several important systematic error properties.

> Long-range or slowly varying systematic errors perturb the pattern in the vicinity of the main beam.

> Small systematic phase errors and systematic amplitude errors give rise to very similar power patterns.

> Strong and very regular systematic errors can give rise to grating-lobe-like peaks (i.e., peaks that mimic the main beam and lie in the side lobe region).

These results are consistent with the Fourier transform relationship that exists between the aperture illumination and the far field. In well-designed arrays, systematic errors are very smooth, and pattern degradation is usually limited to some beam broadening or to elevation of the close-in side lobes.

PROBLEMS

PROBLEM 9-1. A linear phased array is shown schematically in Figure 9-26. When the beam is in the broadside direction, all phasers are set at $\phi_m = 0°$ ($m = 1, \ldots, n$).

a. Calculate a table of values of ϕ_m for scan angles of 15°, 30°, and 60° (off broadside) for $n = 4, 8,$ and 16.

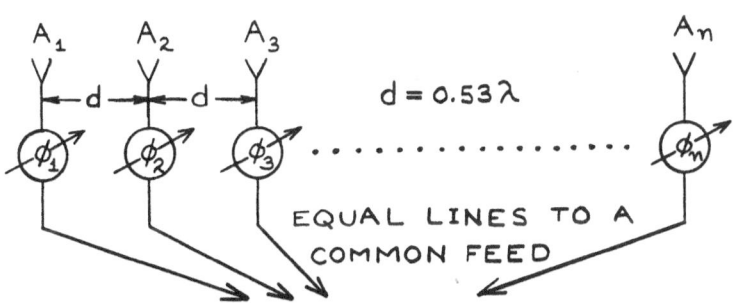

FIGURE 9-26. Illustration for problem 1.

b. A two-bit digital phaser has four positions: 0°, 90°, 180° and 270°. A three-bit phaser has eight positions: 0°, 45°, 90°, 135°, 180°, 225°, 270° and 315°. What would be the tables of phase settings in part (a) of this problem, and what would be the maximal quantization phase errors for a two-bit and a three-bit phaser?

c. Using the results of Problem 9-1a and b, calculate and summarize the effect of quantization phase errors on beam pointing, gain, and side lobe levels as functions of phaser bits and number of elements for uniform linear arrays.

PROBLEM 9-2.

a. Use Eqs. (9-18) through (9-25) to estimate the error amplitude and wavelength λ_e from the data contained in Figure 9-25.

b. How small should ϕ_0 be in order for the error lobes to be at the level of the side lobes, -55 dB?

c. What would be the locations of the error lobes if $\lambda_e = 5 \lambda$?

PROBLEM 9-3. The operation of multiple beam arrays, shown schematically in Figure 9-18, is only hinted in section 9.5. Explain to yourself in detail how each one of the schemes (Fig. 9-18a and 9-18b) produces multiple beams, and how they can be utilized.

CHAPTER 10

Reflector and Lens Antennas

The concept of radiation from an aperture was introduced in Chapter 7. Chapter 8 explained that a narrow, high-gain beam can be produced by a large number of discrete radiating elements that are configured as an array. In this chapter, a similar effect is described—one that can be produced if a large aperture is illuminated by a continuous electromagnetic (EM) field. This approach to beam forming is analogous to that used in optics: reflectors and lenses are used, and geometrical optics is applied extensively in the design and analysis of such antenna systems.

10.1 APERTURE ANTENNA GAIN

With a flat aperture in the x-y plane in which the EM field is polarized (aligned) along the x direction, and satisfying

$$E_x = \eta H_y, \qquad (10\text{-}1)$$

the field in the aperture is (locally) a plane wave. By Eqs. (7-1) and (7-6), the far field that arises from the aperture fields is expressed as follows:

$$E_\theta = \frac{-j}{\lambda} \frac{e^{-jkr}}{r} \left(\frac{1 + \cos\theta}{2} \right) \cos\phi \, F(\theta,\phi)$$

$$E_\phi = \frac{j}{\lambda} \frac{e^{-jkr}}{r} \left(\frac{1 + \cos\theta}{2} \right) \sin\phi \, F(\theta,\phi). \qquad \textbf{(10-2)}$$

The aperture factor $F(\theta,\phi)$ is given by

$$F(\theta,\phi) = \iint\limits_{\text{aperture}} E_x(x',y')e^{\psi(x',y')} \, dx'dy'$$

$$\psi(x',y') = jk(x'\sin\theta\cos\phi + y'\sin\theta\sin\phi) \qquad \textbf{(10-3)}$$

which is analogous to the array factor that was introduced in Chapter 8. Eq. (10-3) was derived in Chapter 7 in connection with radiation from an open rectangular waveguide.

The integral in Eq. (10-3) is a two-dimensional analog of the Fourier integral obtained in Chapter 4 in Eqs. (4-15) and (4-17) when a straight wire antenna was considered. Therefore, if the field in the aperture is known or can be estimated, the far field can be computed by evaluation of a double Fourier integral. As in previous chapters, the aperture factor, in most cases of interest, is a high-gain function characterized by a narrow main beam. An important relationship can be derived from Eqs. (10-2) and (10-3) for the case of $\theta = 0$ (the far field along the z axis and perpendicular to the aperture plane as shown in Fig. 7-1). With $\cos\theta = 1$ and $\sin\theta = 0$,

$$F(\theta,\phi) = F_0 = \iint\limits_{\text{aperture}} E_x(x',y') \, dx'dy', \qquad \textbf{(10-4)}$$

$$E_\theta = \frac{-j}{\lambda} F_0 \cos\phi \frac{e^{-jkr}}{r},$$

$$E_\phi = \frac{j}{\lambda} F_0 \sin\phi \frac{e^{-jkr}}{r}, \qquad \textbf{(10-5)}$$

$$|E|^2 = |E_\theta|^2 + |E_\phi|^2 = \left[\frac{F_0}{\lambda r}\right]^2. \qquad (10\text{-}6)$$

The aperture field may be expressed in the form

$$E_x(x',y') = E_0 e(x',y'), \qquad (10\text{-}7)$$

in which E_0 is the maximal value of E_x, and $e(x',y') \le 1$, for all points (x',y') in the aperture. Clearly, for any illumination

$$|F_0|^2 = \left| \iint\limits_{\text{aperture}} E_x(x',y')dx'dy' \right|^2$$

$$= E_0{}^2 \left| \iint\limits_{\text{aperture}} e(x',y')dx'dy' \right|^2 \le (E_0 A)^2 \qquad (10\text{-}8)$$

in which

$$A = \iint\limits_{\text{aperture}} dx'dy' \qquad (10\text{-}9)$$

is the area of the aperture. The equality sign in Eq. (10-8) is valid when $E_x = E_0$ for all points in the aperture—a condition called uniform illumination. Thus, for any aperture, regardless of its shape, the highest gain is realized when the illumination is uniform. Furthermore, by use of Eqs. (10-6), (10-8), and (10-9) the following is derived for this case:

$$|E|^2{}_{\text{max}} = (E_0 A/r\lambda)^2. \qquad (10\text{-}10)$$

If it is assumed that the aperture has an ideal radiation pattern (i.e., that the radiation is confined to a cone as shown in Fig. 10-1), the law of energy conservation is used to obtain

$$\text{Power} = \frac{1}{2}(E_0^2/\eta) A = \frac{1}{2}(E^2/\eta)S, \qquad (10\text{-}11)$$

in which S is the section of spherical surface subtended by $S = r^2\Omega$. By combination of Eqs. (10-10) and (10-11), the far field and the aperture field are related:

$$\left(\frac{E}{E_0}\right)^2 = \left(\frac{A}{\lambda r}\right)^2 = \frac{A}{r^2\Omega},$$

or

$$\boxed{\lambda^2 = A\Omega} \qquad\qquad \textbf{(10-12)}$$

This result, which is identical to Eq. (8-39), constitutes a relationship between wavelength, aperture size, and beam size and is therefore quite important. Now, by knowing the aperture area and wavelength, the beam width is estimated. For example, a round dish of 3 m diameter that is operated at a frequency of 3 GHz ($\lambda = 0.1$m) has a beam size of

$$\Omega = \frac{\lambda^2}{A} = \frac{0.04}{9\pi} \cong 1.41 \times 10^{-3} \quad \text{(steradian)}.$$

The cone angle, α, for the simple case of a circular cone, is given by

$$\alpha = 2\cos^{-1}\left[1 - \frac{\Omega}{2\pi}\right] \cong 2.4°.$$

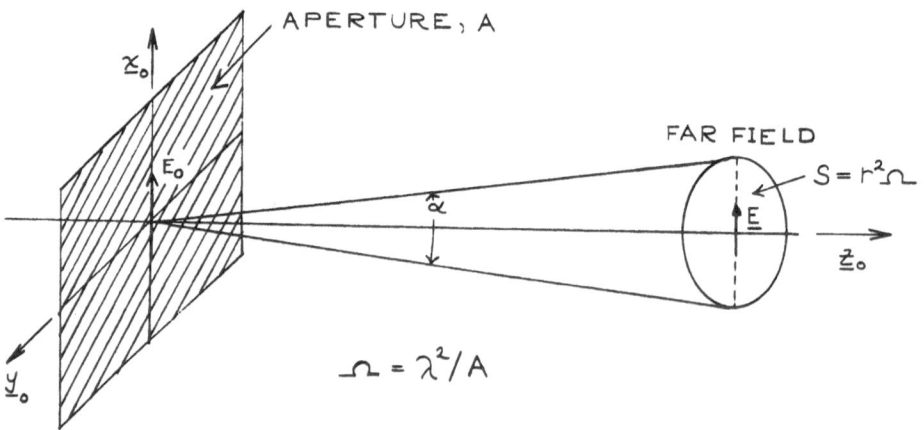

Figure 10-1.   Aperture-beamsize relation.

The circular aperture has a conical beam of circular cross section; however, other aperture shapes have more complicated beam shapes. Nevertheless, Eq. (10-12) remains valid in these cases also, despite the more involved geometry.

The geometrical definition of directive gain is given by Eq. (8-38), that is: $g = 4\pi/\Omega$. If this result is combined with Eq. (10-12), the following is obtained:

$$\boxed{g = 4\pi A/\lambda^2} \qquad (10\text{-}13)$$

This formula was derived by a different method in Chapter 8, in the context of planar arrays. This result is valid for aperture antennas as well and is, therefore, a very general and important relation.

10.2 REFLECTOR GEOMETRY

Most reflectors are based on the geometrical properties of the conic sections, considered here in some detail. The ellipse, shown in Figure 10-2a, is formed by the locus of all points, P, which satisfy the relationship $d_1 + d_2 = 2a$. By use of analytic geometry, it is not difficult to show that the ellipse of Figure 10-2a is described by

$$r^2 = 4f\left(1 - \frac{f}{2a}\right)\left(1 - \frac{z}{2a}\right)z. \qquad (10\text{-}14)$$

The circle, as shown in Figure 10-2b, is a special case obtained when $a = f$ and F_1 and F_2 coalesce. It has the equation

$$r^2 = 2fz - z^2, \qquad (10\text{-}15)$$

which is more familiar in the standard form:

$$f^2 = r^2 + (z - f)^2.$$

The hyperbola, illustrated in Figure 10-2c, is the locus of a set of points that satisfy $d_2 - d_1 = 2a$. Its equation, which is similar to that of the ellipse, is

$$r^2 = 4f\left(1 + \frac{f}{2a}\right)\left(1 + \frac{z}{2a}\right)z. \tag{10-16}$$

If the separation of the foci is increased without limit, both the ellipse and hyperbola approach the form

$$\boxed{r^2 = 4fz} \tag{10-17}$$

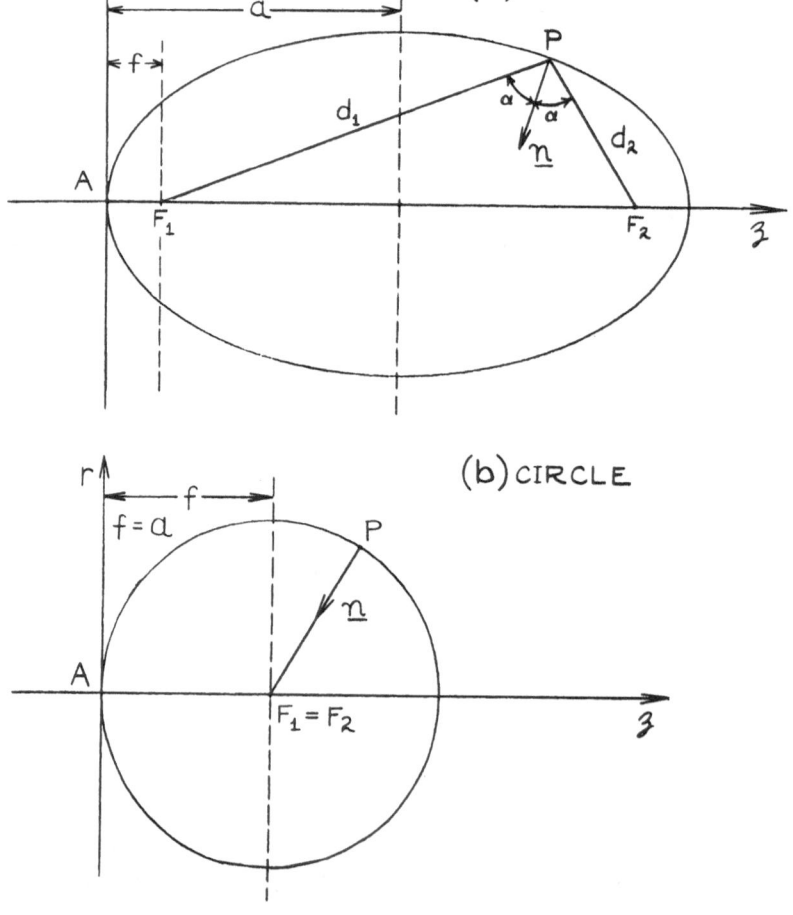

Figure 10-2.   Conic sections.

which is the equation of the parabola, as shown in Eq. (5-18). As seen in Figure 10-2d, the parabola can also be described as the locus of points for which

$$d_1 = d_2, \qquad \textbf{(10-18)}$$

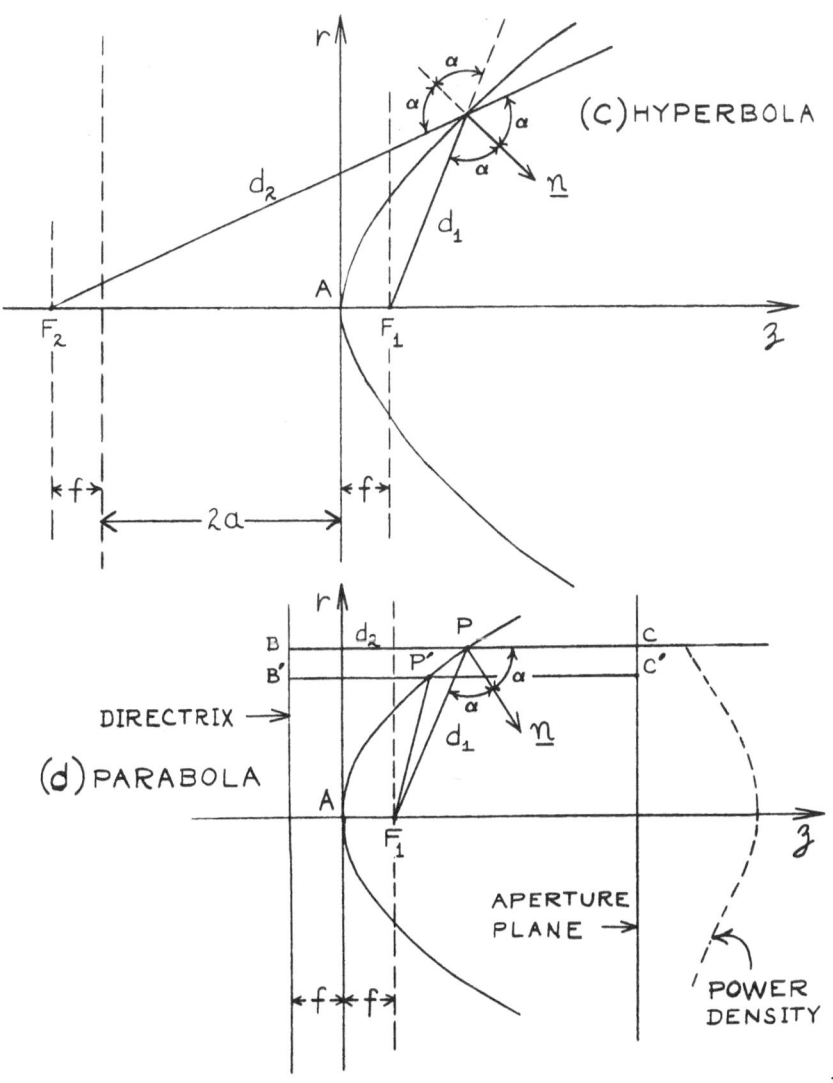

Figure 10-2. (*Cont.*)

in which $d_1 = \overline{F_1P}$ is the distance from the focus to P, and $d_2 = \overline{BP}$ is the distance from P to the directrix—the line perpendicular to the z axis at $z = -f$.

It is not too difficult to show that the normal to a conic section at a point, P, is the bisector of the angle between the lines $\overline{PF_1}$ and $\overline{PF_2}$, which connect the foci to P. For the parabola, F_2 lies at infinity, and $\overline{PF_2}$ is, therefore, parallel to the z axis for all points P on the curve.

The conic sections in Figure 10-2 can be rotated about the z axis to generate surfaces of revolution or translated in a direction normal to the r-z plane to generate cylindrical surfaces. Both types of surfaces are encountered in antenna designs. Chapter 5 described the application of geometrical optics to gently curved conducting surfaces. In a local plane of reflection the angle of reflection of a ray is equal to its angle of incidence. Therefore, rays emanating from one focus of a reflector whose shape is a conic section, are reflected toward the other focus (ellipse, parabola) or appear to emanate from the other focus (hyperbola) following reflection.

Reflectors of elliptical cross section (Fig. 10-2a) are infrequently used in antenna designs.* Of the other shapes, the parabola is by far the most common. When used for transmission, the parabolic reflector is fed from its focus in order to provide a collimated beam of rays along its axis (z axis in Fig. 10-2). As a receiver, the parabolic surface causes the rays that enter parallel to its axis to converge at the focus. Shallow spherical reflector dishes can be viewed as an approximation to a paraboloid (as discussed in Chap. 5). They are occasionally used because of their higher symmetry, which makes beam scanning easier in certain applications. Hyperbolic reflectors are often used as subreflectors. For example, rays that emanate from F_2 and are reflected by the convex side of the hyperbolic section of Figure 10-2c appear to emerge from the focus at F_1. The hyperbolic surface may be oriented such that F_1 lies at the focus of a parabolic reflector, so that the rays reflected by the hyperboloid finally emerge parallel to the parabolic axis. Such secondary reflectors permit the placement of the antenna feed behind or at the center of the primary dish—a more convenient location for many applications. This is shown in Figure 10-8.

Examination of Figure 10-2d and the definition of Eq. (10-18) reveal that

$$\overline{F_1P} + \overline{PC} = \overline{BC},$$

and

$$\overline{F_1P'} + \overline{P'C'} = \overline{B'C'} = \overline{BC},$$

---

*One exception is the Gregorian feed, which uses an ellipsoidal subreflector.

thus, the path length of any ray that passes from the focus to the aperture plane (plane normal to z axis) is constant. Therefore, a spherical wavefront that radiates from F_1 is transformed into a plane wavefront, as illustrated in Figure 10-3. On the other hand, the power flux at the aperture plane is, in general, not constant. For example, the power density in the region bounded by the rays $\overline{PC}$ and $\overline{P'C'}$ is determined by the power density at the segment $\overline{PP'}$, which is inversely proportional to the distance $\overline{F_1P}$. Because $\overline{F_1P}$ varies with location, an isotropic source at the focus gives rise to a tapered power distribution at the aperture plane, as shown by the dashed line in Figure

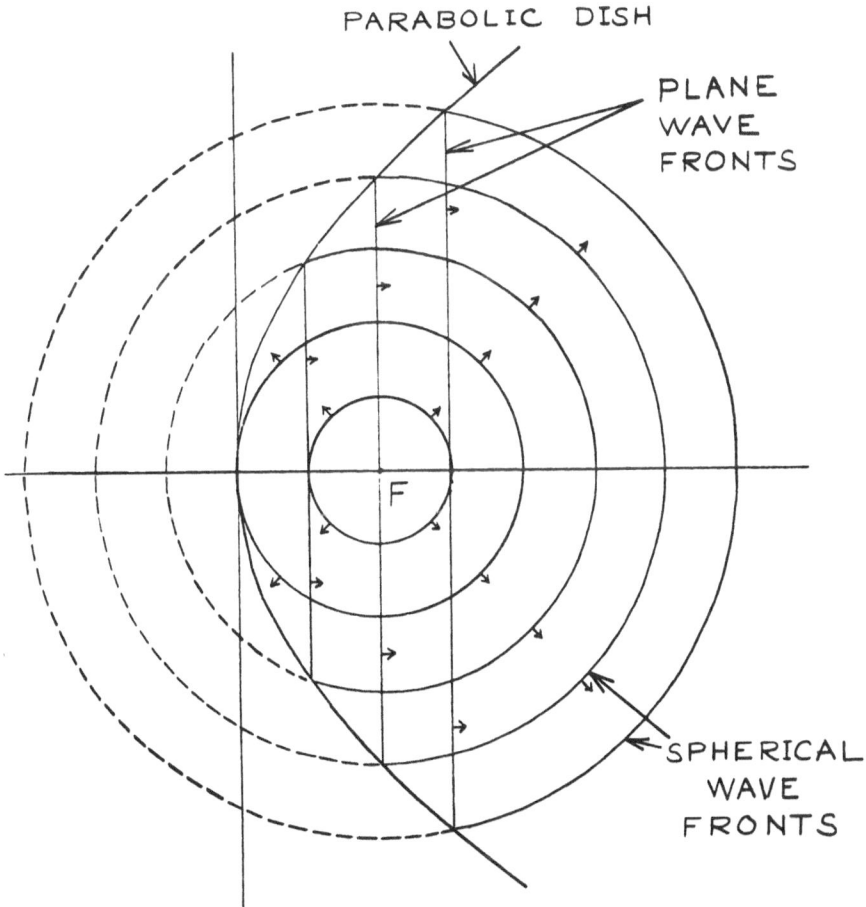

PARABOLIC DISH

PLANE
WAVE
FRONTS

F

SPHERICAL
WAVE
FRONTS

Figure 10-3. Plane wave formed by parabolic reflector.

10-2d. Note that the power density decreases monotonically as the distance from the z axis increases.

The field distribution at the aperture plane can be computed by use of geometrical optics, provided that the reflector size, focal distance, and smallest radius of curvature are large compared to the longest wavelength used. The position chosen for the aperture plane is arbitrary, as long as its distance from the reflector surface is not too large. As discussed in Chapter 5, geometrical optics break down in the vicinity of foci. Because the second focus of the parabolic reflector lies at infinity, geometrical optics are not expected to account for the far field. Instead, Eqs. (10-2) and (10-3) can be used to calculate the far field once the aperture plane distribution has been determined by geometrical optics. Measurements confirm that predictions obtained by this two-step theoretical approach are quite accurate.

10.3 THE FAR FIELD OF A CIRCULAR REFLECTOR

For round reflectors, rotationally symmetrical about an axis (z axis in Fig. 10-2), a simplified expression for Eq. (10-3) may be derived. The geometrical conventions of Figure 10-4 are used in this procedure, which reduces the double integral of Eq. (10-3) to a single integral.

Assuming that the aperture has a radius a, and that the field distribution, E_x, is a function of the distance from the origin, ρ, only, with reference to Figure 10-4 and Eq. (10-3),

$$x' = \rho\cos\phi', \quad y' = \rho\sin\phi',$$
$$x'\sin\theta\cos\phi + y'\sin\theta\sin\phi = \rho\sin\theta\cos(\phi - \phi'),$$
$$dx'dy' = \rho d\rho\, d\phi',$$

and

$$F(\theta,\phi) = \int_0^a E_x(\rho)\rho\, d\rho \int_0^{2\pi} e^{jk\rho\sin\theta\cos(\phi - \phi')}d\phi'. \qquad \textbf{(10-19)}$$

Because the result is expected to be symmetrical about the z axis (independent of ϕ), $F(\theta,\phi) = F(\theta)$, and $\phi = 0$ may be chosen in the integral without loss of generality. The second integral in Eq. (10-19) has the definition

$$\int_0^{2\pi} e^{jx\,\cos\phi}\, d\phi = 2\pi J_0(x), \qquad \textbf{(10-20)}$$

in which $J_n(x)$ is a Bessel function of order n. This well-known and tabulated mathematical function is named after the nineteenth-century German astronomer, Wilhelm Friedrich Bessel. Eq. (10-19) can now be expressed

$$F(\theta) = 2\pi \int_0^a \rho J_0(k\rho\sin\theta)E_x(\rho)d\rho \qquad \textbf{(10-21)}$$

Once $E_x(\rho)$ is specified, this integral can be evaluated quite accurately by various approximation techniques. If the distribution is uniform ($E_x = 1$ for $0 \leqslant \rho \leqslant a$), approximation is unnecessary. Using the Bessel function identity

$$\int_0^a \rho J_0(q\rho)d\rho = \frac{a}{q}J_1(qa),$$

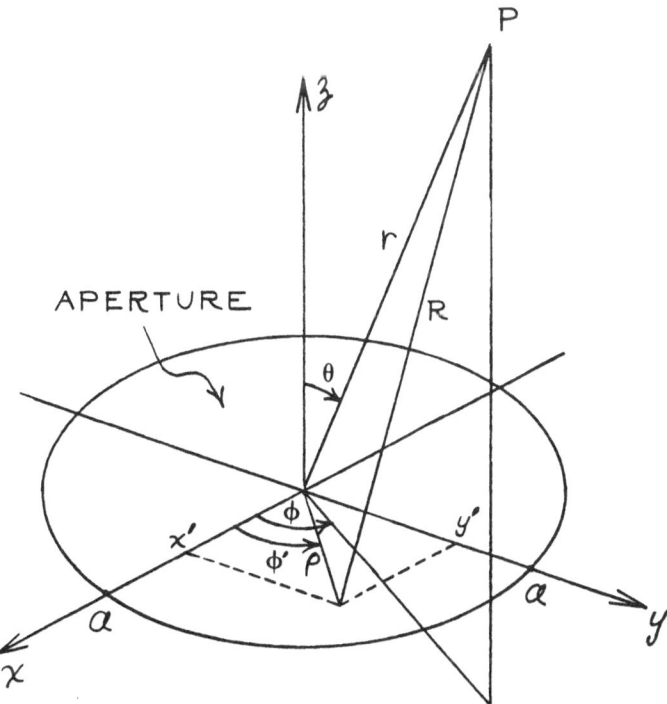

Figure 10-4. Circular aperture geometry.

with $q = k \sin\theta$,

$$F(\theta) = \pi a^2 \frac{2J_1(ka\sin\theta)}{ka\sin\theta} = A \frac{2J_1(u)}{u}, \qquad \textbf{(10-22)}$$

is obtained, in which A is the aperture area, and $u = ka \sin\theta = 2\pi(a/\lambda) \sin\theta$. Eq. (10-22) is analogous to the aperture factor for a straight wire of length, L, with uniform current, for which, referring to Eq. (4-20) or (4-21),

$$F(\theta) = L \sin u/u$$

and

$$u = kL \sin\theta/2$$

(here θ is measured from the normal to the wire). Thus, for the case of a uniformly illuminated circular aperture, the aperture factor is proportional to the aperture area and the function $J_1(u)/u$. This function is similar in appearance to the function $\sin u/u$, however, its side lobes fall off somewhat faster with increasing θ. The uniform wire, as well as the long uniform linear array, produce a first side lobe level of -13.6 dB. For the uniform circular aperture, the first side lobe lies at -17.6 dB, and the radiation pattern has rotational symmetry about z. This radiation pattern is illustrated in Figure 10-5. The width of the main beam (between nulls) is found from the first zero of the Bessel function J_1:

$$\frac{2\pi a}{\lambda} \sin\theta_1 = 1.22\pi \rightarrow \theta_1 \cong 1.22\lambda/2a \quad \text{(radians)}. \qquad \textbf{(10-23)}$$

If a circular hole in an opaque screen is uniformly illuminated, the transmitted light has the far field pattern shown in Figure 10-5, given by Eq. (10-22). The alternating light and dark rings in the pattern that surrounds the main beam are called Airy disks, for the nineteenth-century British astronomer Sir George Airy.

Optical physicists refer to the radiation pattern of Figure 10-5 as a diffraction pattern. These types of patterns are intimately related to the achievable resolution of both antennas and optical instruments such as telescopes. When an antenna is scanned across a distant point source, the variation in signal intensity with scan angle resembles the pattern of Figure 10-5. Clearly, if two sources are separated from each other by an angle less than $2\theta_1$, that is, if both fall within the main beam, or Airy disk, they may not be distinguishable. Furthermore, if one source is much weaker than the other, the side lobes of the stronger source may continue to mask the weaker source for angular separations well in excess of $2\theta_1$. Resolution of sources of comparable strength is enhanced by reducing θ_1, which corresponds to increasing the aperture size or gain. However, if one source dominates the other, resolution is not improved by increasing the gain. Under these circumstances, what is needed is a reduction in side lobe level. Because side lobe levels are, practically speaking, independent of aperture size, they can not be reduced by increasing

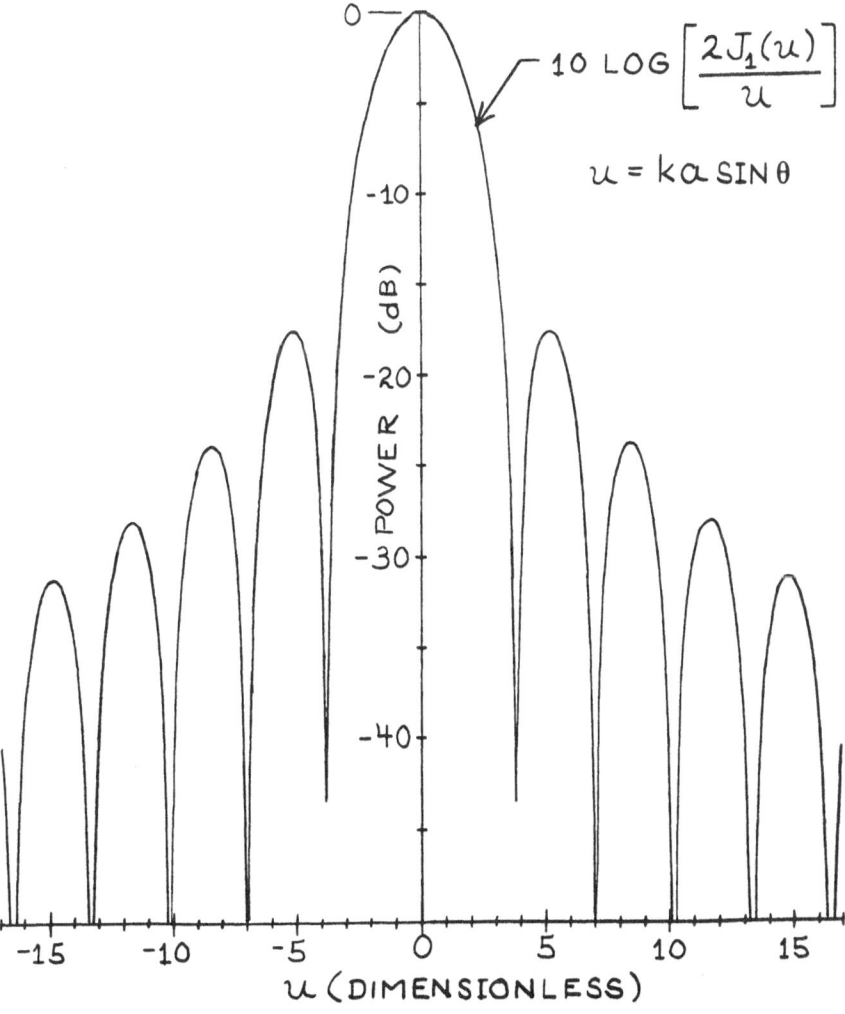

Figure 10-5. Far field pattern for circular, uniformly illuminated aperture.

the gain. However, by changing the aperture illumination function (the field distribution), side lobe levels can be lowered considerably. The echos of radar targets can vary in strength by many orders of magnitude. Good resolution is one of many reasons for designing antennas with low side lobes.

10.4 DESIGN OF VARIOUS REFLECTOR ANTENNAS AND FEEDS

Reflector antennas may be fed by a variety of methods. The particular technique chosen depends on such parameters as frequency range, side lobe level, antenna operational mode (monopulse, beamscanning, etc.), and aperture efficiency. The basic requirements are that a primary source has a phase center located at the focus of the reflector and a good "front-to-back" ratio—i.e., a radiation pattern that illuminates the reflector and does not radiate much into the opposite half-space. Furthermore, the feed structure, which may consist of waveguides or other types of transmission line, must not excessively block the reflector aperture. Some of the most common arrangements are shown in Figure 10-6. For the configurations illustrated, the primary feed is responsible for blockage. Although it is desirable to minimize blockage, the primary feed must have sufficient gain to reduce spillover loss. Figure 10-7 illustrates these conflicting design objectives.

If the feed pattern is highly directive, spillover is minimized, but illumination of the outer area of the reflector is weak. As a result, utilization of the available aperture is poor (aperture efficiency is low). A larger primary feed structure is required to produce the higher directivity, and as a consequence, aperture efficiency is reduced even further by blockage.

Blockage at the center of the antenna also results in higher side lobes. A low-gain feed, such as the two-dipole endfire array in Figure 10-6, presents a small blockage area, but considerable spillover will occur unless the reflector is very deep. However, as noted earlier, a deep dish leads to a strong space taper in the illumination, due to the relatively large dissimilarity in the distances $\overline{OA}$ and $\overline{OB}$ (Figure 10-7). The intensity (i.e., the power density) of the primary spherical wave is inversely proportional to the square of the distance from the source to the reflector surface. Therefore, the ratio of the intensity at B to that at A is, for a parabolic reflector,

$$(\overline{OA}/\overline{OB})^2 = f^2/[r_B^2 + (f - z_B)^2] = 1/[1 + (D/4f)^2]^2. \quad \textbf{(10-24)}$$

The parameter f/D, which is a measure of parabolic reflector depth, has an important role. When f/D is small the dish is deep, and the space taper, given by Eq. (10-24), is large. When f/D is large the reflector is relatively flat, and the illumination is nearly uniform. The most useful range for this parameter

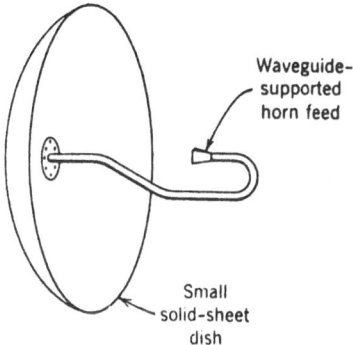

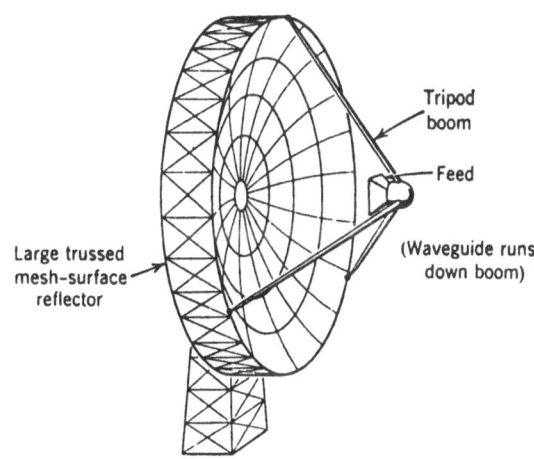

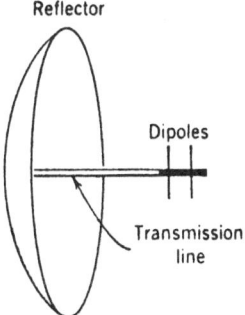

Figure 10-6. Common reflector configurations. (*From L. V. Blake, Antennas, Dedham, MA: Artech House, 1984, pp. 265, 275; copyright ©1984 by Artech House, Inc., reprinted by permission.*)

lies between 0.25 and 0.5. However, for some space communications applications, flatter reflectors ($f/D \approx 1$) have been used.

Detailed design examples for some simple circular parabolic reflector and feed systems are provided in Appendix D. The design process is usually iterative: certain parameters, such as aperture area and illumination, are first chosen to meet the various design objectives and then modified to meet all of the system specifications, if possible. Some nomographs, which represent many of the relationships among the various parameters in graphic form, also appear in Appendix D.

Parabolic cylinder reflectors are quite useful for certain applications. Figure 10-7 can also be interpreted as a cross-sectional view of a parabolic cylinder, in which case the feed at the focus is a line source. The line source might consist of a linear array of horns, waveguides, slots, or other radiating elements that form a long and narrow continuous aperture.

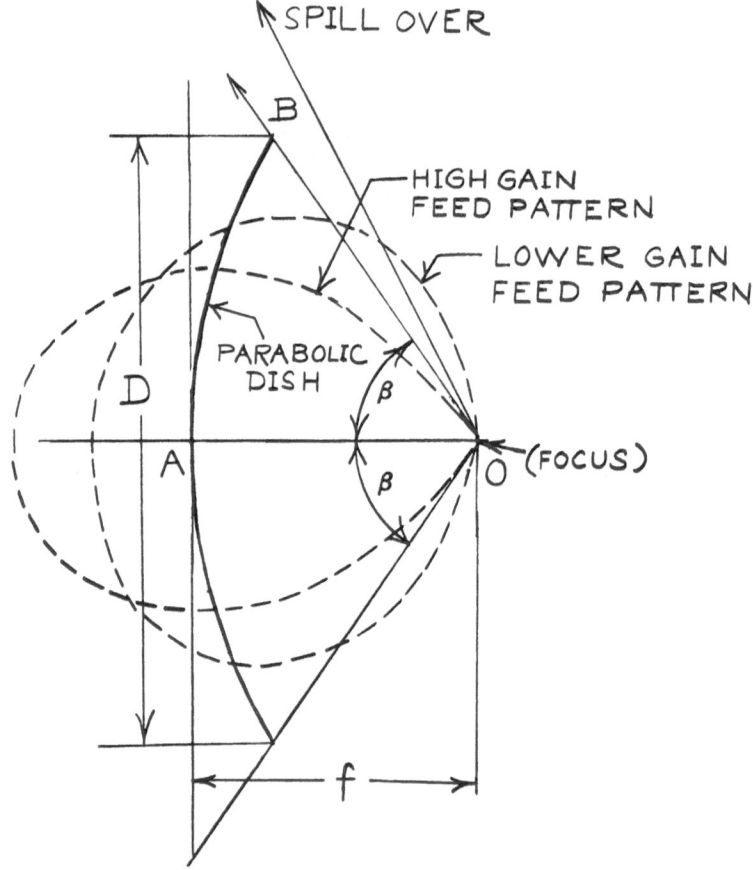

Figure 10-7. Feed gain and spillover loss.

The aperture blockage incurred by large primary feed structures and excessive feed line lengths that would be required in some conventional configurations, have prompted the use of Cassegrain feeds for reflector antennas. This feed geometry, shown in Figure 10-8, was invented in the seventeenth century for use in optical telescopes. The subreflector is one surface of a hyperbola of revolution, whose focus, F_1, coincides with the focus of the main reflector. The other focus, F_2, coincides with the phase center of the feed. A spherical wavefront, which appears to emanate from F_2, is transformed into another spherical wavefront, which appears to radiate from F_1. This wavefront is finally transformed into a plane wavefront when reflected from the main reflector. The Cassegrain geometry facilitates the support of a heavy multichannel feed structure, reduces support structure blockage, and minimizes feed line lengths. The Cassegrain feed is a simple example taken from a whole class of dual (or multiple) reflector antennas, in which the main reflector is fed through one or more subreflectors.

Because some antenna systems do not require a symmetrical pencil beam, a variety of nonsymmetrical configurations that reduce feed (or subreflector) blockage and simplify mechanical design can be used. One example, the offset parabolic reflector, is illustrated in two views in Figure 10-9. The main reflector is formed from a piece of a paraboloid that has an elliptical parallel projection onto the aperture plane. The main beam is fan shaped—wide in the elevation plane and narrow in azimuth. This type of beam is often used

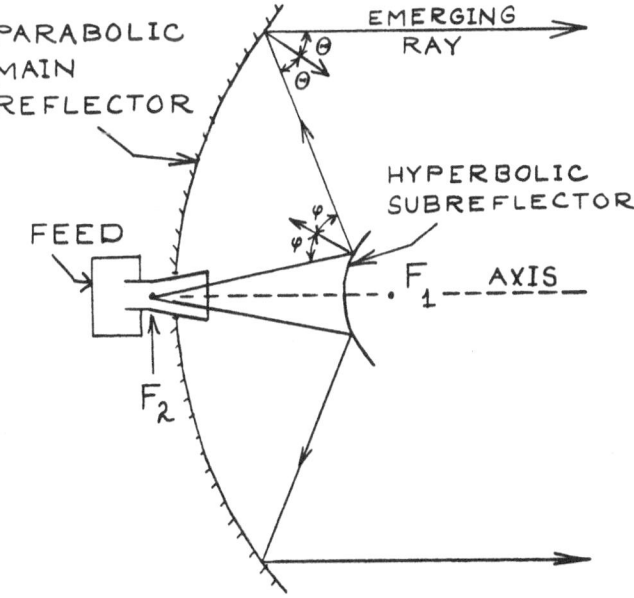

Figure 10-8. Basic Cassegrain geometry.

by air defense and traffic control radars, which must detect low-flying aircraft. The offset paraboloid antenna can be configured to eliminate aperture blockage, as seen in the side view in Figure 10-9. Note also that the wide side of the feed horn is aligned with the narrow side of the main reflectors to provide good illumination and minimize spillover. The main beam shape in the elevation plane is not symmetrical; however, this is of little consequence in many applications.

Offset parabolic reflectors often use subreflectors in Cassegrain fashion. Such a configuration can offer mechanical and electrical advantages. Clearly, the use of a pair of reflectors (main and sub) allows the designer additional degrees of freedom for obtaining the desired phase and amplitude distribution in the aperture plane. For example, by deviating from the parabola and hyperbola surface shapes, families of illuminations with very high gain for space telecommunications and specially shaped beams for covering irregular areas can be generated.

For large reflector antennas, wind loads pose a threat to operation and maintenance. To alleviate this problem, many reflectors have surfaces constructed of metallic grids or meshes. A metal wire or thin rod aligned with the incident wave polarization will scatter most of the incident energy. An array of closely spaced wires or rods (about $\lambda/4$ spacing is adequate) acts, for all practical purposes, as a solid reflector. The polarization perpendicular to the grid wires is not strongly affected. If two linear polarizations or circular polarization is required, the reflector surface should be made of mesh. This strategy reduces weight and wind loads with practically no adverse electrical effects. Some examples of commercially available grid-reflector antennas are displayed in Figure 10-10.

Another commercially available reflector antenna is pictured in Figure 10-11. This unit is a mobile earth station that supports satellite communications.

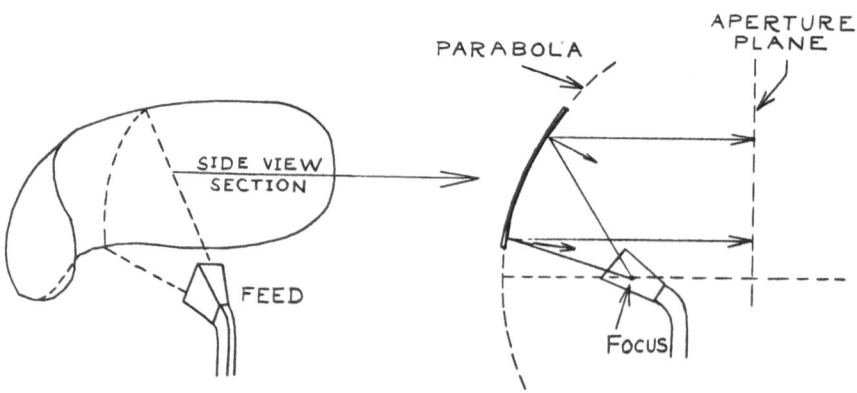

Figure 10-9. Two views of offset parabolic reflector.

Figure 10-10. Commercial grid reflector antennas. (*Photos courtesy of Andrew Corporation.*)

The 4.5-m dish can be hydraulically pointed and has an advertised gain of ~44 dB at 4 GHz (C-band) and 3-dB beam width of 1.2°. This antenna can be deployed and made operational within 15 min., according to the manufacturer.

The 9.3-m mechanically steerable antenna (Fig. 10-12) is a dual-reflector type that features a Gregorian feed. The Gregorian feed geometry is quite similar to the Cassegrain feed geometry; however, a concave ellipsoidal subreflector is used instead of the Cassegrain hyperboloid subreflector. This antenna has a gain of about 51 dB at 4 GHz and 3-dB beam width of 0.51 degrees. The manufacturer claims an efficiency of 77 percent. Although the Gregorian feed geometry is less compact than that of the Cassegrain, it is

Figure 10-11.   Commercial 4.5-m mobile dish antenna. (*Photo courtesy of Andrew Corporation.*)

believed that the Gregorian provides better side lobe performance in large reflectors.

Another type of reflector antenna, which minimizes noise and interference from other antennas, is the horn reflector or cornucopia shown in Figure 10-13. The horn reflector is frequently used for telecommunications in heavy traffic areas (i.e., locations where many closely spaced communications links pose "crosstalk" problems). It minimizes the antenna noise temperature (see Chap. 13) by reducing side lobes that would illuminate the Earth's surface and alleviates crosstalk by reducing side lobes in the horizontzal plane, where competing sources are located.

Figure 10-12. 9.3-m reflector antenna with Gregorian feed. (*Photo courtesy of Andrew Corporation.*)

In all of the reflector antennas described thus far, the primary feed is located at the exact focus of the main dish, or appears to be at the focus if subreflectors are used. In some cases deviations from this "ideal" condition are caused by fabrication and assembly errors; in other instances feed displacement from the focal point is intentional. A qualitative illustration of the effects of off-focus feeding is provided by ray tracing, as shown in Figure 10-14. If the feed is displaced along the reflector axis, rays remain symmetrically distributed but are no longer parallel to the axis. When the feed is "too far" from the reflector, the reflected rays converge to form a focus-like area of high-field strength at a certain distance from the antenna. Such a configuration can obviously offer an advantage in certain applications. If the feed is moved in the focal plane (normal to the reflector axis), the emerging rays, to a first approximation, are still parallel but no longer lie along the axial direction, as shown in Figure 10-14b. Such a feed displacement can provide limited beam scanning, but only within a narrow cone, because if the displacement is too large, beam broadening and gain loss are excessive. When only narrow angle scanning is required, such as for conical scan tracking applications, it is much easier to move the feed or subreflector than the main reflector. Large angle scanning is required for radar surveillance and must be performed by movement of the whole antenna assembly.

Beam scanning using feed or subreflector displacement is less restricted if

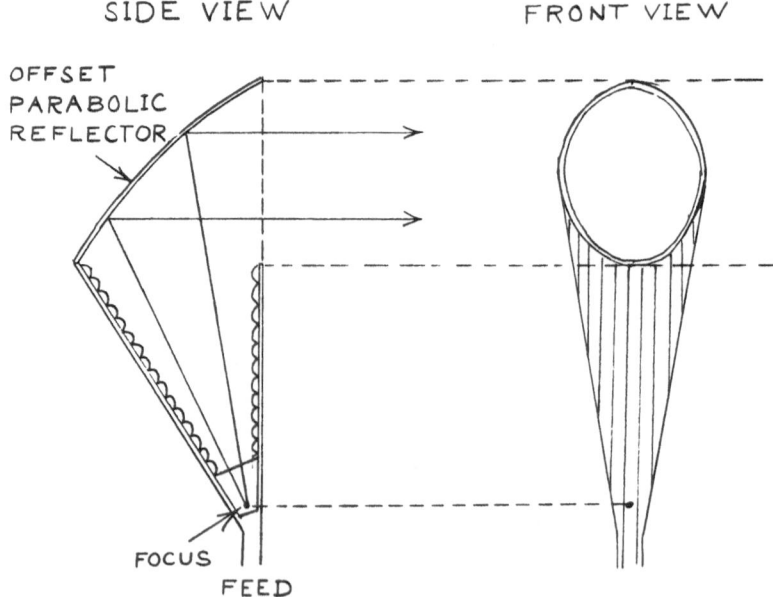

Figure 10-13. Schematic of a hog horn antenna.

the main reflector is spherical rather than parabolic. As shown in Figure 5-8, the sphere has a caustic instead of a focus. In a spherical reflector, a point source positioned near the tip of the caustic gives rise to an aperture field distribution whose phase is not constant (spherical aberration). As a result, the beam will be wider and have lower gain than that produced by a parabolic dish of the same diameter. However, because of symmetry the feed excursion and resultant beam scan can be larger than with a parabolic dish for comparable beam degradation. The properties of both the sphere and paraboloid are combined in the parabolic torus, a reflector that produces a high-gain

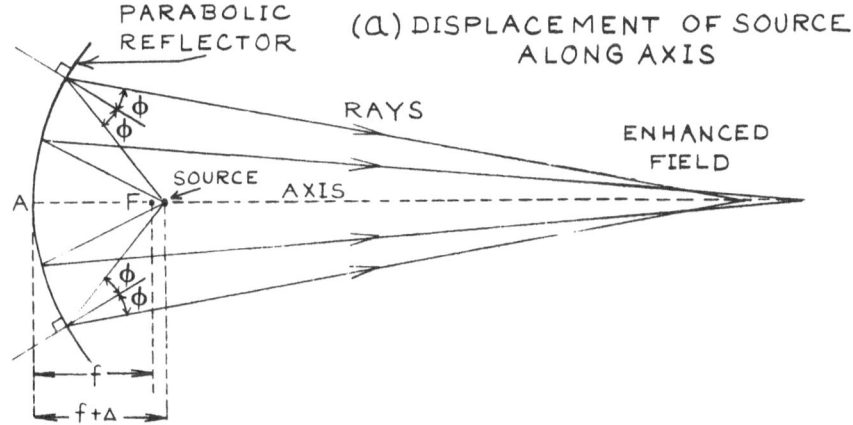

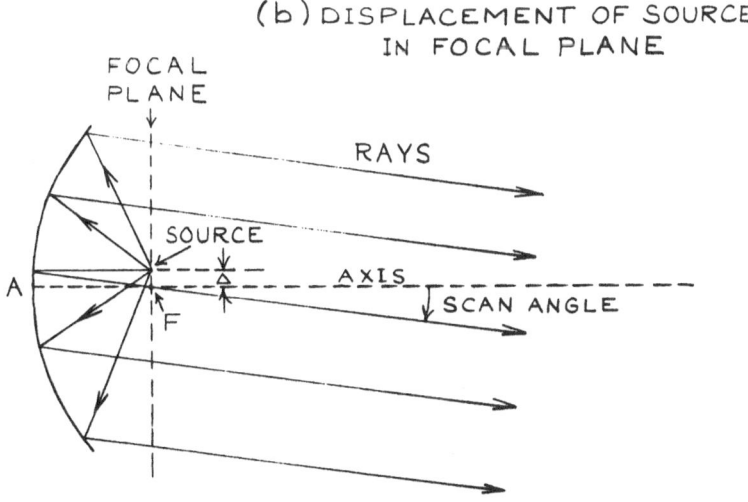

Figure 10-14. Effect of source displacement for parabolic reflector.

beam in the elevation plane, while allowing feed displacement scanning over an appreciable angle in azimuth.

Corner reflectors are another class of antennas (also discussed in Chap. 14). The example shown in Figure 10-15 is not really an aperture antenna, but actually an array. A corner reflector that has infinite reflecting planes and dihedral angle $\beta = \pi/m$ ($m = 1, 2, 3, \ldots$) can be rigorously analyzed by the method of images discussed in Chapter 4. The feed source (an array or

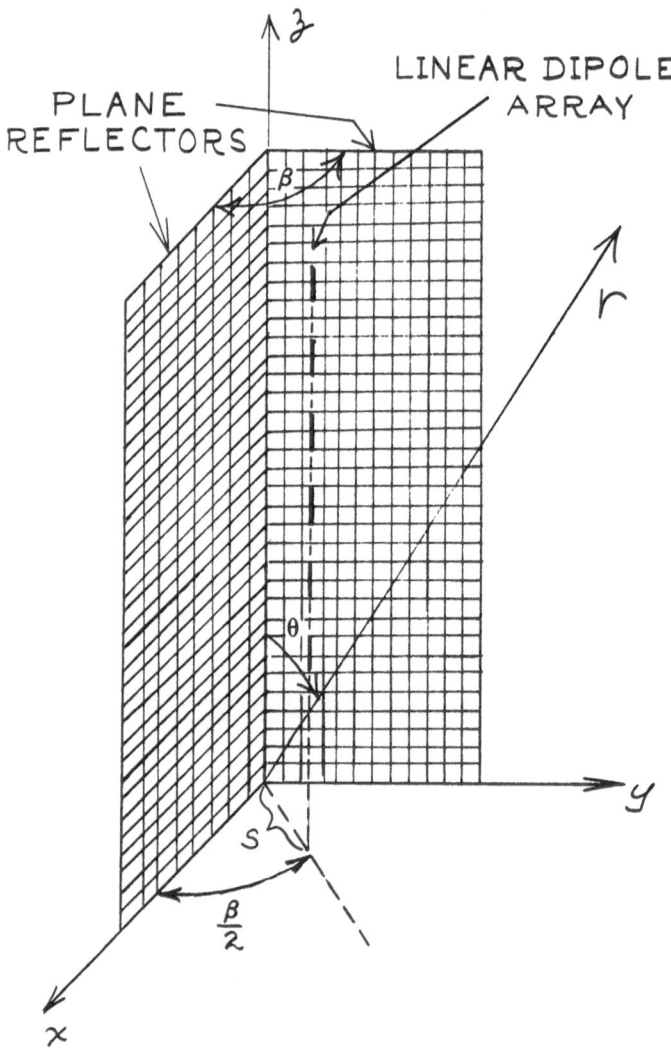

Figure 10-15.   Corner reflector with B = 90°.

a single element) combined with $2m - 1$ image sources forms an array whose radiation pattern can be calculated by use of the array methods discussed in Chapter 8. (If the angle β is not π/m, the radiation pattern cannot be rigorously determined by this simple approach, but an approximate solution can be found.) Analysis of practical corner reflectors with finite reflecting planes is more difficult, particularly if the plane dimensions are not very large compared to a wavelength.

Corner reflectors are inexpensive, moderate-gain antennas, and their simplicity is advantageous. They can be used instead of parabolic reflectors when extreme accuracy is not a requirement. As is the case for other reflectors, the corner reflector surface may consist of wire grid or mesh to reduce wind load.

Gain, beam width, and radiation resistance of corner reflectors depend on the angle, β, and the distance, S, of the source from the vertex (it is assumed that the source is located in the bisecting plane). In the $\theta = 90°$ plane, the practical maximal gain is about 12.5 dB for $\beta = 90°$, 14.5 dB for $\beta = 60°$, and 16.5 dB for $\beta = 45°$. A variety of radiation pattern shapes can be obtained by varying S.

10.5 LENS ANTENNAS

The lens, illustrated in Figure 10-16, represents another method by which a collimated (parallel) beam of rays may be produced over an aperture. Like the parabolic reflector, the lens converts a spherical wavefront, centered on

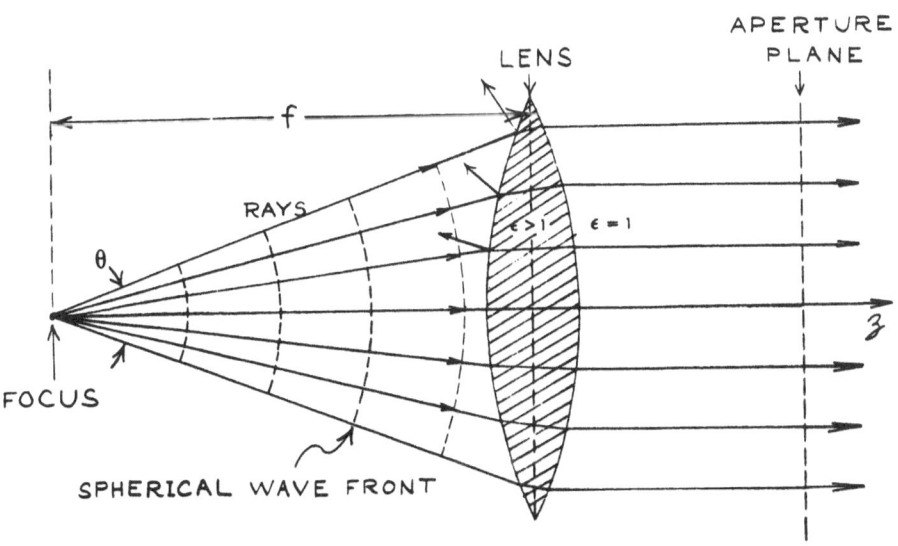

Figure 10-16. Collimated beam production by lens.

the focus, into a plane wavefront at the aperture plane. Conversely, an incident plane wave is transformed into a spherical wave that converges toward the focus. Although the lens arrangement is not subject to the blockage incurred by most reflectors, a certain amount of reflection at the dielectric interface cannot be avoided. Theoretically, the curved surfaces of a lens are parabolic in shape (e.g., parabolic cylinders, paraboloids of revolution); however, for practical lenses these surfaces are often approximated by circular sections. This approximation is quite satisfactory as long as θ is small (paraxial rays) and is easier to fabricate. Artifical dielectrics can be fabricated by embedding small metallic particles ($a/\lambda \ll 1$, in which a is a typical particle dimension) in a lightweight dielectric matrix. Artificial dielectrics can have a relatively high dielectric constant and thereby save both space and weight. Reflection losses at the interfaces can be reduced by application of a quarter wavelength coating of light dielectric with constant ε, such that $\varepsilon = \sqrt{\varepsilon_1 \varepsilon_2}$, in which ε_1 refers to the lens and ε_2 refers to the surrounding space. (An analgous method for matching transmission lines is described in Chapter 2—quarter wave transformer.)

Lens weight can also be reduced by *zoning,* as shown in Figure 10-17. The zones are cut in such a way that the phase difference between rays passing through adjacent zones is 2π radians. Disadvantages of zoning include increased losses through scattering and shadowing. Dielectric lenses are attractive for millimeter wavelength applications because lens size becomes more manageable.

A lens effect can also be produced by a dielectric layer of constant thickness and variable refractive index. Consider the ray geometry of Figure 10-18. If

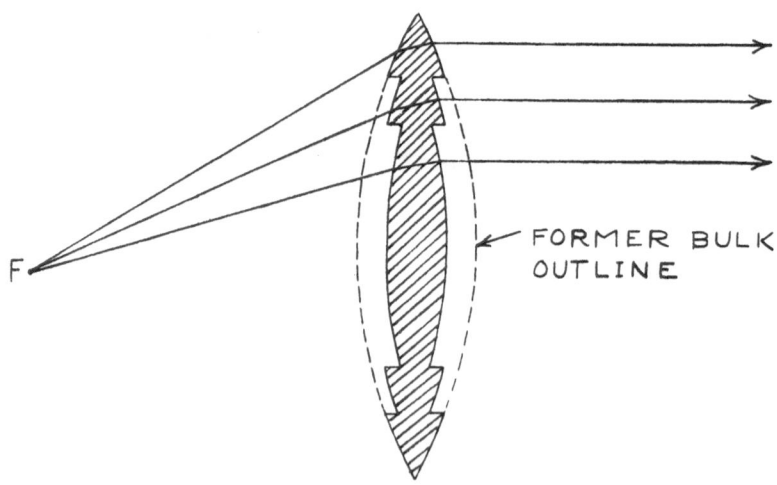

Figure 10-17.   Lens zoning.

it is assumed that r is much less than the focal length, f, the phase at point A of a wave that originates at F is given by

$$\psi_A = k\left[n_1\sqrt{r^2 + f^2} + n_2d\right] \approx k\left[n_1\left(f + \frac{r^2}{2f}\right) + n_2d\right]. \quad (10\text{-}25)$$

In order for ψ_A to be independent of r, the refractive index in the layer must satisfy

$$n_2 = n_0\left[1 - \left(\frac{r}{a}\right)^2\right] \quad (10\text{-}26)$$

with $n_1/2f = n_0d/a^2$. Thus, the index of refraction varies in parabolic fashion, with a maximum of n_0 at the center ($r = 0$). An artificial delay effect of this type can be achieved by the configuration shown in Figure 10-19. In this case the lens consists of many elements, each of which has a receiving antenna, A, and delay line, B, attached to a transmitting antenna, C. If a parabolic delay—a delay that is proportional to the refractive index of Eq. (10-26)—is introduced across the plane of the lens elements, the desired lens effect is obtained. The beam produced by this arrangement may be scanned if the phase delay, B, is electronically adjustable. Addition of a linear phase slope ($\psi_s \propto r$) to the beamforming phase will steer the beam off axis with minimal aberration. Such an antenna is an interesting phased array—optical feed hybrid with a large, unblocked aperture. By fine-tuning the phase delays, a desired aperture illumination can be achieved to minimize side lobe levels and optimize aperture efficiency at any scan angle. The price of such an antenna system is very high of course. This type of antenna has been successfully incorporated in some air defense systems.

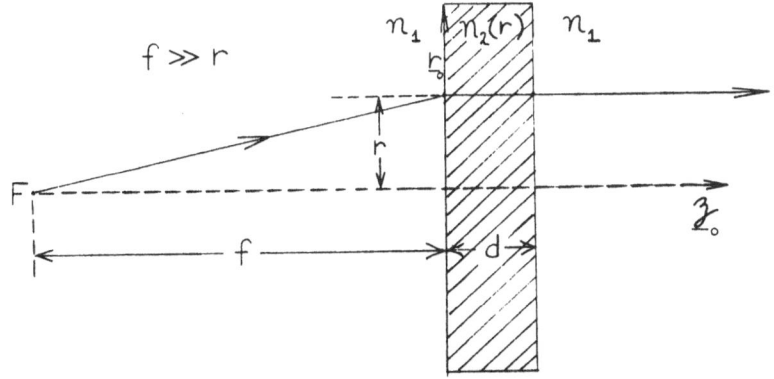

Figure 10-18. Variable refractive index lens.

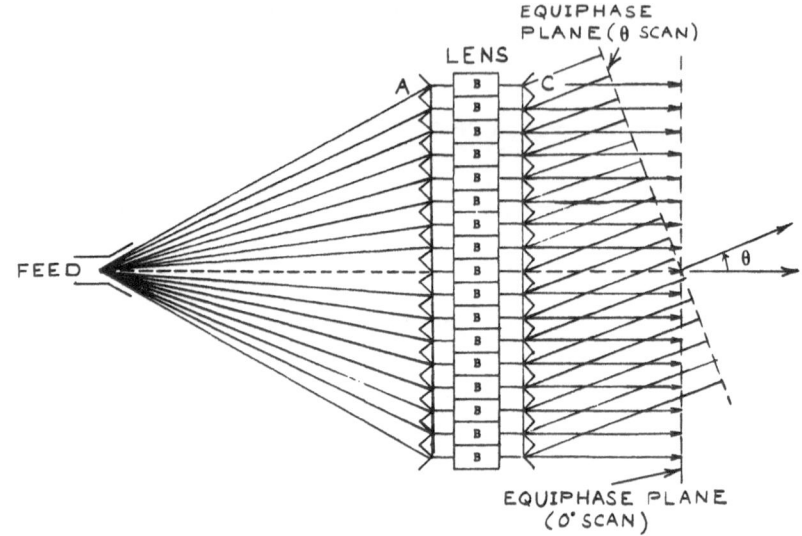

Figure 10-19.   Delay line lens.

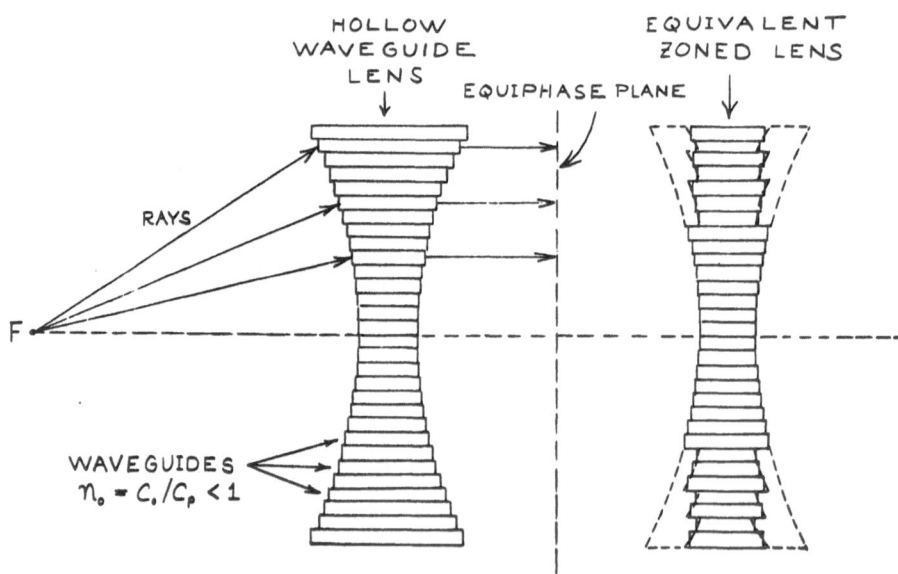

Figure 10-20.   Hollow waveguide lens configuration.

The delay type lenses "straighten out" a spherical wavefront by introducing a delay that decreases with increasing separation from the lens center. The same effect can be produced by introducing a phase advance as the distance from the center of the lens increases. This may be accomplished by use of hollow metallic waveguides, in which (as in Chap. 6) the phase velocity, c_p, exceeds c_0, the velocity of light in free space. This principle is the basis for the design of metallic lenses, which consist of many lightweight waveguides packed together, as shown schematically in Figure 10-20. The spherical wave excites a waveguide mode at the left of each waveguide, and this mode propagates to the other end and is re-radiated. The metallic waveguide lens design offers the possibility of building large, very rugged, lightweight lenses. As in dielectric or delay line lenses, there are losses caused by reflections at each interface. Metallic lenses are usually advantageous at the lower end of the microwave spectrum, where dielectric lenses would have to be either very heavy or structurally weak. The Radant Lens™, described in section 9.4 in the context of scanning arrays, can be seen as a hybrid; it combines some elements of a waveguide lens with some elements of a delay line lens.

PROBLEMS

Problem 10-1. Reconstruct all the necessary intermediary steps that were omitted in deriving Eqs. 6, 10, 12, 14, 16, 17, 23, 24 and 25 of this chapter.

Problem 10-2.

a. Estimate the maximum gain, beam width, and the location (in degrees) of the first two side lobes of a circular reflector antenna with $D = 8$ m over the band from 0.8 to 1.3 GHz. See Figure 10-21.

b. Find the focal distance, f, and the ratio, f/D, for a parabolic reflector as shown, when $t = 0.5$, 1.0, and 1.6 m.

Problem 10-3.*

A Cassegrain antenna with given D and f_1 and fed by a circular horn is shown in Figure 10-22. Try to minimize aperture blockage and maximize efficiency.

---

*Advanced problem.

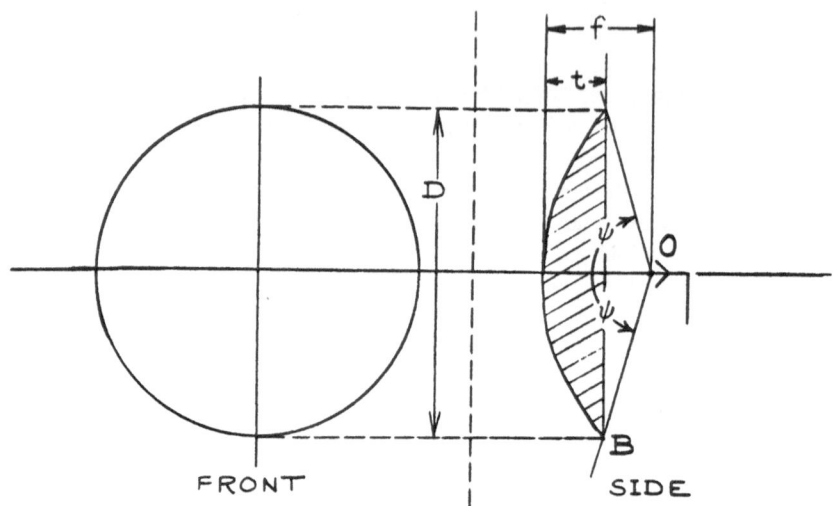

Figure 10-21. Illustration for problem 2.

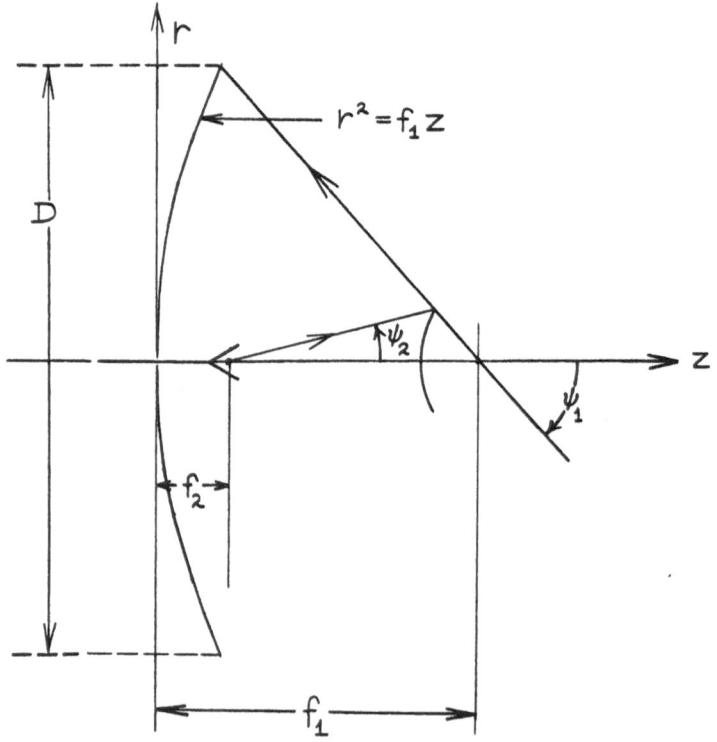

Figure 10-22. Illustration for problem 3.

a. Find the diameter and equation of the hyperbolic subreflector as a function of f_2.
b. Find ψ_2 as a function of f_2.
c. Show how the selection of f_2 affects the radii of the horn and and the subreflector and how to find a trade-off between efficiency and blockage.

PROBLEM 10-4.* Study Appendix D carefully. Then apply the method to the following.

a. A circular parabolic reflector, as shown in Figure 10-21 with $D = 3$ m and $f/D = 0.35$, is to be used for a communication channel from 3.8 to 4.4 GHz. The feed is a circular horn whose radiation curves can be approximated by the H plane curve in Figure D-3. (A is now the horn's diameter). Choose a horn with a diameter such that its radiated power at the edge, B, will be 10 dB or more below its peak for all frequencies in the band.
b. Using the curves of Figures D-1 and D-2, find the edge taper, the efficiency, the gain, the beam width, and the level of the two first sidelobes at the higher and the lower ends of the band.

---

*Advanced problem.

Miscellaneous Types of Traveling Wave and Broad-Band Antennas

The preceding chapters covered the most common types of antennas—simple wire antennas, small aperture antennas, antenna arrays, and large aperture antennas, both reflectors and lenses. There are many other antennas that do not clearly fit into any of these categories. Some very useful antennas include traveling wave, surface wave, and broad-band antennas, which are described in this chapter.

11.1 TRAVELING WAVE AND SURFACE WAVE ANTENNAS

Chapter 4 described a wave of the form $\exp(-jknz)$ that travels along a wire—a simple type of traveling wave antenna. Various wire configurations carrying traveling waves are used at high frequencies (HF) (3–30 MHz) and the lower very high frequencies (VHF—up to 100 MHz) to form fairly high-gain antennas. For the higher frequencies, such as ultrahigh frequencies (UHF) and microwaves, it is feasible to use other structures to guide traveling waves while they radiate their energy. One such structure, the dielectric rod, was described briefly in Chapter 6. The physical principle involved in guiding a wave along an open surface, even a flat plane, for example, can be explained in simplified fashion using, for example, the half-space $y > 0$ that lies above the $y = 0$ plane, as shown in Figure 11-1. As noted in Chapter 5 and else-where, the electromagnetic (EM) fields must satisfy wave equations, such as Eqs. (5-4). A field component, u, satisfies such a wave equation, and it is assumed, for the sake of simplicity, that u does not depend on z. Therefore, u satisfies the two-dimensional wave equation,

$$\frac{\partial^2 u}{\partial x^2} + \frac{\partial^2 u}{\partial y^2} + k^2 u = 0 \ (y > 0) \tag{11-1}$$

where $k = \omega\sqrt{\varepsilon\mu} = 2\pi/\lambda$. At the surface, $y = 0$, it is assumed that the field must also satisfy an impedance type boundary condition, which can be expressed

$$\frac{\partial u}{\partial y} = -kZu \ (\text{at } y = 0), \tag{11-2}$$

in which Z, the surface impedance, is constant. The object then is to find a solution, $u(x,y)$, that behaves like a traveling wave along the surface and decays exponentially in the direction perpendicular to the surface. The solution should have the form,

$$u(x,y) = e^{-jknx} e^{-k\gamma y}, \tag{11-3}$$

with $\gamma > 0$. If this expression is differentiated twice with respect to x and y and the results are substituted into Eq. (11-1), then

$$k^2(\gamma^2 - n^2 + 1) = 0. \tag{11-4}$$

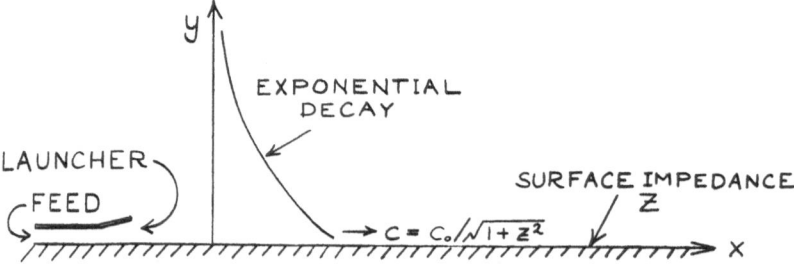

Figure 11-1. Surface mode schematic.

Furthermore, differentiating once with respect to y and substituting the result into Eq. (11-2) yields

$$\gamma = Z. \qquad (11\text{-}5)$$

Now, Eqs. (11-4) and (11-5) can be solved for n,

$$n = \sqrt{1 + Z^2}, \qquad (11\text{-}6)$$

so that the solution represented by Eq. (11-3) is completely characterized by Z. Recall that the refractive index, n, relates the propagation velocity c to c_0, the speed of light in free space:

$$c = c_0/n. \qquad (11\text{-}7)$$

According to Eqs. (11-2) through (11-7), if an impedance condition such as in Eq. (11-2) can be created along a plane, surface waves may exist. (Surface waves are bound to the plane and travel along it at a phase velocity less than that of light.) If the surface impedance, Z, diminishes for $x > x_0$, the surface wave of Eq. (11-3) speeds up, and it separates from the supporting surface $y = 0$. When Z goes to zero, the surface wave is radiated into the quarter-space $x > 0$, $y > 0$. The beam width of such an antenna depends on Z and on its rate of change.

Real physical structures that support electromagnetic surface waves are shown in Figure 11-2. That of Figure 11-2a is a dielectric layer over a conducting plane. This structure can be viewed as a waveguide, similar to the

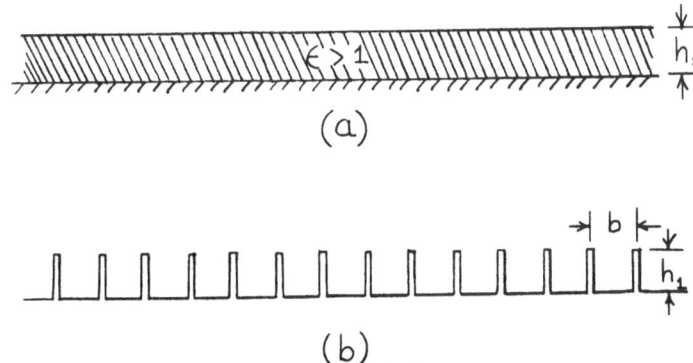

Figure 11-2. Surface wave support structures.

plane dielectric waveguide and the dielectric rod discussed in Chapter 6. The thickness, h_1, dielectric constant, ε, and wavelength, λ, must satisfy certain relationships for the surface mode to exist, and if h_1 is sufficiently large, more than one mode may exist simultaneously. Each surface mode has its own velocity and decay rate.

The corrugated conducting plane of Figure 11-2b is more complicated, but once again the parameters h_2, λ, and b must satisfy certain relationships for a surface mode to exist.

A surface wave can be launched on such a structure by a small horn, as illustrated in Figure 11-1. The guiding structure, which is several wavelengths long, provides the gain and narrowing of the beam in an endfire mode (in a direction tangent to the guiding surface). The flatness of the structure is advantageous for airborne applications that require conformality. An example of a streamlined design is shown in Figure 11-3.

The surface waves that propagate in the type of antenna described above are "slow," having phase velocities less than c_0. As described in Chapter 6, "fast" traveling waves ($c > c_0$) exist inside hollow metallic waveguides. As in Chapter 7, short ($\sim\lambda/2$) waveguide slots may or may not radiate, depending on their location. In Figure 11-4, a different type of waveguide slot, which may be many wavelengths long, is shown. If the slot were cut straight along the centerline, it would not radiate. However, when the slot is offset to one side, a tangential electric field is excited across the slot, and the strength of

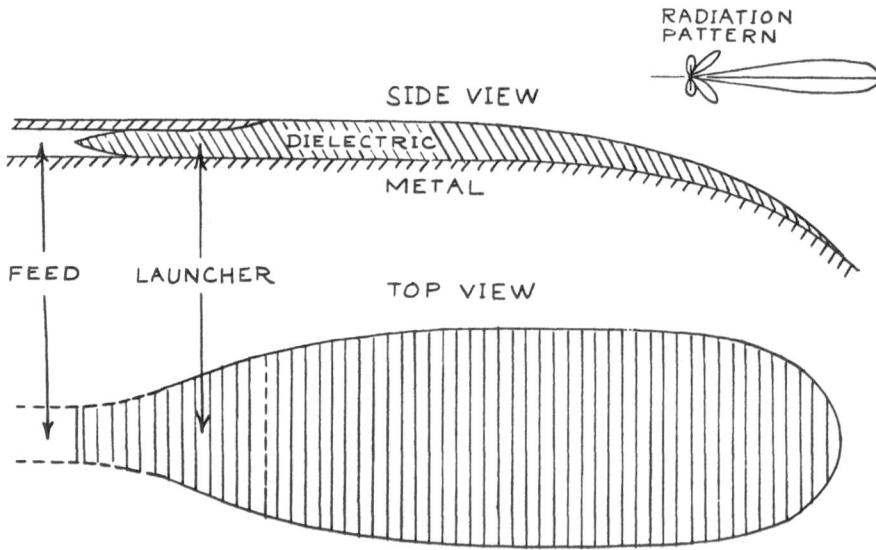

Figure 11-3.   Surface wave antenna.

the field is roughly proportional to $\sin(\pi x/a)$. If the slot is shaped by varying x, the field in the slot receives a taper that may be used to control side lobe levels, as described in Chapter 8. The phase of the field in the slot is determined by the dominant (TE_{10}) mode, which exists in the waveguide. The radiation from this waveguide slot has a main lobe in the direction θ, determined by the ratio of λ to λ_g, as shown in Figure 11-5. The slot, which lies along the z axis in this side view, is excited by the TE_{10} mode, which has the indicated z dependence, as described in detail in Chapter 6 and Eqs. (6-3) through (6-9). Radiation that emerges from the slot at points A and B is mutually reinforcing in that direction for which the phase lag along path AB' is equal to that along path AB. Planes of constant phase (wavefronts) are thereby formed along BB' or any parallel line such as CC'. The condition for reinforcement may be expressed as

$$(2\pi d \cos\theta)/\lambda = 2\pi d/\lambda_g$$

or

$$\boxed{\cos\theta = \lambda/\lambda_g} \qquad \textbf{(11-8)}$$

As this relation indicates, the beam angle depends on both the free space wavelength, λ, and the waveguide wavelength, λ_g. Antennas of this type are often called leaky wave antennas because of the continuous leak of energy

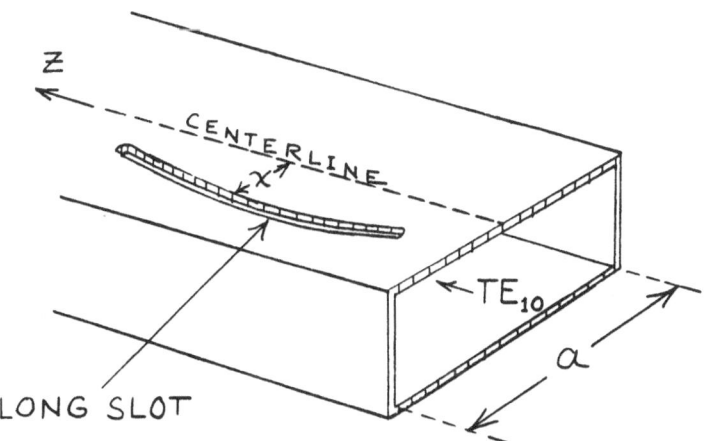

Figure 11-4. Long slot traveling wave antenna.

from the feeding transmission line to the outside. The feed structure of Figure 11-4 is a rectangular waveguide for which, from Eqs. (11-8) and (6-8):

$$\cos\theta = \sqrt{1 - (\lambda/2a)^2}. \qquad (11\text{-}9)$$

A limited amount of beam scanning can be realized by changing the frequency. Eq. (11-8) also holds for other types of waveguides; however, the relationship expressed by Eq. (11-9) applies only for the TE_{10} mode in rectangular waveguides.

The radiating aperture of a waveguide-fed, leaky wave antenna does not have to be cut in the broadwall nor does it have to be continuous. The structure illustrated in Figure 11-6 has an aperture that consists of many closely spaced holes along the narrow wall of a waveguide. By varying the diameter of the holes, the desired aperture illumination may be obtained, and the phase relation of Eq. (11-8) remains valid.

Ideally, no energy should remain in the feed structure beyond the end of the radiating structure of the leaky wave antenna. In practical antennas of this type, as much as 10% of the total incident EM energy is not radiated, but continues through the waveguide. If precautions are not taken, this energy will be partially reflected at the end of the waveguide and travel back across the antenna, giving rise to an additional lobe at an angle of $\pi - \theta$ radians. For this reason, the waveguide (or other traveling wave feed) must be terminated with a matched load to absorb any residual energy. As in Chapter 2, this load should have an impedance equal to the line's characteristic impedance, because the line will then behave as though it were semi-infinite (reflectionless).

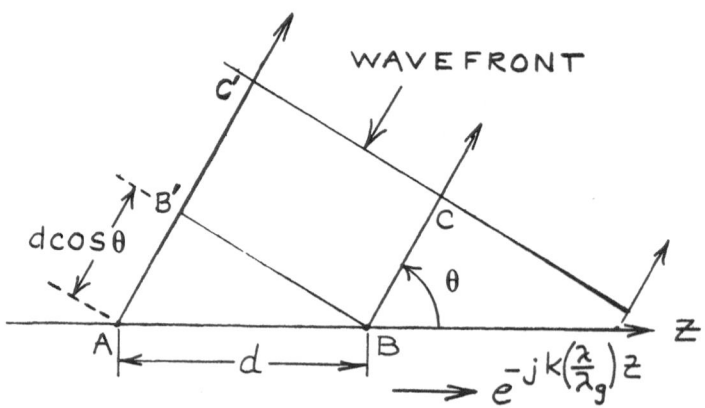

Figure 11-5. Formation of wavefront.

Leaky wave antennas, which may also be built using stripline and coax, have many applications. Their gain and beam width depends on the normalized length of the radiating structure. The few examples described here represent a large group of traveling wave antennas, including both slow and fast wave types.

11.2 BROAD-BAND ANTENNAS

Another important class of antennas are those designed to operate over very wide frequency bands. The antennas described thus far can be characterized by certain length parameters that must be kept within specific bounds in relation to the wavelength in order for the antenna to function properly. Some electrically small antennas (short dipoles, small loops) can operate over a fairly wide band, in the sense that their radiation patterns do not vary sharply with frequency. However, the impedance of these antennas is very low and changes rapidly with frequency, which makes it difficult to feed them efficiently. Helix and "fat" (relatively thick transverse dimensions) dipole antennas are not very frequency sensitive, but when very wide ratios (10:1 or more) between the highest and lowest frequencies must be accommodated by the same antenna, new principles must be used.

Geometrical structures that can be specified in terms of angles only, without reference to length, provide antenna configurations that are useful when

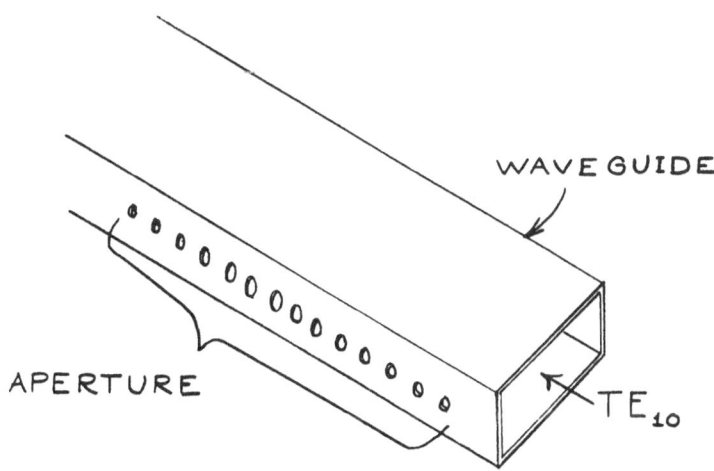

Figure 11-6. Leaky waveguide antenna.

bandwidth requirements are quite large. These structures are the logarithmic spiral and spiral of Archimedes, either of which may be plane or conical.

A plane logarithmic spiral is a curve characterized by a single angle, α, which is the angle between the radius vector and the tangent to the curve at any point on the curve, as shown in Figure 11-7. For the planar log-spiral, this angle is a constant. In polar coordinates, the equation of the planar log-spiral is,

$$r = r_0 e^{a(\phi - \phi_0)}, \tag{11-10}$$

where a is constant and related to the angle, α, by $a = 1/\tan\alpha$ and (r_0, ϕ_0) is an arbitrary reference point on the spiral. Figure 11-7 shows pieces of two planar log-spirals. One passes through A', B', and C' with $r = OB'$ for $\phi = 0$; the other passes through A'', B'', and C'', with $r = OB''$ for $\phi = \pi$. Each spiral can be generated from the other by rotation about the origin. Note that as $\phi \to \infty$, $r \to \infty$, and as $\phi \to -\infty$, $r \to 0$ for each spiral. This antenna geometry may be implemented by forming the spirals from wires fed by a transmission line. For infinite spirals there is no length scale, and (theoretically) the structure

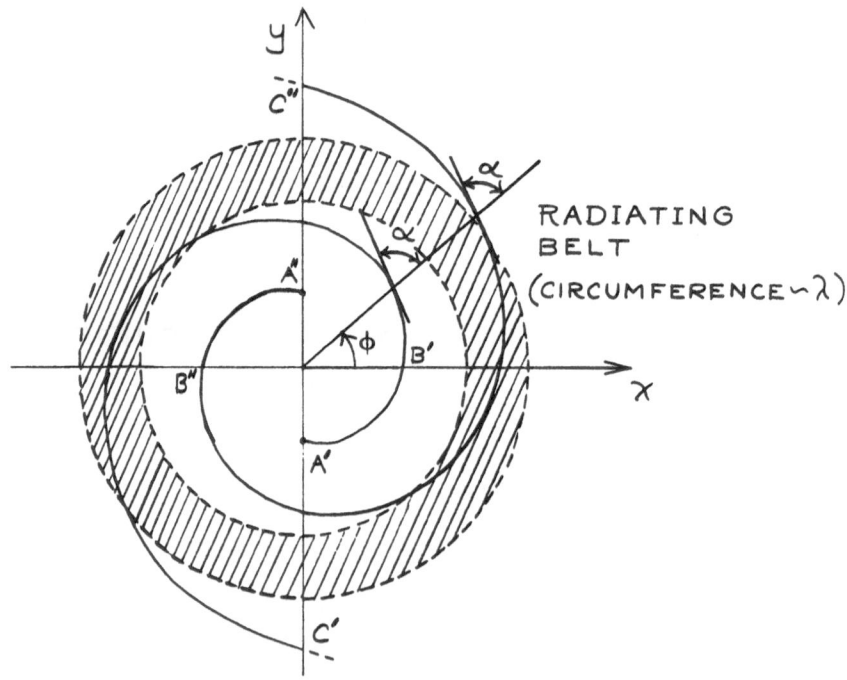

Figure 11-7.   Log-spiral antenna.

is frequency independent. A practical spiral antenna, however, has both inner and outer terminations that set the upper and lower frequency limits. The feed, which may be a balanced transmission line, is connected at the innermost points—A' and A'' in Figure 11-7. At any given frequency within the operational limits of the spiral, most of the energy is radiated from an annular region whose circumference is about one wavelength; the remaining structure contributes little. Thus, when the frequency increases, the active portion of the antenna moves inward, and when the frequency decreases, the active part of the antenna moves outward. Antenna parameters, such as impedance and gain, do not change much. Log-spiral antennas can cover frequency ratios in excess of 10:1. Their radiation pattern is circularly symmetrical about the z axis (perpendicular to the x-y plane) and has a wide lobe in the z direction. The polarization is circular when viewed along the z axis.

Four-armed spiral antennas can be fed in ways that use other radiation modes. For example, a "difference" mode has a torus-like radiation pattern with a null along the z direction and is produced if the radiation ring circumference is approximately 2λ. A schematic cross-sectional cut through such a pattern is shown in Figure 11-8. A feed network somewhat similar to that used for monopulse, as described in Chapter 9, can simultaneously sense both the sum and difference patterns. As illustrated in Figure 11-8, the ratio of sum to difference power is θ dependent, and the four-arm spiral may therefore be used for wideband direction finding.

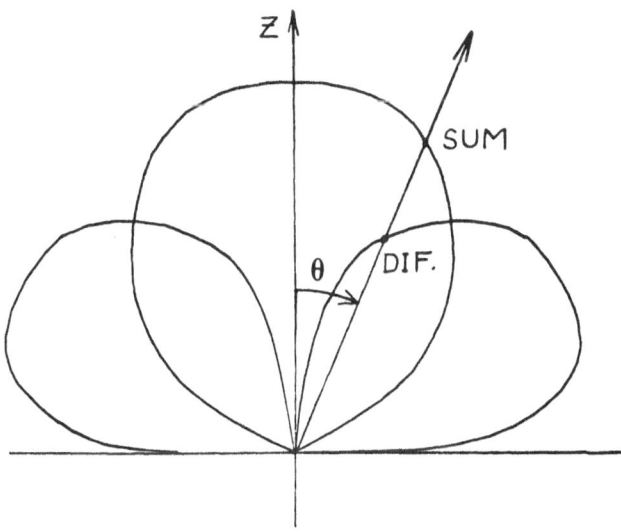

Figure 11-8. Four-armed spiral pattern for direction finding.

To provide a unidirectional pattern, planar spirals are sometimes mounted over a cavity or box filled with absorbing material. This method wastes 50% of the available power. However, it is one of the few ways of obtaining a single lobe with a radiation pattern that is frequency independent over a wide band.

Planar spirals may be formed by metal strips on a dielectric substrate in lieu of thin wires. If the width of the strips is equal to the width of the interstrip spaces, the structure is called self-complementary.

Another form of spiral geometry, the Archimedes' spiral, has the mathematical form

$$r = r_0 + b(\phi - \phi_0). \tag{11-11}$$

A single arm of this spiral, with $r_0 = OA$ for $\phi_0 = 0$, is shown in Figure 11-9. Note that Eq. (11-11) is the approximate form for Eq. (11-10) when the value of the exponent is small compared to unity. Spiral antennas are often constructed in accordance with Eq. (11-11), because the linear spread of the arms permits more turns per unit area than the logarithmic geometry. Archimedes' spiral antennas have properties that are quite similar to those of the log-spiral antennas.

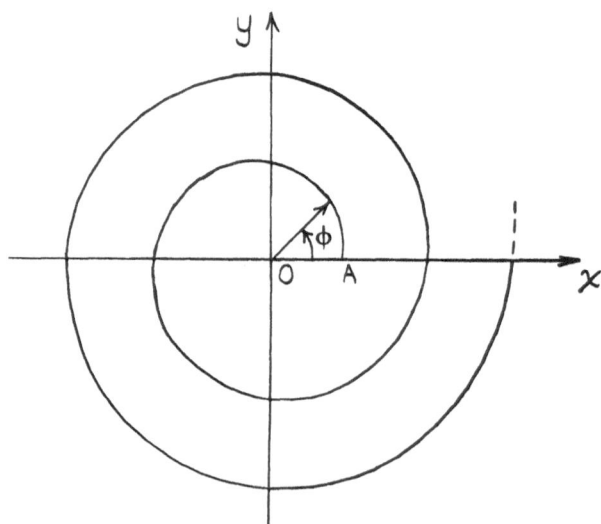

Figure 11-9. Archimedes's spiral.

The spiral illustrated in Figure 11-10 is wrapped around the conical surface as described in spherical coordinates by $\theta = \theta_0$ where $\theta_0 > \pi/2$. The planar spiral is a special case for which $\theta_0 = \pi/2$. As in the planar spiral, the feed is attached at the smaller radius, and the antenna may have two, four, or more spiral arms. The conical and flat spiral antennas are distinguished mainly by their different directivities along the z direction. For the conical spiral illustrated in Figure 11-10, the main lobe is directed in the z direction, and a lobe of much lower power lies along the $-z$ direction. The front-to-back lobe ratio is a function of the cone angle, θ_0. Thus, a good front-to-back ratio can be achieved without using absorbers if a larger structure is acceptable.

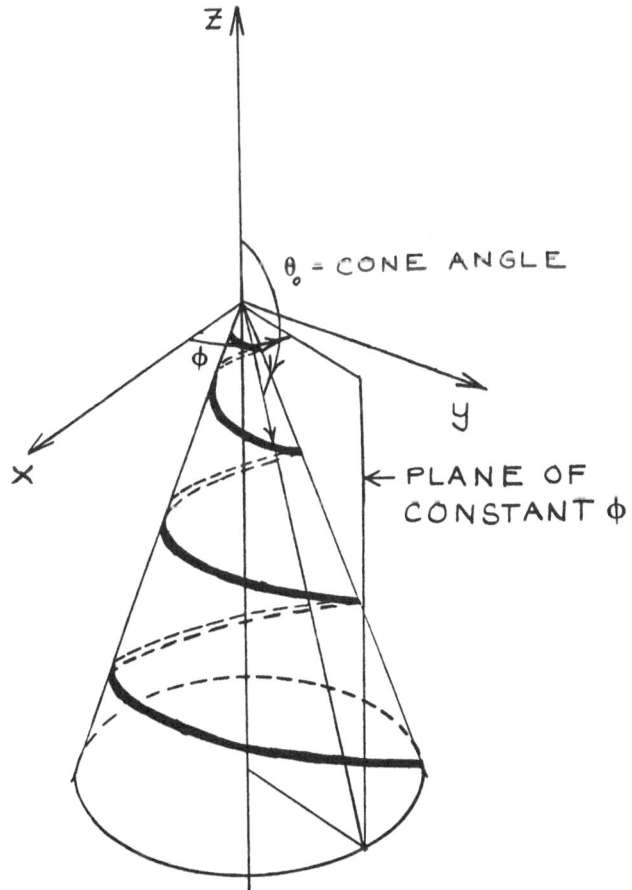

Figure 11-10. Conical spiral.

The design of spiral antennas is not greatly assisted by theoretical analysis, because realistic spiral structures are not easily modeled, even by crude approximations. As is the case with many antennas, design is ultimately an empirical process.

Another class of broad-band antenna and array structures is log-periodic. The broad-band characteristic is achieved by shaping the structure such that the angle, θ, is a periodic function of log r. For example,

$$\theta = \theta_0 \sin[b \log(r/r_0)] \tag{11-12}$$

is such a function. Clearly, when $r > r_0$, the period is stretched out, and for $r < r_0$, the period is increasingly compressed. A segment of the function specified by Eq. (11-12) appears in Figure 11-11. A structure that has this log-periodic property is similar to the log-spiral in that for a given frequency most of the radiation originates from a limited section of the antenna. Because the active section will move up or down the antenna as the frequency changes, the broadband characteristic occurs. Many different types of log-periodic shapes

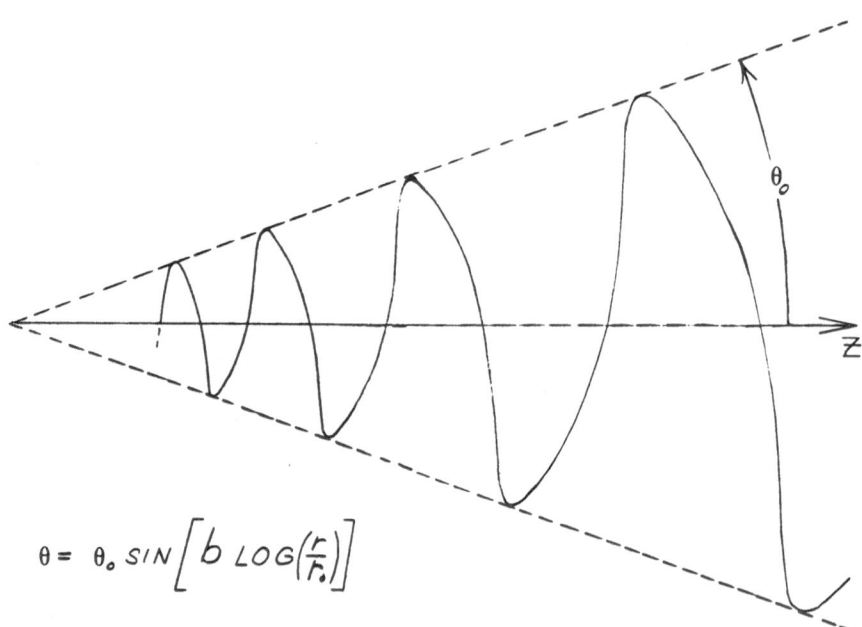

$$\theta = \theta_0 \, SIN\left[b \, LOG\left(\frac{r}{r_0}\right)\right]$$

Figure 11-11.   Log-periodic function.

have been tested, including metallic shapes and wire structures (Fig. 11-12). The structure of Figure 11-12b consists of two log-periodic branches, whose "vee" configuration provides forward directivity. Wire structures are more commonly used at low frequencies; metallized shapes are usually employed at the higher frequencies. Steady radiation patterns and input impedance over a 3:1 frequency range are typical, and good performance over ratios as high as 10:1 has been achieved.

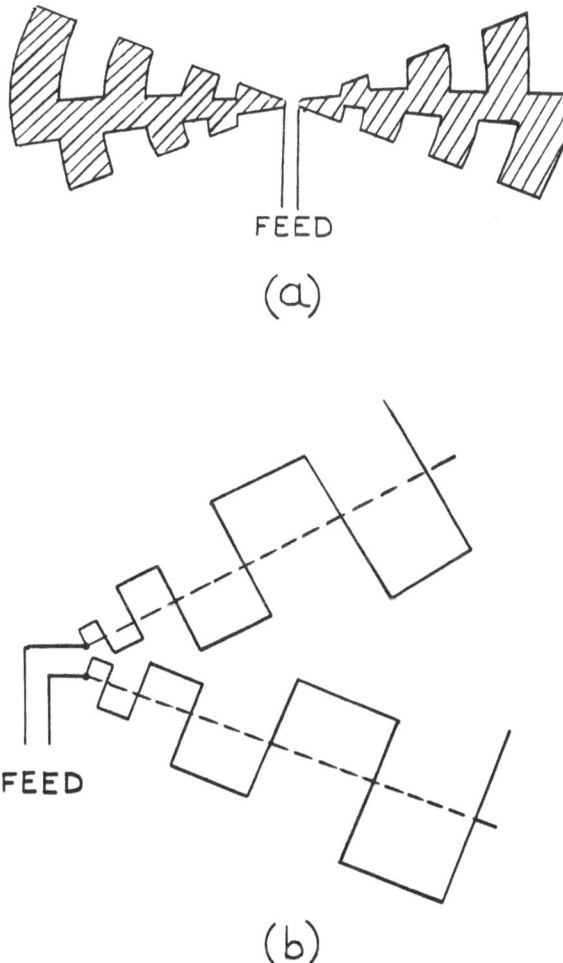

Figure 11-12. Log-periodic antenna structures.

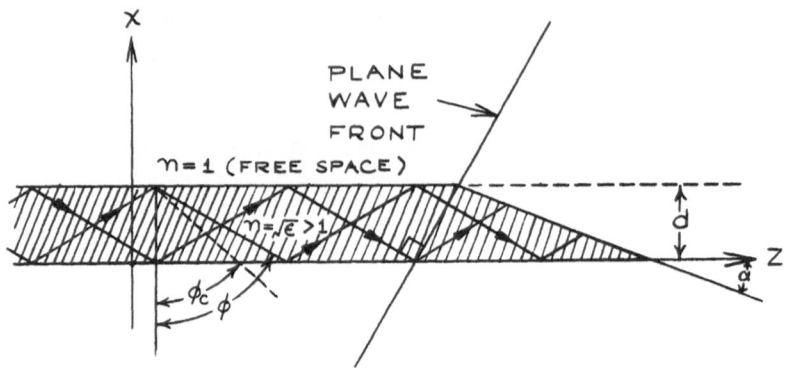

Figure 11-13.   Illustration for problem 1.

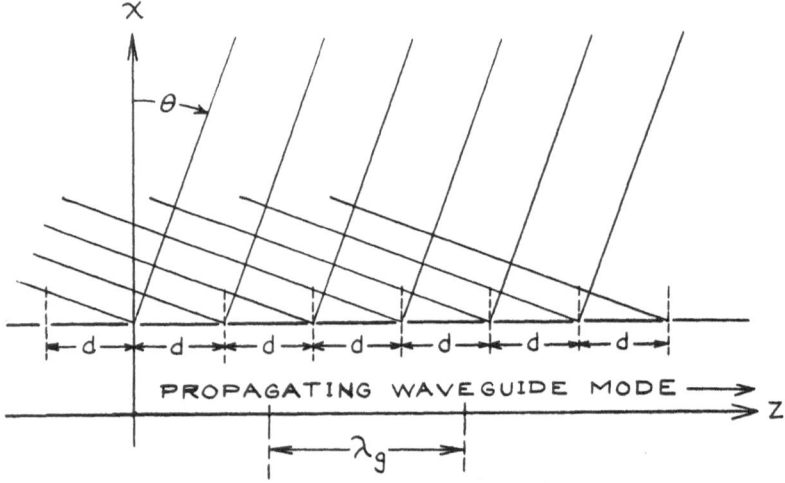

Figure 11-14.   Illustration for problem 2.

PROBLEMS

PROBLEM 11-1. * A plane dielectric layer ($\varepsilon > 1$) of thickness, d, is backed by a perfect conductor and is terminated by a wedge region defined by $\alpha > 0$ as shown in Figure 11-13.

---

*Advanced problem.

a. A guided mode consists of two plane waves, which propagate at an angle, ϕ (with respect to x), in the plane layer and are continuously reflected at the boundaries $x = 0$ and $x = d$. Using Snell's law and the boundary conditions at the interfaces $x = d$ and $x = 0$, show that the phase velocity of the mode (in the z direction) must be smaller than the free space velocity c_0 and that as a result such a mode cannot radiate energy into the upper half space $(x > d)$.

b. Using geometric optical considerations, show that when such a mode enters the wedge region, it must radiate energy into the upper half space for any wedge angle, α. Show that when α is very small, the energy will "leak out" over an extended region in the wedge.

PROBLEM 11-2. A waveguide feeds a leaky wave antenna that consists of many holes separated by a distance, d, from each other as shown in Figure 11-14. The propagating mode has a phase velocity c_g and wavelength λ_g related by $c_g = \lambda_g f$, $c_g > c_0$, and $\lambda_g > \lambda_0$.

a. Use simple interference arguments to show that radiation from the holes form a beam in the direction θ, which is independent of d. Assume that the waveguide is rectangular, with $a = 0.9$ in., and the propagating mode is TE_{10}.

b. Find the range of θ for the frequency band $8 \leqslant f \leqslant 12$ GHz.

PROBLEM 11-3. A planar spiral antenna is to operate over the range from 2 GHz to 8 GHz. What should be the outer and inner radii of the structure?

PROBLEM 11-4. Show how Eq. (11-11) is an approximation of Eq. (11-10) and the relationship between the parameter b and α.

CHAPTER 12

Radomes and Windows

Antennas and arrays must often be protected from environmental effects (wind, ice, sun, dust, etc.) and, in some instances, enemy surveillance. The structures that provide such protection are known as electromagnetic windows and radomes. Ideally, these structures should not degrade the antenna electrical performance; however, in practice some compromises must be accepted. In some cases the electrical perturbations that arise from a radome are minimal and can be ignored or easily accounted for. In many other applications, radome effects are substantial and must be considered carefully in the overall antenna system design. Often the combined efforts of electrical, structural, aeronautical, and materials engineers are necessary to realize performance objectives. An understanding of the design of radomes and windows is an important part of antenna engineering.

12.1 THE UNIFORM DIELECTRIC WALL

Radomes can be classified according to their location (airborne, shipboard, ground based, etc.) and structure (thin wall, sandwich, solid wall, compensated wall, etc.). As a first example, the interaction of a plane wave with a uniform dielectric wall of thickness, d, and dielectric constant, ε, is analyzed. It is assumed that ε is complex with a small imaginary part. The plane wave has an angle of incidence, θ, and a polarization that may be resolved into components perpendicular and parallel to the plane of incidence. At the first air-dielectric interface, the reflection coefficient is

$$\Gamma = (y_1 - y_2)/(y_1 + y_2), \qquad \textbf{(12-1a)}$$

in which y_1 is the air's admittance,

$$y_1 = \begin{cases} \cos\theta \ \text{(perpendicular polarization)} \\ 1/(\cos\theta) \ \text{(parallel polarization)}, \end{cases} \qquad \textbf{(12-1b)}$$

and y_2 is the dielectric's admittance:

$$y_2 = \begin{cases} \sqrt{\varepsilon - \sin^2\theta} \ \text{(perpendicular polarization)} \\ \varepsilon/\sqrt{\varepsilon - \sin^2\theta} \ \text{(parallel polarization)}. \end{cases} \qquad \textbf{(12-1c)}$$

Recall that Γ was given for a general interface by Eqs. (5-13). At the second interface the positions of air and dielectric are reversed with respect to the first interface, and the reflection coefficient is therefore $-\Gamma$. In general, it can be shown that the wall will cause partial reflection and partial transmission of the wave. The ratios of the electric field amplitudes of the reflected and transmitted waves to the amplitude of the incident wave are denoted by R and T, respectively. To calculate these ratios, one must account for multiple reflections that occur within the dielectric wall. Derivation of the results are not given here but can be performed as an exercise.

$$\begin{aligned} R &= (2j\Gamma\sin\psi)/D, \\ T &= (1 - \Gamma^2)/D, \end{aligned} \qquad \textbf{(12-2)}$$

in which,

$$D = (1 - \Gamma^2)\cos\psi + j(1 + \Gamma^2)\sin\psi,$$

and

$$\psi = kd\sqrt{\varepsilon - \sin^2\theta}.$$

If ε is purely real (lossless dielectric), it is not difficult to show that $R^2 + T^2 = 1$, which is consistent with power conservation. From Eqs. (12-2), it is seen that reflection will not occur when $\psi = m\pi$, where m is an integer. Strictly speak-

reflection will not occur when $\psi = m\pi$, where m is an integer. Strictly speaking, this condition can only be satisfied for a lossless dielectric. It corresponds to wall thicknesses given by

$$d = m\lambda/2 \, (\sqrt{\varepsilon - \sin^2\theta}). \qquad (12\text{-}3)$$

This is the multiple half wave condition for total transmission.

For the polarization component parallel to the plane of incidence, there is also no reflection if the angle of incidence is equal to the Brewster angle,

$$\theta_B = \tan^{-1} \sqrt{\varepsilon}, \qquad (12\text{-}4)$$

as discussed in Chapter 5, Eq. (5-17). For $\theta = \theta_B$, $R = 0$, $T = e^{-j\psi_B}$, and $|T|^2 = 1$. Therefore, at the Brewster angle, the parallel polarization component is totally transmitted, regardless of wall thickness. The perpendicular polarization component is partially reflected, unless Eq. (12-3) is also satisfied.

Because the wave slows down during its passage through the dielectric wall, its phase experiences a delay that exceeds that which would have occurred if the wall were absent. This insertion phase delay (IPD) is, in general, a function of ε, d, and θ.

The thin wall for which $\psi \ll \pi$, or $d \ll \lambda/2\sqrt{\varepsilon}$ is an important special case. For this special case, $\cos\psi \cong 1$, $\sin\psi \cong \psi$, and, from Eqs. (12-2),

$$R \cong 2j\Gamma\psi/(1 - \Gamma^2),$$

$$T \cong 1 - j\psi(1 + \Gamma^2)/(1 - \Gamma^2),$$

$$|T|^2 \cong 1 - \left(\frac{2\pi d}{\lambda}\right)^2 (\varepsilon - \sin^2\theta)[(1 + \Gamma^2)/(1 - \Gamma^2)]^2. \qquad (12\text{-}5)$$

Although this approximation breaks down in the vicinity of $\theta = 90°$ ($\Gamma \to 1$), it can be safely used for small-to-moderate incidence angles. Figure 12-1 provides a graphical summary of some electrical properties of a solid wall.

12.2 THE EFFECTS OF RADOMES

Thin wall structures can form effective radomes over a range of frequencies that extends up to the mid-microwave region. For example, a tough, thin, plastic membrane may be supported by air pressure or stretched over a supporting frame. The supporting structure will cause blockage and scatter energy in various directions. Blockage and scattering lead to gain reduction and increased side lobe levels; formulas for estimating these effects have been derived. For many applications, the thin membrane, supported either by air pressure or by a structure, is very satisfactory, because the electrical effect of the membrane itself is usually quite small and is easily estimated by use of Eqs. (12-5).

For many applications, a thin wall radome is unsuitable. Examples include airborne radomes, which require very strong aerodynamically shaped struc-

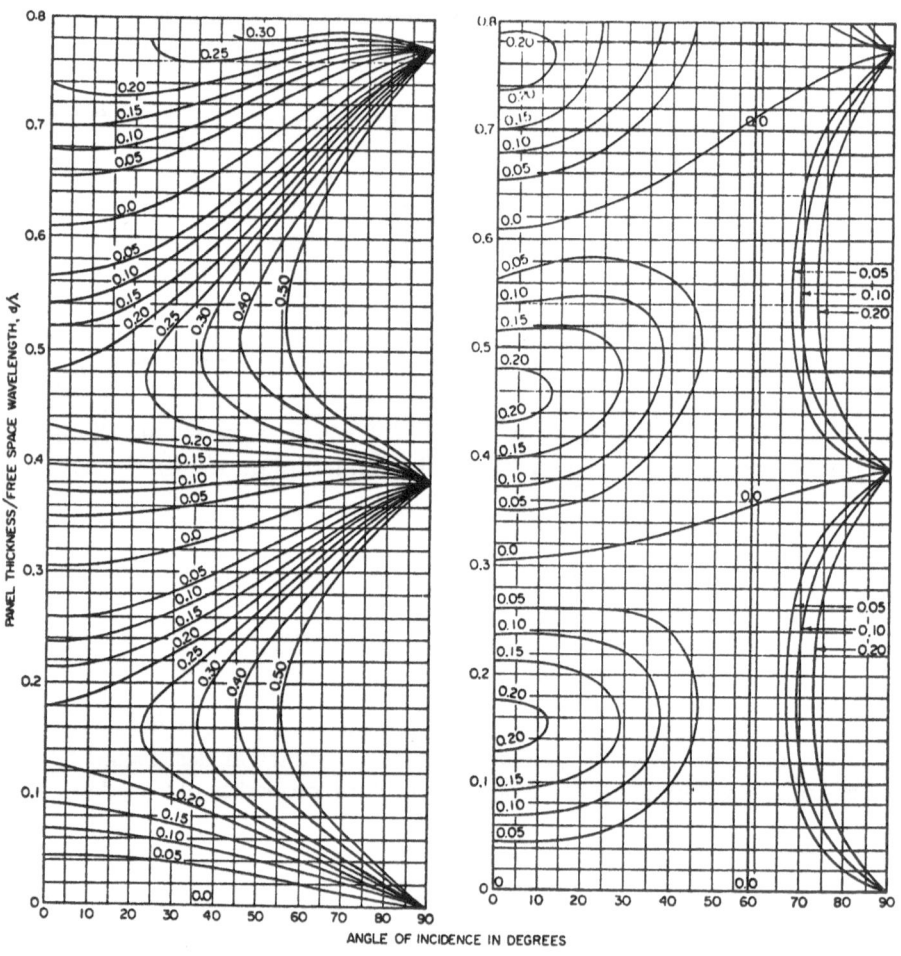

ANGLE OF INCIDENCE IN DEGREES

(a) PERPENDICULAR   (b) PARALLEL
     POLARIZATION   POLARIZATION

Figure 12-1. Contours of constant reflected power; dielectric constant = $2.7E_0$. (*From R. C. Johnson and H. Jasik, Antenna Engineering Handbook, 2nd ed., New York: McGraw-Hill, 1984, pp. 32-9, 32-10, 32-14, 32-15; copyright © 1984 by McGraw-Hill Book Company, reprinted by permission.*)

tures; radomes for antennas that operate at very short wavelengths comparable to the thinnest practical membrane; and large radomes, for which the required supporting structures would cause unacceptable degradation of electrical performance. Dielectric shell radomes are needed for applications such as these. Some effects that arise from shell structure are illustrated in Figure

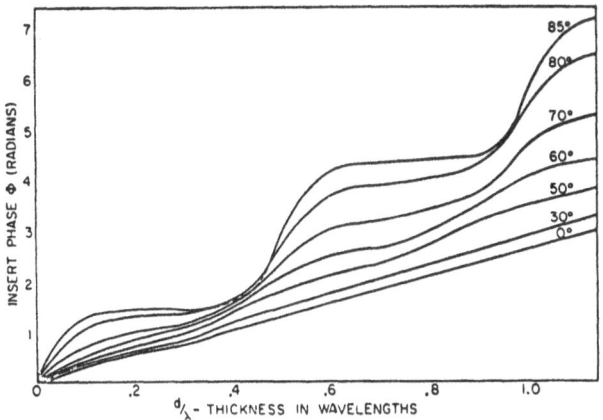

(c) PERPENDICULAR POLARIZATION

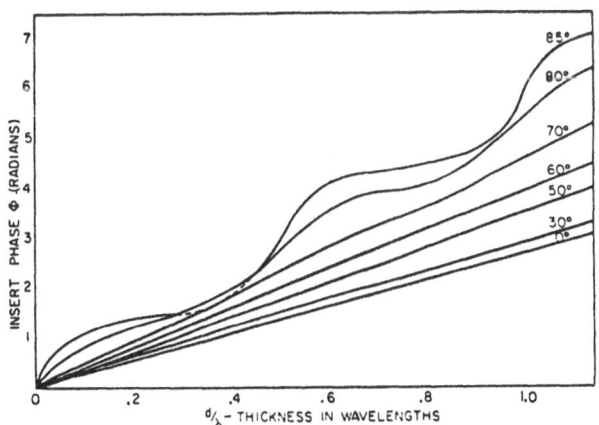

(d) PARALLEL POLARIZATION

Figure 12-1. Insertion phase delay for wall with $E = 2E_0$.

12-2. Experience indicates that geometrical optics (discussed in Chap. 5) frequently provides a satisfactory mathematical framework for calculating radome effects. Therefore, electromagnetic radiation emitted by the antenna can be represented by a bundle or beam of parallel rays, which emerge from the antenna aperture. Upon striking the radome shell, these rays are partially reflected and partially transmitted with a strength that, because of radome curvature, depends on the local angle of incidence and polarization geometry. Movement of the beam, by either mechanical or electronic scanning, causes each ray-radome intersection point to change. The trajectory and polarization component strengths of each ray, therefore, depend on beam position. Furthermore, because the radiated power occupies a finite frequency bandwidth, the ratio of shell thickness to wavelength can vary significantly. With so many factors influencing the ray-radome interaction, any structure must represent a compromise. When mechanical-structural design factors are considered, the engineering problem becomes even more complex. The electrical impact of a radome can be summarized as follows.

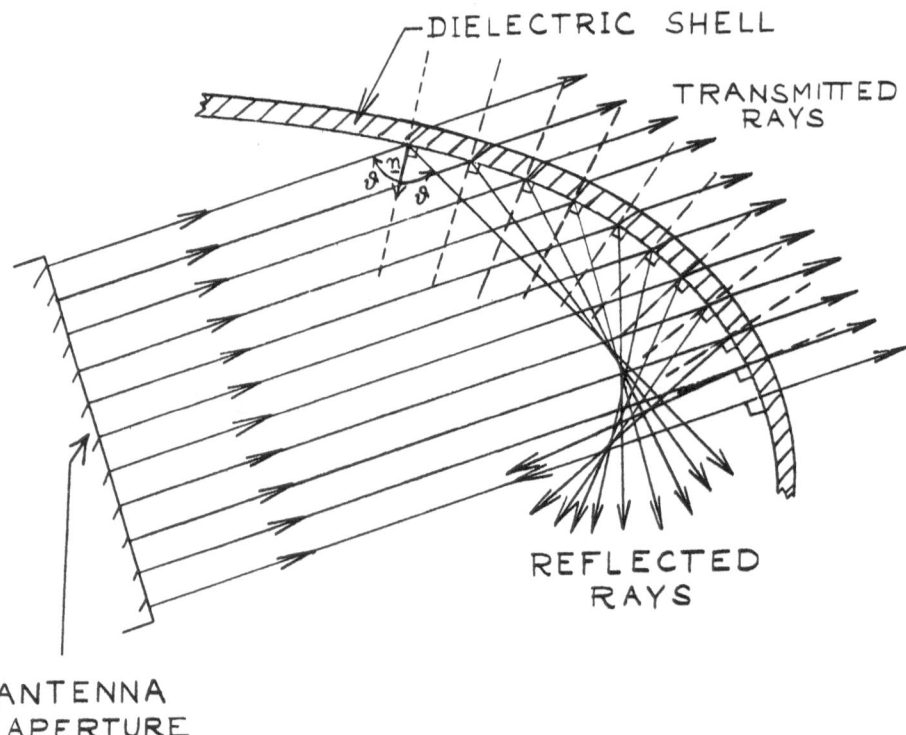

Figure 12-2.   Radome ray trace geometry.

Radome losses caused by wall absorption, antenna aperture blockage, and scattering reduce antenna gain.

If the radome is built up in sections (i.e., geodesic dome) or has other structural periodicity, it may cause systematic error lobes (see section 9.9) to appear in the antenna pattern.

Power reflected by the radome, which travels in unintended or undesirable directions, appears as increased side lobes or reflection lobes. Such reflections can be especially troublesome for airborne radars if they illuminate the ground. Power reflected from the ground (clutter) may completely obscure targets of interest.

The insertion phase delay and transmission coefficient of the transmitted beam vary from one point on the radome to another. These perturbations of aperture phase and amplitude distribution cause beam defocusing, the effects of which include beam broadening, gain reduction, and side lobe degradation. In addition, refraction of the beam by the radome may cause a change in beam direction, which is called boresight shift or pointing error.

The radome can cause depolarization, which may shift the transmitted polarization or enhance the cross polarization (i.e., the polarization component that is perpendicular to the desired one). In the case of circular polarization, the axial ratio is increased.

12.3 RADOME CONSTRUCTION

For many narrowband microwave applications, a half-wave (or a multihalf wave) wall radome provides an adequate solution. After determining the required frequency and incidence angle ranges, and the ε of available materials, a wall thickness, d, is chosen in accordance with Eq. (12-3). The wall thickness must give adequate mechanical strength. The transmission and loss at a given frequency can be found with the help of Eq. (12-2).

A single solid wall radome structure often does not provide the designer with enough controllable parameters to satisfy the conflicting requirements. In some instances, multilayer shell structures furnish the needed flexibility. Some of the structures that are commonly used are shown in Figure 12-3. Figure 12-3a represents a single dielectric layer to which a coating has been applied for protection from the elements. However, the thickness and material properties of the coating may be used as design parameters in order to enhance certain electrical characteristics. This structure, sometimes called a *half sandwich*, can be a broadband structure. In Figure 12-3b, the A-sandwich is formed by two strong, high-ε skins that enclose a weaker, low-ε core material. The C-sandwich has three skins and two cores (Fig. 12-3c). Sandwiches have a high strength-to-weight ratio, which is especially desirable in airborne radome applications. The reflection from the alternating interfaces of these structures interfere with each other. When the reflection coefficient of the structure is shown as a function of frequency and angle of incidence, a pattern of pass-bands and stop-bands emerges, which can be quite intricate and difficult to predict. Sandwiches can be tailored to satisfy a variety of requirements. For example, in airborne radar applications, extremely low reflections (typically 0.1% power or −30 dB) over a wide range of incidence angles are often desirable for a narrow bandwidth signal. For electronic counter measures, reasonable power transmission (e.g., better than 80%) over a very wide (multi-octave) frequency band may be the design objective. Other applications can require good transmission in two or more widely separated but

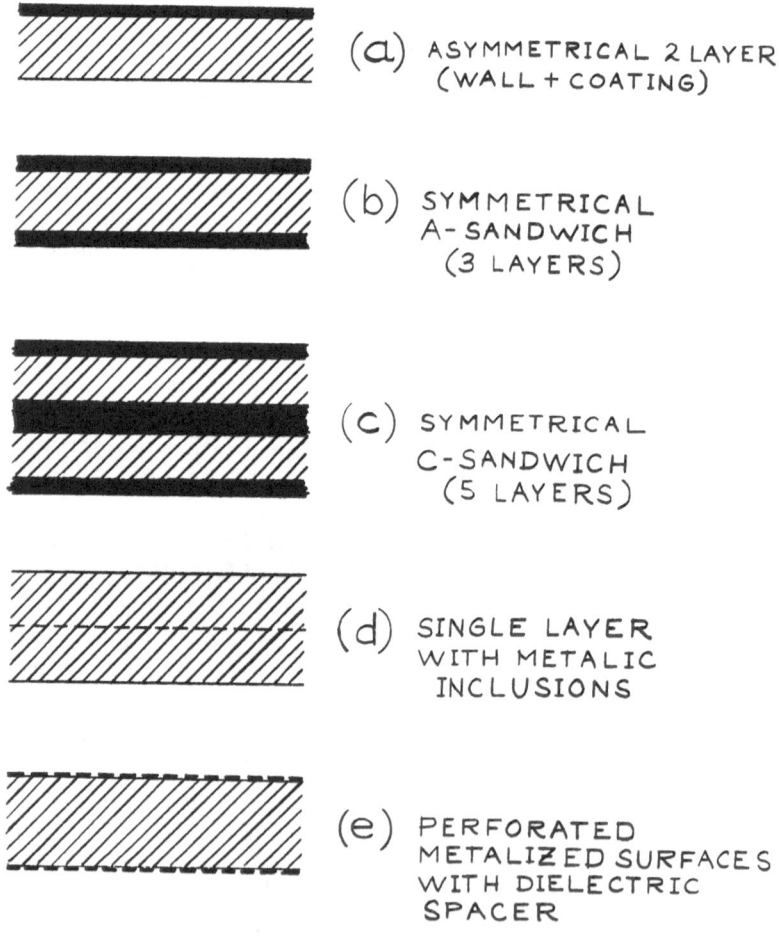

(a) ASYMMETRICAL 2 LAYER (WALL + COATING)

(b) SYMMETRICAL A-SANDWICH (3 LAYERS)

(c) SYMMETRICAL C-SANDWICH (5 LAYERS)

(d) SINGLE LAYER WITH METALIC INCLUSIONS

(e) PERFORATED METALIZED SURFACES WITH DIELECTRIC SPACER

Figure 12-3.   Radome shell structures.

narrow frequency bands. For each of the objectives described, a tailored radome design is needed to ensure good system performance.

Figure 12-3 illustrates another class of radome structures with one or more dielectric layers combined with metallic structures, which may take the form of embedded wires, grids, particles, or metallized, perforated surfaces. Metallic inclusions (embedded objects) are reactive (i.e., they may increase or decrease the apparent dielectric constant of the medium). If the inclusions are small compared to the wavelength and uniformly fill the bulk, the dielectric constant is artificially increased. If the embedded metallic structure is medium to large in size (e.g., long wires), it can be treated as a localized reactance in a generalized network approach. Perforated metallized surfaces are also known as frequency-selective surfaces. The shapes, sizes, and patterns of the metallized areas determine the reflectivity or transmissivity of the surfaces over frequency bands (which are usually narrow) and certain ranges of incidence angle. Synthesis and fabrication of such radomes are not simple, and this is an active area of research and development.

The E-3A airborne warning and communications system (AWACS) radome is shown prominently in Figure 12-4. The E-3A is one of the most sophisticated airborne radar surveillance systems. The *rotodome* atop the aircraft has the exterior shape of an oblate ellipsoid of revolution, with major and minor diameters of 30 ft and 6 ft, respectively. The rotodome houses a large phased array microwave antenna system, covered by a radome of semi-ellipsoidal form. To satisfy requirements of mechanical load, weight, aerodynamics, and ultralow reflections and side lobes, designers conceived a tapered C-sandwich in which the thicknesses of the two cores and three skins

Figure 12-4. E3A (AWACS) aircraft and rotodome. (*Courtesy of The MITRE Corporation.*)

vary continuously from point to point. The E-3A radome is the first five-layer variable thickness radome to be incorporated in an operational radar system.

The list of materials used for radomes is beyond enumeration, but the common materials include the following. A lamination of fabric and resin, which, when cured, becomes a nearly homogeneous solid, provides good mechanical and electrical properties. Fiberglass is a commonly used fabric material and is abundant, inexpensive, and very strong. Quartz fabric is strong, has low ohmic losses and withstands very high temperatures, but is quite costly. Kevlar is not as expensive as quartz fabric, has unexceptional electrical properties, but is very strong. The resins in use include polyester, various epoxies, polyimides, and polybutadiene compounds. In addition, various plastics have been used for specific applications. Sandwich cores can be constructed from light paper or from fiberglass and resin honeycomb structures and from foaming plastics or ceramics that become hard. Modern ceramics are candidates for use in radomes and windows incorporated in high-speed missiles and spacecraft. Indeed, the search for new structural materials that are transparent to electromagnetic radiation never ceases.

PROBLEMS

PROBLEM 12-1.   A circular parabolic reflector antenna with diameter, D, focal length, f, and gain, G, is to be protected by the radome shown in Figure 12-5. The specifications are:

> Frequency: 2.7 GHz to 3.0 GHz
> Azimuth scan: 360°
> Elevation scan: 0° to 90°
> Antenna: $D = 6$ m, $f = 2.1$ m, vertical polarization, prime
> focus feed, $G = 34.5$ dB at 3 GHz, maximum side
> lobes -32 dB (below main beam)

a.   Estimate the size of a radome (R and H) such that the surface area will be minimal, and calculate the range of angles of incidence.

b.   Using a radome consisting of a space frame and membrane, where the frame blocks 6% of the aperture area, estimate the reduction in antenna gain due to the frame effect alone.

c.   Assume that the membrane is a plastic material of 1 mm thickness, $\varepsilon = 3.2$, $\tan\delta = 0.01$. Estimate the reduction in antenna gain due to this effect.

d.   What is the total radome effect on antenna gain. Would the antenna diameter have to be increased to maintain the 34.5-dB gain? If yes, by how much?

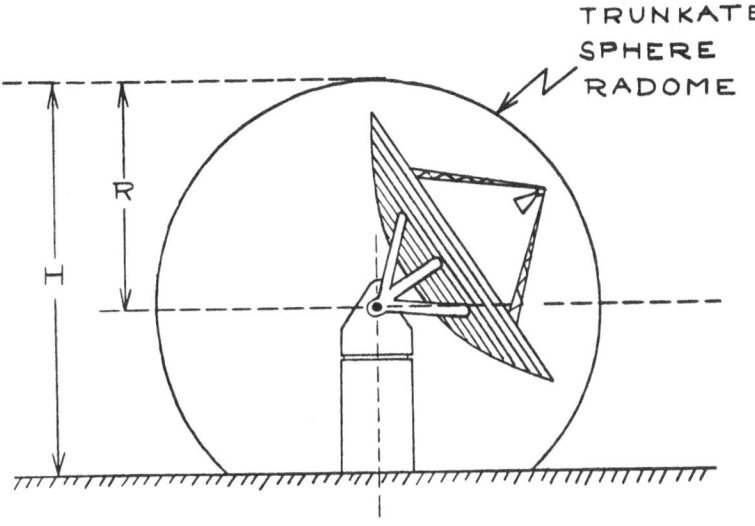

Figure 12-5. Illustration for problem 1.

e.* Assume that all the power that is scattered by the radome is scattered isotropically. Estimate the effect this would have on the side lobe level.

f. Using a λ/2 wall radome to avoid the effect of the space frame, assume ε = 4.0 and a specific weight of 1800 kg/m³. What would be the weight of such a radome?

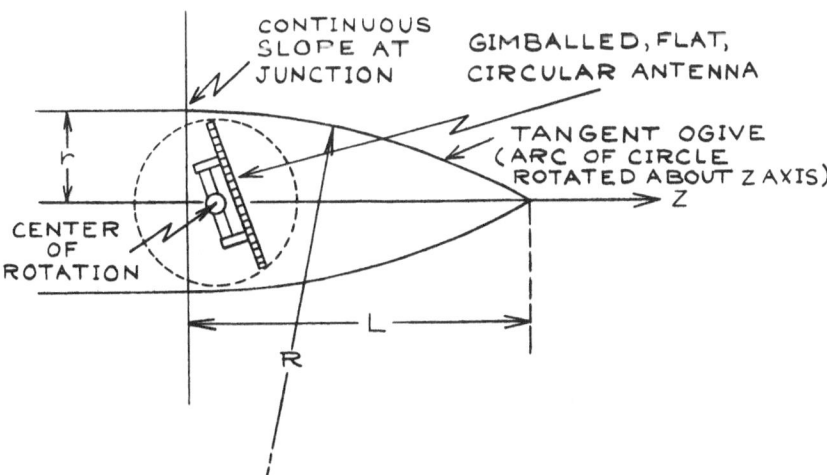

Figure 12-6. Illustration for problem 2.

PROBLEM 12-2.   The radome of a high-speed vehicle has the shape of a tangent-ogive that is generated by rotating an arc of a circle about the z axis as shown in Figure 12-6. Let r = 15 cm, L = 80 cm, and let the antenna frequency range be from 9.0 GHz to 10.0 GHz.

    a. Find R and write the equation of the radome surface.
    b. Calculate the nominal wall thickness for a $\lambda/2$ ceramic radome with ε = 5.4.
    c.* The antenna is a vertically polarized, circular flat plate, mounted on a gimbal to scan a conical space from 0° to 45° about the z axis. Find the range of angles of incidence at 0° and at 45° scan.
    d.* Estimate the antenna gain, without a radome, given the measurements and frequencies. Make a rough estimate of the radome effect on antenna gain.

PROBLEM 12-3.   A 3-m reflector antenna is to operate at 15 GHz in a very severe environment. A conical protective radome is chosen as shown in Figure 12-7. The material is a fiberglass polyester laminate with ε = 3.5, $\tan\delta$ = 0.01.

    a. What is the thickness, d, for a $\lambda/2$ wall?
    b. If a minimal thickness of 2 mm is required for protection from flying ice and hail, should d = 2 mm be chosen rather than a $\lambda/2$ wall?

---

*Advanced problem.

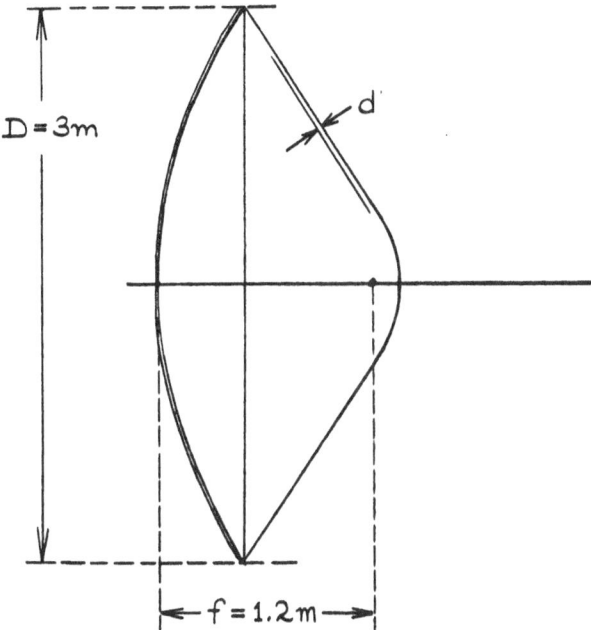

Figure 12-7. Illustration for problem 3.

CHAPTER 13

Reciprocity, the Friis Formula, and Antenna Noise Temperature

This chapter considers three topics that are germane to all antennas, regardless of their shape, size, or operational mode. The reciprocity theorem states that the power patterns of any antenna are identical for transmitting and receiving and is of fundamental importance in both theory and practice. The Friis formula is implicitly based on the reciprocity theorem. It allows determination of the gain or effective area that an antenna must have in a communications link. The antenna noise temperature represents the noise power collected by the antenna from its environment and is of great importance when very weak signals must be received.

13.1 RECIPROCITY

Antennas are used for both transmission and reception, and there is a close relationship between antenna transmitting and receiving characteristics. This relationship can be clarified with the aid of electrical network theory.

A network with two accessible ports is illustrated in Figure 13-1a. If the network is linear and there are no internal sources, the voltages and currents at the two ports are related by

$$V_1 = Z_{11} I_1 + Z_{12} I_2,$$

$$V_2 = Z_{21} I_1 + Z_{22} I_2, \qquad \textbf{(13-1)}$$

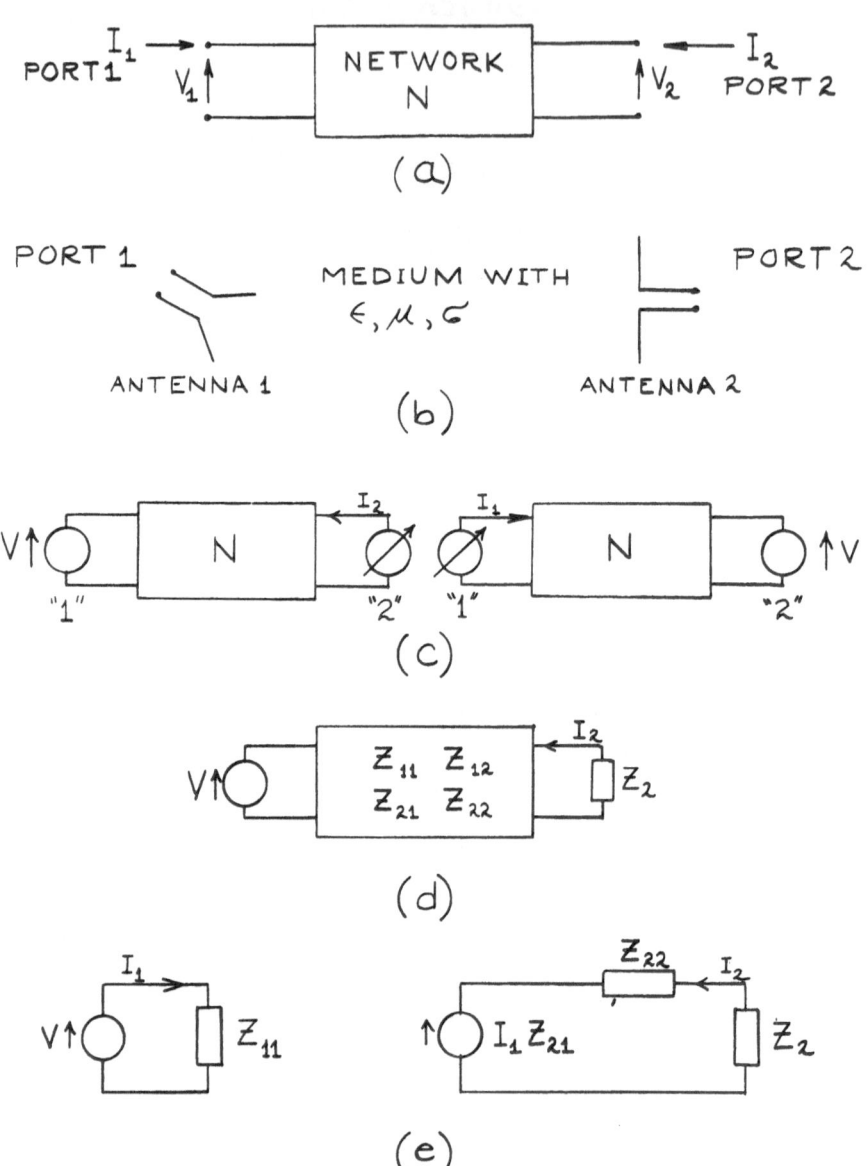

Figure 13-1. Network representations.

or alternatively by

$$I_1 = Y_{11} V_1 + Y_{12} V_2,$$

$$I_2 = Y_{21} V_1 + Y_{22} V_2, \qquad (13\text{-}2)$$

in which the coefficients (impedances or admittances) are independent of time, the voltages, and the currents. In general, the coefficients are complex and depend on frequency and the physical parameters of the network.

An example of a two-port network is an arbitrary pair of antennas in a general setting, as shown in Figure 13-1b. The network represented by Eq. (13-1) or (13-2) is said to be reciprocal if

$$Z_{12} = Z_{21}$$

or

$$Y_{12} = Y_{21}. \qquad (13\text{-}3)$$

If these relationships hold, reciprocity can be shown: a current, I_2, due to a voltage, V, at port 1 is equal to the current, I_1, due to a voltage, V, at port 2, as illustrated in Figure 13-1c. The voltage source and ammeter are assumed to be ideal, which means they have no internal impedance. The relationship of Eq. (13-3) and the consequent reciprocity do not imply symmetry. That is, the network, in general, does not appear the same at ports 1 and 2. For example, the input impedance at port 1 is not necessarily equal to that at port 2. Also, the circuit elements represented by the impedances Z_{11}, Z_{22}, and Z_{12} may be quite complicated.

For the system of two antennas shown in Figure 13-1b, antenna 1 is considered as a transmitter, represented by a voltage source, V, and antenna 2 as a receiver, represented by a load, Z_2, as illustrated in Figure 13-1d. The power delivered to the load is proportional to $|I_2|^2$ and can be calculated by use of Eq. (13-1) and $V_2 = -I_2 Z_2$, with the result

$$|I_2|^2 = V^2 |Z_{21}|^2 / |Z_{11} (Z_{22} + Z_2) - Z_{12} Z_{21}|^2. \qquad (13\text{-}4)$$

For the case of two antennas, coupling is usually very weak—the current, I_2, has virtually no effect on antenna 1, so that

$$V \cong Z_{11} I_1, \qquad (13\text{-}5)$$

in which Z_{11} is approximately equal to the input impedance of antenna 1. Under these conditions the network representation of Figure 13-1e is appropriate, and it is not difficult to show that

$$|I_2|^2 \cong V^2|Z_{21}|^2/|Z_{11}(Z_{22} + Z_2)|^2. \tag{13-6}$$

This expression is similar to Eq. (13-4) when $Z_{11}Z_{22} >> Z_{12}Z_{21}$. Here the coupling terms are small compared to the self-terms, Z_{11} and Z_{22}. Examination of Eq. (13-6) reveals that the power delivered to antenna 2 is proportional to $|Z_{21}|^2$, which is the coupling term from antenna 1 to antenna 2. Similarly, if antenna 2 were the transmitter and antenna 1 the load, the power delivered to antenna 1 would be proportional to $|Z_{12}|^2$.

For the antenna geometries shown in Figure 13-2, antenna 1 of Figure 13-2a is a transmitter and antenna 2 has been rotated through an angle θ

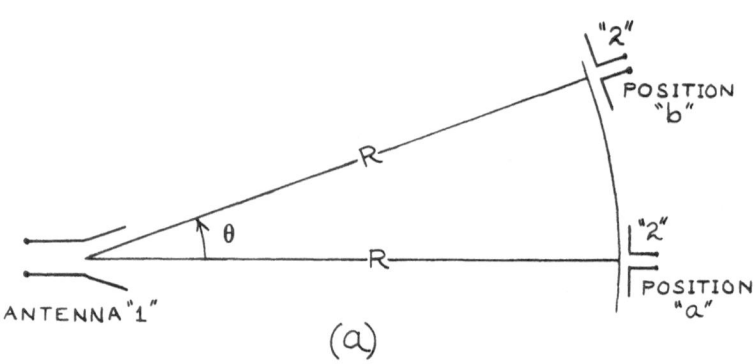

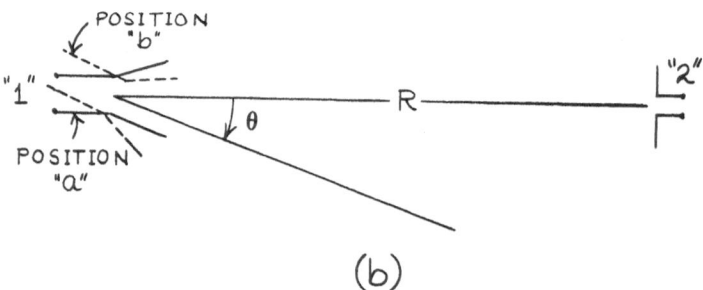

Figure 13-2.   Equivalent antenna geometries.

with respect to the axis of antenna 1, while keeping the separation, R, constant. The ratio of the powers received in positions a and b is, according to Eq. (13-6),

$$W_{2b}/W_{2a} = |Z_{21}|_b^2/|Z_{21}|_a^2. \tag{13-7}$$

If antenna 2 is used as a stationary transmitter and antenna 1 is rotated by an angle, θ, about its axis, the ratio of the powers received by antenna 1 at positions a and b is expressed similarly:

$$W_{1b}/W_{1a} = |Z_{12}|_b^2/|Z_{12}|_a^2. \tag{13-8}$$

For a reciprocal system, $Z_{12} = Z_{21}$ holds for both positions a and b. Thus, Eqs. (13-7) and (13-8) indicate that the normalized power radiation pattern is identical, whether antenna 1 is used as a transmitter while antenna 2 is rotated along a sphere of radius, R, or antenna 2 is used as a stationary transmitter while antenna 1 is rotated about its axis. In arriving at this conclusion, the only assumptions were reciprocity by Eq. (13-3), linearity, and weak coupling between the antennas. When these conditions are satisfied, the antenna separation and each antenna's individual characteristics have no bearing on the result. However, if the coupling between the antennas involves the ionosphere, reciprocity may not exist, because the ionized ionosphere is an example of a medium that may have unidirectional properties in the presence of the Earth's magnetic field. Reciprocity also can break down if other antennas in the area are simultaneously radiating at the same frequency, so that some of their energy is coupled into antennas 1 and 2. In most situations, however, reciprocity applies and the following assertion is valid:

> The normalized total power patterns of any antenna are identical for transmit and receive.

This result lies at the heart of receiving antenna design and pattern measurements. Usually, the antenna undergoing measurement is used as a receiver and is mounted on a gimballed (rotatable) pedestal while the transmitter is stationary and mounted on a tower.

Although the arguments that led to the pattern equivalence principle involved only elementary network theory, they are nevertheless quite subtle. If the line of reasoning is still unclear, a review of this section is suggested.

13.2 THE FRIIS FORMULA

The derivation of the following general antenna relationship refers to Figure 13-2a. The power collected by antenna 2 is proportional to its effective area, A_{eff}. If the power density in the immediate vicinity of antenna 2 is denoted ʰʸ P_2, then the total power received, W_2, is given by

$$W_2 = A_{eff2} P_2. \qquad \text{(13-9)}$$

P_2 can be expressed in terms of W_1, the total power accepted at the input of antenna 1, the antenna separation, R, and the gain of antenna 1, g_1:

$$P_2 = W_1 g_1/4\pi R^2. \qquad \text{(13-10)}$$

However, gain is proportional to effective area as in Eq. (8-33)

$$g_i = 4\pi A_{effi}/\lambda^2; \; i = 1,2. \qquad \text{(13-11)}$$

The results can be substituted into Eq. (13-9) to yield

$$\boxed{\frac{W_2}{W_1} = \frac{A_{eff1}A_{eff2}}{\lambda^2 R^2} = \frac{g_1 g_2 \lambda^2}{(4\pi R)^2}} \qquad \text{(13-12)}$$

This result is known as the Friis transmission formula. It allows calculation of the power transferred from antenna 1 to antenna 2 in terms of these antennas' gains (or areas), their separation, and the wavelength. The assumption that Eq. (13-11) can be used to describe both transmission and reception (for both antennas) is implicitly based on the reciprocity theorem as discussed earlier. Although certain losses are neglected in this derivation, these and other effects can be considered to derive a more general formula not described here.

13.3 ANTENNA NOISE TEMPERATURE

The Friis formula allows determination of the gain or effective area that an antenna must have in order to receive W_2 watts (at antenna 2) when the radiated power is W_1 watts (at antenna 1). However, the received power, in a practical communications link, cannot be arbitrarily small, because a signal that is weaker than the ambient electrical noise level is not ordinarily intelligible. Many sources of noise contribute to the ambient noise level. However, the discussion here is limited to noise that enters the antenna along with the desired signal. If the antenna ohmic losses are neglected, this type of noise is characterized by the antenna noise temperature discussed below.

Any body with a temperature greater than absolute zero (0 K, $-273°C$, or $-459.4°F$) radiates electromagnetic energy over a wide spectrum of frequencies. This is obvious for hot objects that visibly glow but is equally true for objects that "feel" cold. The radiation or emission spectrum of a body depends on its surface temperature; a few examples of such spectra are shown in Figure 13-3.

An antenna receives this thermal radiation from bodies in its environment and delivers it to the receiver where it is manifested as noise. In that part of

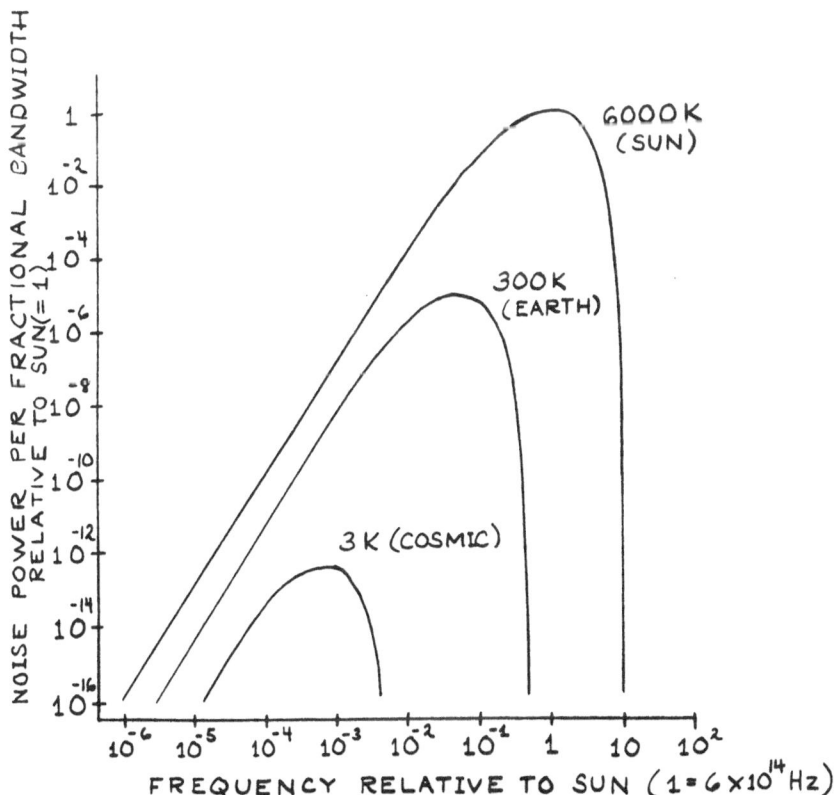

Figure 13-3. Blackbody radiation spectra for various temperatures.

the electromagnetic spectrum that concerns the antenna engineer ($\lambda = 10^3$ m to 10^{-3} m), the intensity of thermal radiation is roughly proportional to temperature and is a potentially important factor in antenna performance.

The antenna noise temperature represents the noise power collected by the antenna and is the average of the temperature of everything in the antenna field of view, weighted by antenna gain:

$$T_a = \frac{1}{4\pi} \iint_{\substack{\text{unit} \\ \text{sphere}}} g(\theta, \phi) \, T(\theta, \phi) \, d\Omega, \qquad \textbf{(13-13)}$$

where $d\Omega = ds/r^2 = \sin\theta \, d\theta \, d\phi$. Clearly, for an isotropic antenna ($g(\theta, \phi) = 1$) with no ohmic losses, T_a reduces to the actual temperature averaged over all angle space (4π steradians). If the antenna has significant ohmic losses represented by the loss factor, $L \geq 1$, the antenna noise temperature given by Eq. (13-13) must be modified in accordance with

$$T_a' = \frac{T_a}{L} + T_\ell \left(1 - \frac{1}{L}\right), \qquad \textbf{(13-14)}$$

in which T_ℓ is the temperature of the lossy material of the antenna, and T_a' is the corrected antenna noise temperature. T_a might be regarded as an external contribution and T_ℓ as an internal contribution.

To relate the antenna noise temperature to the noise power that enters the front end of the receiver, it is assumed that the antenna is replaced by a resistor, R_N, at temperature T_a, at the end of the feed line, as shown in Figure 13-4. The entering noise power P_N is found by using the well-known result for the noise power generated by a resistor in a bandwidth, B,

$$P_N = k \, T_a \, B, \qquad \textbf{(13-15)}$$

in which $k = 1.38 \times 10^{-23}$ W-s/K—Boltzmann's constant.

Of course, the antenna noise, as calculated by Eq. (13-13) or (13-14) and (13-15), is only part of the overall noise that limits system performance as characterized by the system noise temperature. However, antenna noise is an important noise source in systems that must detect very weak signals, such as communications from distant spacecraft. As with Eq. (13-13), the antenna noise temperature depends on both the full antenna pattern and the surrounding environment of the antenna. The standard or nominal temperature of the Earth's surface is 300 K, which is approximately room temperature. The temperature of the sky depends on the part viewed: The sun is very hot—much hotter, in the microwave region, than the 6000 K in the visible

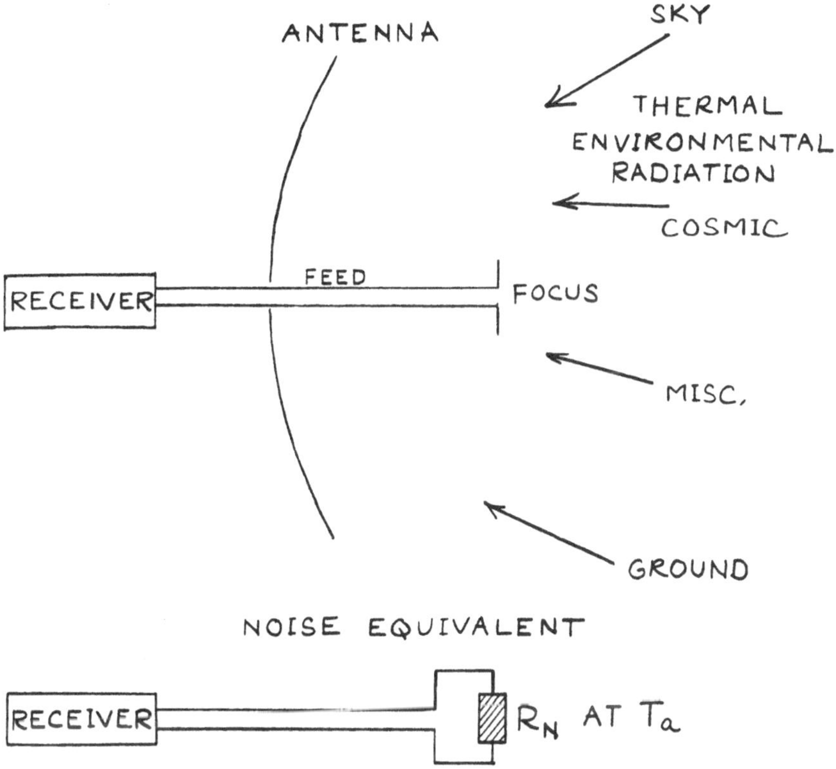

ANTENNA

SKY

THERMAL
ENVIRONMENTAL
RADIATION

COSMIC

RECEIVER

FEED

FOCUS

MISC,

GROUND

NOISE EQUIVALENT

RECEIVER R_N AT T_a

Figure 13-4. Equivalent circuit for antenna noise.

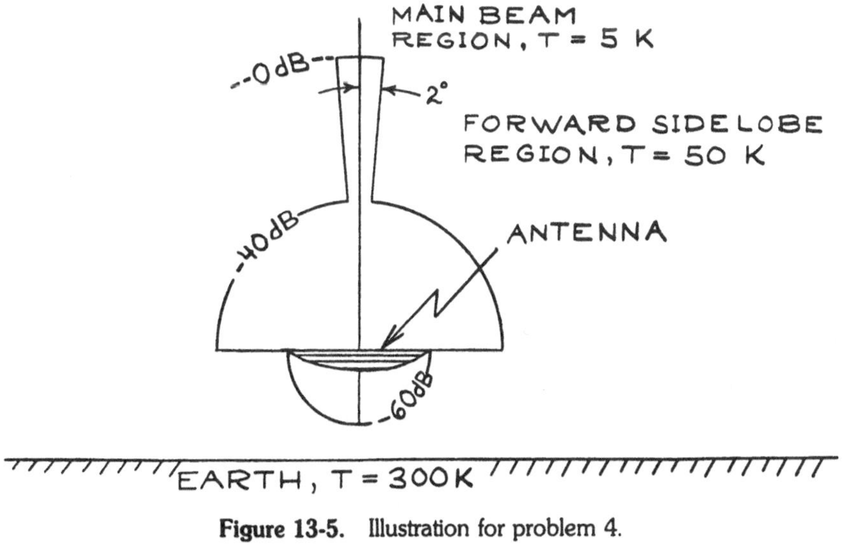

MAIN BEAM
REGION, T = 5 K

-0dB-- 2°

FORWARD SIDELOBE
REGION, T = 50 K

-40dB

ANTENNA

-60dB

EARTH, T = 300 K

Figure 13-5. Illustration for problem 4.

region of the spectrum. Certain stars and galaxies emit enough radiation to constitute a measureable noise source. Indeed, even the darkest areas of the sky are not completely quiet, because all space is permeated by the primordial cosmic background radiation that is characterized by a temperature of ~3 K.

Because the entire radiation pattern, including far out and even backward side lobes, contributes to the antenna noise temperature, precautions must sometimes be taken to ensure that the antenna does not strongly illuminate the relatively hot earth. For this reason, some very sensitive satellite communications and radioastronomy antennas are designed to suppress backward radiation. One design is the completely enclosed horn-reflector antenna, illustrated in Figure 10-13. In general, low side lobe levels help reduce the antenna noise temperature, which is another benefit in addition to those already mentioned.

PROBLEMS

PROBLEM 13-1.   Derive Eqs. (13-4), (13-6) and (13-12) in detail.

PROBLEM 13-2.   A spacecraft is designed to send data to Earth from distances of up to 100 million km. It is reasonable to assume that the solar cells can provide the spacecraft with 20 W of transmitted power, and its deployable antenna has an effective area of 50 m². The receiver on Earth needs a signal of 10^{-14} W to be able to detect the transmission. What effective area and what gain does the antenna on Earth need at $f = 3$ GHz and at $f = 30$ GHz?

PROBLEM 13-3.   A point-to-point communication link between two mountains 50 km apart uses a frequency of 6 GHz and two identical circular reflector antennas of 3 m diameter. Assume that the effective aperture area is 50% of the physical area and that a received power of 10^{-10} W is needed to maintain good service. What is the necessary transmitted power?

PROBLEM 13.4.   An antenna points into a "quiet" region in the sky, where $T = 5$ K. It has an idealized radiation pattern as shown in Figure 13-5. Assuming the average temperature of the sky in the forward side lobe region to be 50 K and of the earth 300 K, calculate the antenna noise temperature. How much of an improvement would be obtained by reducing the back radiation to -80 dB? How much improvement would be obtained by reducing the average forward side lobes to -50 dB?

CHAPTER 14

The Radar Equation and the Radar Cross Section

This chapter considers topics of central importance in the theory of radar: the radar equation (14.1) and the radar cross section (14.2). Many concepts covered in previous chapters form the necessary background for understanding these subjects.

Radar (radio detection and ranging) is a technology that has found a wide variety of applications in both military and civilian fields. Much of the versatility of radar derives from its ability to "see through" clouds and darkness in a remote sensing role.

Radar uses electromagnetic waves, usually in the ultrahigh frequency (UHF) or microwave regions, to "illuminate" an area of interest. The energy scattered from the objects in the illuminated field is detected and processed to provide information such as location, size, velocity, and other target characteristics. Antennas are used for both the transmission and reception of radar waves. In addition, the scattering process incorporates many concepts described in previous chapters.

14.1 THE RADAR EQUATION

Figure 14-1 illustrates in schematic form a general radar configuration. The transmitting antenna illuminates a distant target that scatters the incident energy in all directions. In general, the energy is scattered anisotropically (i.e., more energy is reflected in certain directions than in others). A second antenna collects a portion of the scattered energy, referred to as the target's "echo." The angle, β, between the lines that join the transmitting antenna, target, and receiving antenna is called the bistatic angle. For various practical reasons, the great majority of radars are "monostatic," with the transmitter and receiver colocated ($\beta = 0$, $R_1 = R_2$). The theory of the more general bistatic case, however, is not more difficult to understand.

The total transmitted power is denoted by W_1. The power density at the target, apart from propagation losses, is simply the product of the isotropic power level and the gain of antenna 1, g_1:

$$P_t = g_1 W_1/4\pi R_1^2; \quad (\text{W/m}^2). \tag{14-1}$$

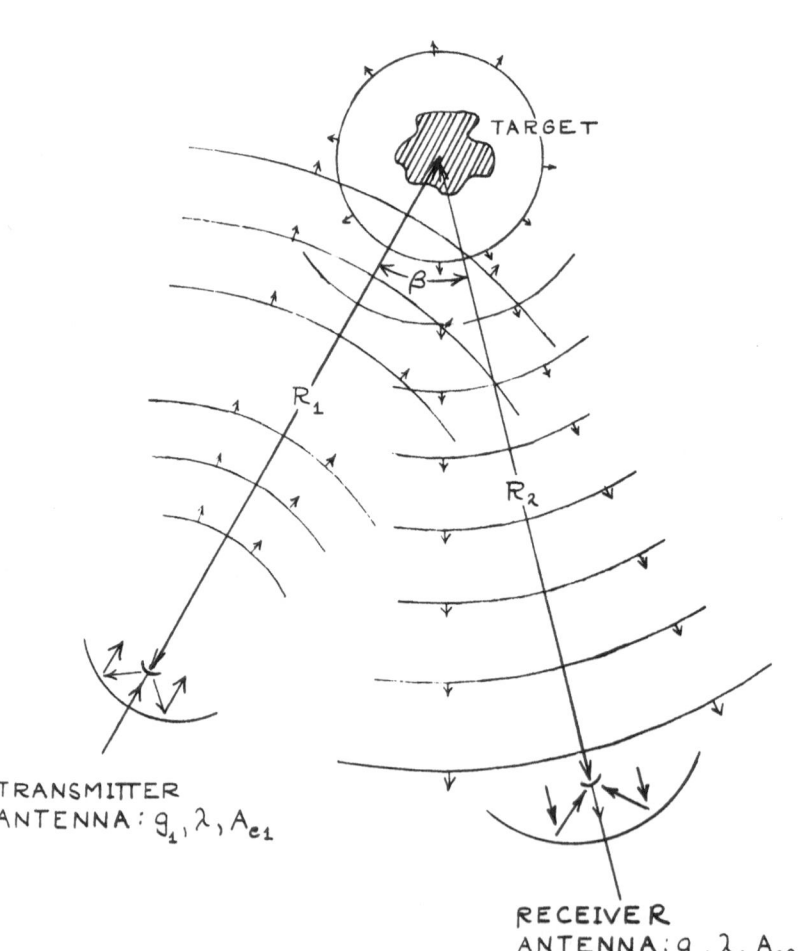

Figure 14-1.   Basic radar geometry.

The amount of power absorbed by the target depends upon its absorption cross section, A_a, which has units of area. With the total absorbed power denoted by W_a, then

$$W_a = P_t A_a \quad \text{(W)}. \qquad \textbf{(14-2)}$$

If the target is a lossless dielectric or a perfect conductor, the total scattered power is equal to W_a. For realistic targets, the total scattered power, W_s, is less than W_a, and can be expressed by

$$W_s = P_t A_s \quad (A_s \leq A_a) \quad \text{(W)}, \qquad \textbf{(14-3)}$$

in which A_s is the scattering cross section of the target. If the power were scattered isotropically, the scattered power per unit solid angle would be

$$\langle W_s \rangle = W_s/4\pi = P_t A_s/4\pi; \quad \text{(W/steradian)}. \qquad \textbf{(14-4)}$$

Using Eq. (14-4), the scattering cross section can also be defined as

$$\sigma_{av} = A_s = 4\pi\langle W_s \rangle/P_t = 4\pi \left[\begin{array}{c} \text{average scattered power} \\ \text{per unit solid angle} \\ \text{power density at target} \end{array} \right], \qquad \textbf{(14-5)}$$

which has units of area. Because the amount of power scattered by the target is a function of direction, β, (as referenced to the direction of incidence) the radar cross section (RCS) is defined in analogy with Eq. (14-5),

$$\sigma(\beta) = 4\pi \left[\begin{array}{c} \text{power scattered per unit} \\ \text{solid angle in } \beta \text{ direction} \\ \text{power density at target} \end{array} \right] \qquad \textbf{(14-6)}$$

The RCS also has units of area. In analogy with Eq. (14-3), the power scattered in the direction of the receiver can be expressed in terms of the RCS as

$$W_s' = P_t\sigma(\beta) = W_1 \, g_1\sigma(\beta)/4\pi R_1^2 \quad \text{(W)}, \qquad \textbf{(14-7)}$$

and the power density at the receiver as

$$P_r = W_s'/4\pi R_2^2 = W_1 \, g_1\sigma(\beta)/(4\pi R_1^2)(4\pi R_2^2) \quad \text{(W/m}^2). \qquad \textbf{(14-8)}$$

In all of these derivations propagation losses have *not* been included. The power collected by the receiving antenna, W_2, is proportional to its effective area, A_{e2}:

$$W_2 = P_r A_{e2} = W_1 g_1 \sigma(\beta) A_{e2}/(4\pi)^2 R_1^2 R_2^2 \quad \text{(W)}. \quad \textbf{(14-9)}$$

If the relation between gain and effective area is used (discussed in previous chapters), $g = 4\pi A_e/\lambda^2$ and the ratio of the received to transmitted power becomes

$$\frac{W_2}{W_1} = \frac{g_1 g_2 \sigma(\beta)\lambda^2}{(4\pi)^3 R_1^2 R_2^2} = \frac{A_{e1} A_{e2} \sigma(\beta)}{4\pi\lambda^2 R_1^2 R_2^2} \quad \textbf{(14-10)}$$

For the very common case of monostatic radar, $\beta = 0$ and $R_1 = R_2 \equiv R$. The RCS for $\beta = 0$ is called the backscatter RCS, denoted as σ. Thus, for monostatic radar,

$$W_2/W_1 = g_1 g_2 \sigma \lambda^2/(4\pi)^3 R^4 L. \quad \textbf{(14-11)}$$

The loss factor, $L > 1$, introduced here takes into account any propagation losses. Note that the receiving and transmitting antennas are not necessarily the same. If $W_2 = W_{min}$ represents the smallest detectable signal, the monostatic radar maximum range can be expressed by use of Eq. (14-11):

$$R_{max} = \left[\frac{W_1 g_1 g_2 \sigma \lambda^2}{(4\pi)^3 W_{min} L} \right]^{1/4} \quad \textbf{(14-12)}$$

From Eq. (14-12), which is one form of the radar range equation, it is clear that antenna gain and the RCS play a crucial role in radar theory and design. If the same antenna is used for both transmitting and receiving, which is very common in practice, then $g_1 = g_2 = g$. The maximum range in that case is in proportion to $\sqrt{g}$, the square root of the antenna gain (or, equivalently, the square root of the antenna area.) The influence of other factors, such as transmitted power, W_1; minimal detectable power, W_{min}; the radar cross section σ; and the losses, L, are all proportional to their fourth root. Thus, a radar antenna is one of the most important and cost determining components of a radar system.

14.2 THE RADAR CROSS SECTION

Because the RCS of a body depends on its shape, size, orientation (with respect to the incident energy), and material properties, it is a difficult function to compute. However, there are some basic configurations that can be used. Consider first the RCS of a perfectly conducting sphere of radius r. If $r \gg \lambda$, geometrical optics, as illustrated in Figure 14-2, can be used. Also, the incident field is assumed to be a plane wave, which is tantamount to requiring that the target lie in the far field region of the transmitter. As shown in Figure 14-2a, there is rotational symmetry about an axis that passes through the center of the sphere and is parallel to the incident wave vector. For convenience, it is assumed that the incident power density is 1 W/m². By the law of reflection, energy that is scattered in the β direction must be reflected from surface elements on the sphere that are perpendicular to the cone of the half angle, $\beta/2$, as shown in Figure 14-2b. The energy incident on the narrow annular region on the sphere surface, which lies between the conical sections defined by $\beta/2 + \delta/2$ and $\beta/2 - \delta/2$ ($\delta \ll 1$), is given by the product of the incident power density and the projected area of the annulus:

$$W_s(\beta) = 1 \left(\frac{W}{m^2}\right) \left(2\pi \, r \sin\frac{\beta}{2}\right) (r\delta) \cos\frac{\beta}{2}$$

$$= \pi r^2 \, \delta \, \sin\beta. \tag{14-13}$$

The corresponding cone of reflected rays centered on the half angle β occupies a solid angle given by

$$\Delta\Omega = \Delta s/r^2 = (1/r^2)(2r\pi\sin\beta)(2r\delta) = 4\pi\delta\sin\beta. \tag{14-14}$$

Thus, the power given by Eq. (14-13) is scattered into the solid angle given by Eq. (14-14). By use of the definition of the RCS in Eq. (14-6), the result obtained is

$$\boxed{\sigma(\beta)_{sphere} = \pi r^2 = \text{constant}} \tag{14-15}$$

The RCS of a conducting sphere that is very large compared to a wavelength ($r \gg \lambda$) is equal to its projected area. It is independent of wavelength, polarization, and bistatic angle, β.

The result of Eq. (14-15) is an approximation that improves as r/λ increases. It is valid for all values of β, save for a small cone about $\beta = \pi$ within which diffraction predominates.

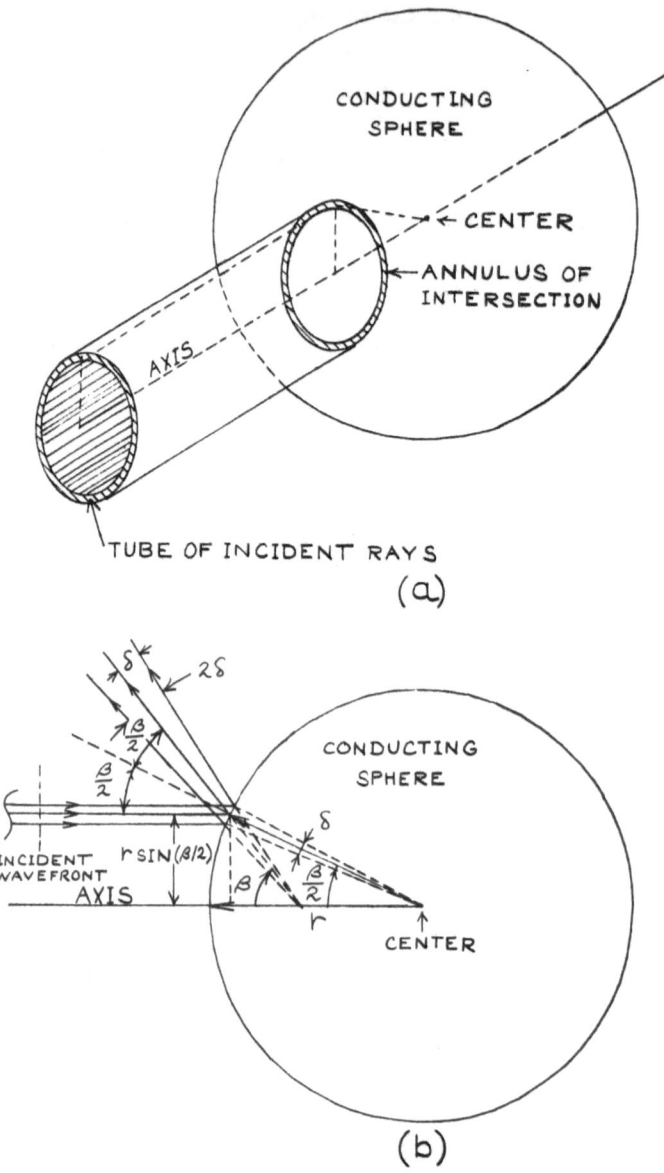

Figure 14-2.   Ray construction for RCS of conducting sphere.

Because of its simplicity, the sphere is one of the few shapes for which the RCS can be computed rigorously for all wavelengths, polarizations, and bistatic angles. The result, illustrated in Figure 14-3, is for $\beta = 0$, the backscatter RCS. Clearly, a sphere's scattering characteristics change drastically as r/λ becomes small. Indeed, in the Rayleigh region geometrical optics is invalid, and the backscatter RCS depends strongly on wavelength:

$$\sigma = 80 \; \pi^3 \; V^2/\lambda^4 \cong 2.5 \times 10^3 \; V^2/\lambda^4, \qquad \textbf{(14-16)}$$

in which $V = 4\pi r^3/3$ is the sphere's volume. Eq. (14-16) also provides a good approximation for general small scatterers, even if their shapes are not spheres. Thus, in the Rayleigh region, the scattering cross section decreases rapidly as the frequency decreases. This digression emphasizes that Eq. (14-15) should not be used in those regimes of r/λ where it does not apply.

Geometrical optics fails if the condition $r \gg \lambda$ is not satisfied in the case of a spherical scattering body. Another case for which geometrical optics cannot be used, regardless of the scatterer's size, is that of scattering by a flat con-

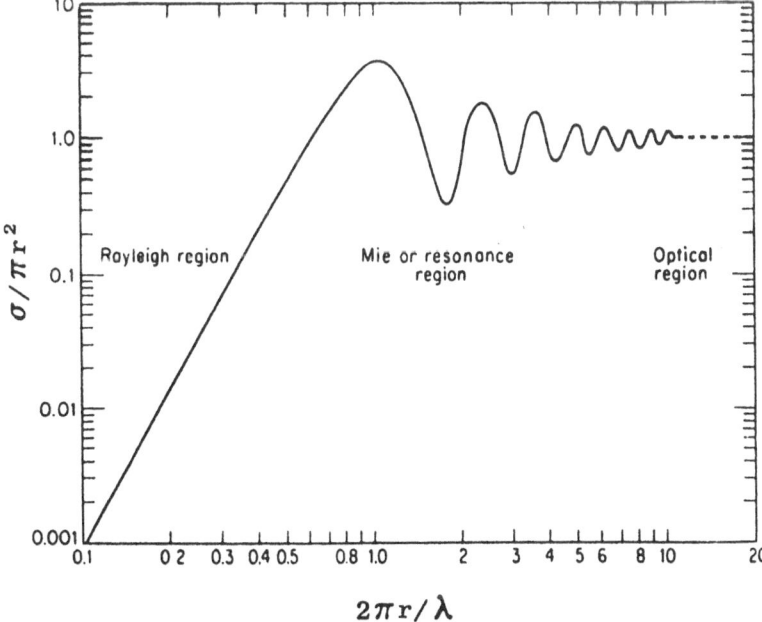

Figure 14-3. Radar backscatter cross section for a sphere. (*From M. I. Skolnik, Introduction to Radar Systems, New York: McGraw-Hill, 1980, p. 34; copyright © 1970 by McGraw-Hill Book Company, reprinted by permission.*)

ducting plate, as shown in Figure 14-4. The tube of parallel rays intercepted by the flat plate is redirected by reflection. However, the rays remain parallel, (i.e., they continue as a *finite* collimated beam). Chapters 5 and 10 explained that the far field produced by a collimated beam cannot be predicted by geometrical optics. However, the field scattered by the flat area of the plate can be described by use of the boundary condition of Eq. (3-22), namely

$$\mathbf{n} \times \mathbf{H} = \mathbf{J}_s, \tag{14-17}$$

in which $\mathbf{H}$ is the magnetic field at the conductor's surface, $\mathbf{n}$ is the unit normal at the surface, and $\mathbf{J}_s$ is the induced surface current. As illustrated in Figure 14-4, the incident magnetic field is that of a plane wave, which gives rise to a reflected magnetic field that has the same tangential component. Referring to Eq. (5-2), the incident field has the form

$$\mathbf{H}_{in} = \mathbf{H}_o \, e^{-j\mathbf{k}\cdot\mathbf{r}},$$

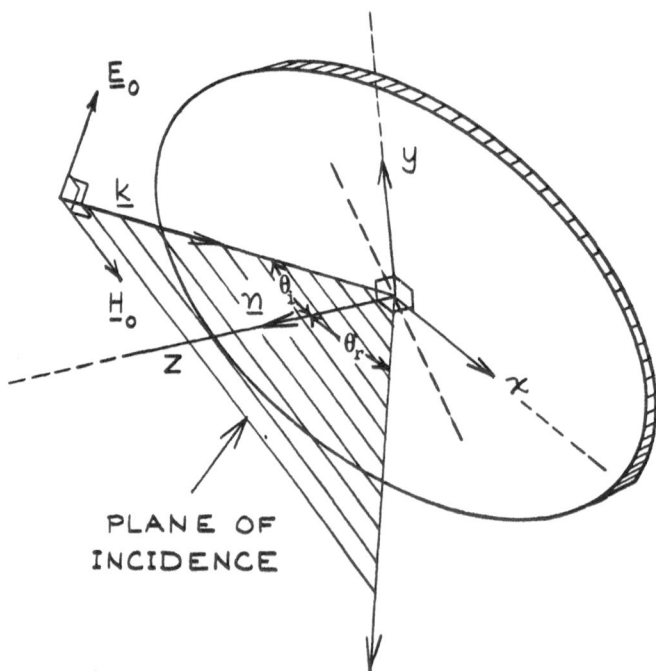

Figure 14-4.   Scatterning geometry for flat conducting plate.

and the unit normal at the plate surface, in this case, is $\mathbf{n} = \mathbf{z}_o$, and so the surface current is given by

$$\mathbf{J}_s = 2(\mathbf{z}_o \times \mathbf{H}_o)\, e^{-j\mathbf{k}\cdot\mathbf{r}}, \qquad (14\text{-}18)$$

in which the factor of 2 is due to reflection. The wave vector $\mathbf{k}$ and $\mathbf{H}_o$ must be known, and $\mathbf{r}$ is evaluated at the point of reflection. For the special case of normal incidence, $\mathbf{k} = -k\mathbf{z}_o$, $\mathbf{k}\cdot\mathbf{r} = 0$, and for convenience, $\mathbf{H}_o = \mathbf{x}_o H_o$. The surface current is then $\mathbf{J}_s = 2H_o\mathbf{y}_o$ on the plate and $\mathbf{J}_s = 0$ elsewhere.

All edge effects have been entirely neglected. The current has a uniform distribution over the finite area of the plate, which can be identified with an aperture. Chapters 7 and 10 described how the far field arising from an aperture distribution is proportional to the double Fourier integral of the current distribution. All of the mathematical tools required have been developed to calculate the scattered field and the associated RCS; however, the details are not provided here. This method is often referred to as *physical optics* or the *Kirchhoff integral* procedure. For a square conducting plate with sides of length b ($b \gg \lambda$), the calculation of the backscatter RCS yields

$$\sigma_b(\theta) = \frac{4\pi b^4}{\lambda^2}\left[\frac{\sin(kb\sin\theta)}{kb\sin\theta}\right]. \qquad (14\text{-}19)$$

The angle θ here is the same as θ_i in Figure 14-4; it is the angle of incidence with respect to the normal $\mathbf{n}$ of the plate. At normal incidence, $\theta = 0$, and

$$\sigma_b = \frac{4\pi A^2}{\lambda^2} = A g_{max} \qquad (14\text{-}20)$$

in which $g_{max} = 4\pi A/\lambda^2$.

The form $\sin u/u$ has appeared several times previously in this text, in connection, for example, with the far field of a linear current distribution. For a large plate, $kb = 2\pi b/\lambda$ is very large, and the function has a narrow and strongly peaked main beam centered on $\theta = 0$. The RCS is very large in the vicinity of $\theta = 0$. For example, a square plate with $b = 1$ m has a physical area of 1 m^2; however, at normal incidence its RCS is approximately 140 m^2 at 1 GHz ($\lambda = 30$ cm), and about 1.26×10^4 m^2 at 30 GHz ($\lambda = 1$ cm). On the other hand, for those directions that correspond to $kb\sin\theta = m\pi$, or $\theta_m \cong m\lambda/2b$, Eq. (14-19) predicts no backscatter at all. This result is somewhat misleading, because in this approximation the diffraction contribution from the edges has been neglected. Nevertheless, it is clear that the flat plate concentrates the scattered energy which, for an irregularly shaped object, would be scattered in all other directions. The creation of a bright spot or "glint" by playing with a mirror is similar to this effect. Thus, the RCS does not usually correspond to the physical projected area of a scatterer.

Because real radar targets are often combinations of elementary shapes, the discussion of the RCS can be extended to a few well-known composite structures. Among these, the corner reflector is particularly interesting and important. A structure that acts as a corner reflector for waves that travel parallel to the x-y plane is illustrated in Figure 14-5a. The plane surfaces of the reflector lie in the x-z and y-z planes, respectively. Using geometrical optics, it is shown, as in Figure 14-5b, that rays that enter the corner reflector undergo double reflection and emerge in a direction antiparallel to their direction of incidence. The corner reflector of Figure 14-5 is referred to as two-dimensional, because only rays that lie in planes perpendicular to the reflector axis (z axis) can experience the perfect reversal that characterizes this class of structures.

Although geometrical optics can accurately predict the field's behavior in the vicinity of the reflector, physical optics must be used to calculate the far field. The backscatter RCS can once again be shown to be proportional to $4\pi(A/\lambda)^2$, where A is the effective aperture area. The computation of the effective aperture area is somewhat involved and is not shown here. However, the result, for the backscatter RCS, is

$$\sigma \cong (4\pi/\lambda^2)\left[2\sqrt{2}\ ab\ \sin\left(\phi + \frac{\pi}{4}\right)\right]^2, \qquad \textbf{(14-21)}$$

in which the azimuthal angle, ϕ, and the parameters a and b are as indicated in Figure 14-5a. The backscatter RCS of the corner reflector is large over a large portion of the range $0° < \phi < 90°$, in contrast to the flat plate backscatter RCS, which is sharply peaked at normal incidence only. Thus, the two-dimensional corner reflector is highly visible or *retrodirective* in the azimuthal plane ($\theta = 90°$) for incidence directions in the quarter plane $0° < \phi < 90°$.

The corner reflectors of Figure 14-6 are three-dimensional generalizations of that shown in Figure 14-5. Each is characterized by a vertex at which the x-y, x-z, and y-z planes meet, although their overall shapes are different. The backscatter cross section of a corner reflector, in general, is again given by a formula similar to Eq. (14-20),

$$\sigma \cong 4\pi(A_{eff}/\lambda)^2. \qquad \textbf{(14-22)}$$

A_{eff} is the aperture area found by projecting that part of the corner reflector surface that is reflecting rays toward the source on a plane parallel to the

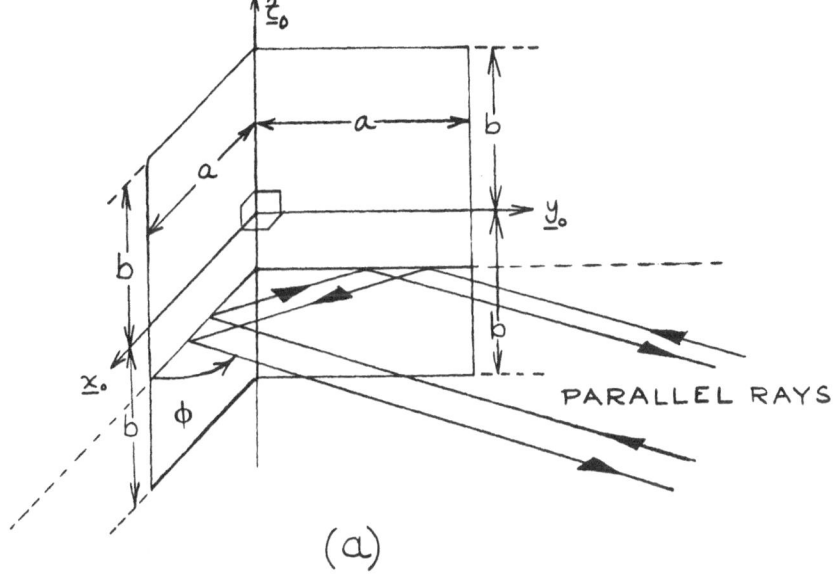

(a)

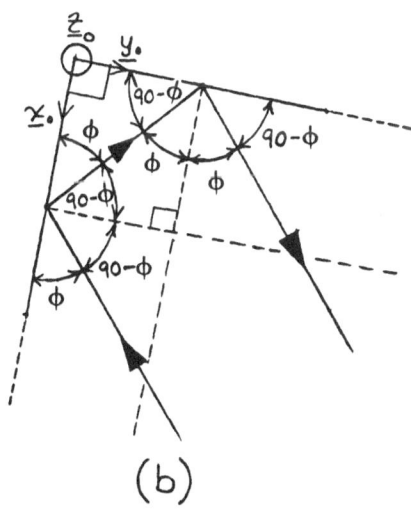

(b)

Figure 14-5. Two-dimensional corner reflector.

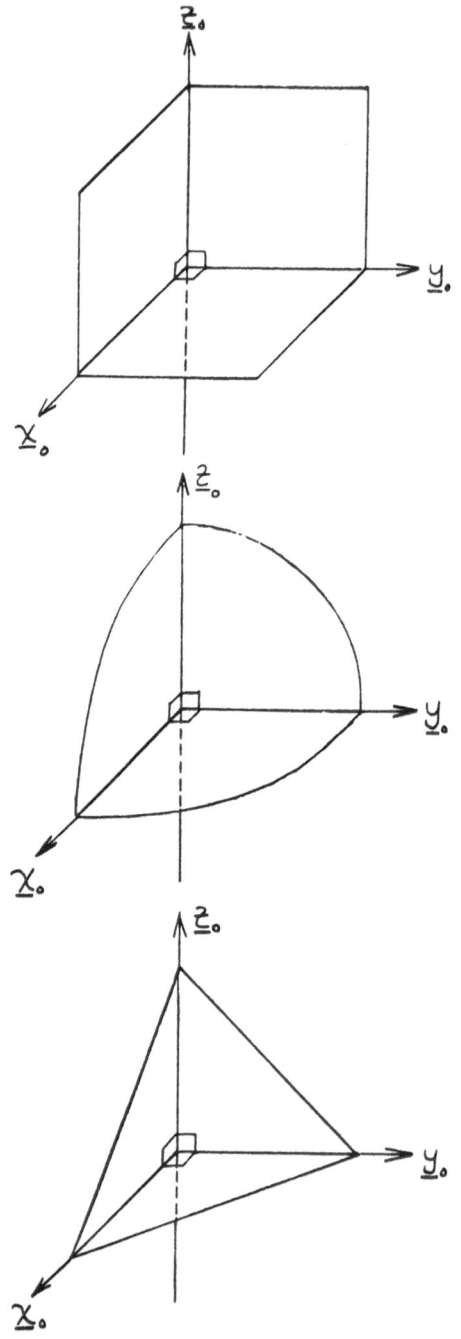

Figure 14-6.   Cube corner reflectors.

incident wavefront. The various types of corner reflectors shown in Figure 14-6 provide a large backscatter cross section over a wide range of angles within a cone that is centered on the symmetry axis of the reflector (this is an axis of three-fold rotational symmetry). Reflectors of this type are used to test radars and provide good radar visibility for commercial aircraft. The Apollo astronauts, in fact, placed an optical corner reflector on the moon as part of a highly accurate laser ranging system. Constructed environmental features such as buildings, ships, bridges, and cities usually contain many corner shapes and are therefore very visible to radar.

Many radar targets can be interpreted as collections of several simpler targets. The scattered fields from the target's components may interfere, giving rise to an RCS that fluctuates with observation angle. Consider the geometry of the simple example shown in Figure 14-7a. Bodies A and B are separated by a distance, d, and are illuminated by a plane wave that arrives from a direction, ϕ, as measured from the line that joins the bodies. The directly backscattered field is proportional to $\sqrt{\sigma}$, because σ is proportional to the backscattered power. At the point of observation, the backscattered fields have a phase difference of

$$\psi = 2(kd\cos\phi) = (4\pi d\cos\phi)/\lambda. \qquad \textbf{(14-23)}$$

Thus, if interactions, such as multiple reflections between A and B, are neglected the total backscattered field is expressed by the law of cosines, as illustrated in Figure 14-7b:

$$\sigma_t = \sigma_A + \sigma_B + 2\sqrt{\sigma_A\,\sigma_B}\,\cos\psi. \qquad \textbf{(14-24)}$$

If the two scatterers are identical $\sigma_A = \sigma_B = \sigma$, and

$$\sigma_t = 2\,\sigma\,(1 + \cos\psi), \qquad \textbf{(14-25)}$$

which is similar to the power pattern for an array of two sources. Thus, the RCS of two similar scatterers will be four times larger than that of each one alone in those directions for which $\psi = 2\,\pi m$, where $m = 0,\ \pm 1,\ \pm 2\ldots$, or $\cos\phi = m\lambda/2d$. However, in those directions for which $\psi = (2m + 1)\pi$, where $m = 0,\ \pm 1,\ \pm 2\ldots$, and $\cos\phi = (\lambda/4d)(2m + 1)$, the interference between the two components causes the RCS to vanish.

A complicated target, such as an airplane has many scattering centers, such as engines or other protruding parts, flat areas on wings and stabilizers, corner reflectors, and curved convex parts. Therefore, an aircraft's RCS is a complicated function of wavelength, observation direction, and polarization. The RCS can be characterized by a single number (e.g., $10\ m^2$) that represents an average value. However, unless how the average is calculated is specified, the meaning is ambiguous. The measured RCS fluctuates wildly as the direction of incidence changes. The average is usually taken over sectors of specified solid angle. Even this averaged RCS may vary by orders of magnitude when viewed "head-on," from the side, or from below.

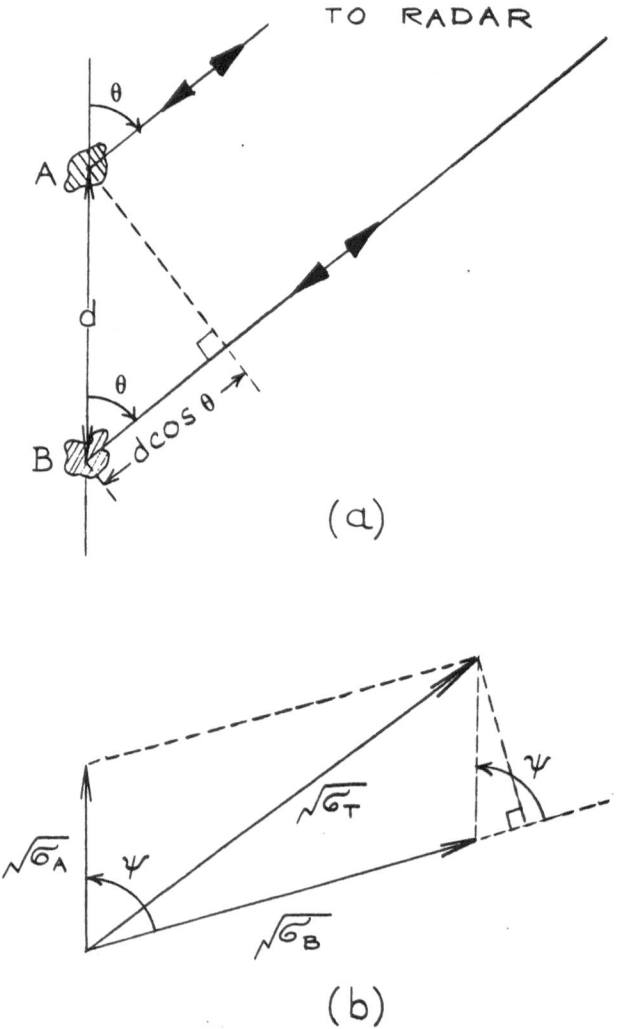

Figure 14-7.   Backscatter RCS for two targets.

PROBLEMS

PROBLEM 14-1.* Calculate the backscatter RCS and the backscatter beam width of a smooth circular metal plate with a 20-cm diameter at $f_1 = 3$ GHz and $f_2 = 30$ GHz. Compare these results to the RCS and beam width of a sphere with the same diameter at these frequencies. Use the results of Chapter 10 for circular apertures.

PROBLEM 14-2. A CW Doppler radar for police operations transmits 1 W at 30 GHz. Assume that some vehicles have an average RCS of as little as 2 m^2. What would be the maximum detection range with a 0.6-m-diameter circular reflector antenna if the needed received power is 10^{-10} W?

PROBLEM 14-3. Two identical spherical metallized balloons of 3-m diameter are used to calibrate a radar.

    a.   What is the backscatter RCS of each balloon at $f_1 = 1$ GHz and at $f_2 = 6$ GHz?

    b.   The balloons float at a distance of 10 m from each other. Calculate their combined RCSs when the values of ϕ in Figure 14-7a are 90° and 30° for the two frequencies f_1 and f_2 in part a above.

    c.   The distance from the balloons to the radar is 10 km. The radar antenna is circular and has a 4-m diameter. Can the two targets be resolved at f_1 and f_2?

---

*Advanced problem.

The Algebra of Complex Numbers and Its Application to Waves

A.1 DEFINITIONS AND ALGEBRAIC OPERATIONS

Complex algebra provides a convenient framework within which to treat wave phenomena.

The imaginary unit, j, is defined by

$$j^2 = j \times j = -1 \text{ or } j = \sqrt{-1}. \qquad \text{(A-1)}$$

The square root of any negative number, $-y^2$, can be expressed as a product of the real number, y, and the imaginary unit, j, and is called an imaginary number:

$$\sqrt{-y^2} = \sqrt{-1}\,\sqrt{y^2} = jy. \qquad \text{(A-2)}$$

The sum of a real number, x, and an imaginary number jy, is a *complex number*, usually denoted by z:

$$z = x + jy. \qquad \text{(A-3)}$$

The rules of algebra for real numbers also apply to complex numbers:

$$z_1 \pm z_2 = (x_1 + jy_1) \pm (x_2 + jy_2) = (x_1 \pm x_2) + j(y_1 \pm y_2)$$

$$z_1 \cdot z_2 = (x_1 + jy_1)(x_2 + jy_2) = (x_1 x_2 - y_1 y_2) + j(x_1 y_2 + x_2 y_1)$$

$$\frac{z_1}{z_2} = \frac{x_1 + jy_1}{x_2 + jy_2} = \frac{(x_1 + jy_1)(x_2 - jy_2)}{(x_2 + jy_2)(x_2 - jy_2)}$$

$$= \frac{x_1 x_2 + y_1 y_2}{x_2{}^2 + y_2{}^2} - \frac{j(x_1 y_2 - x_2 y_1)}{x_2{}^2 + y_2{}^2}. \tag{A-4}$$

It is often useful to plot complex numbers in the x-y plane. The x coordinate represents the real part, and the y coordinate represents the imaginary part. When illustrated in this fashion, complex numbers are seen to obey the rules

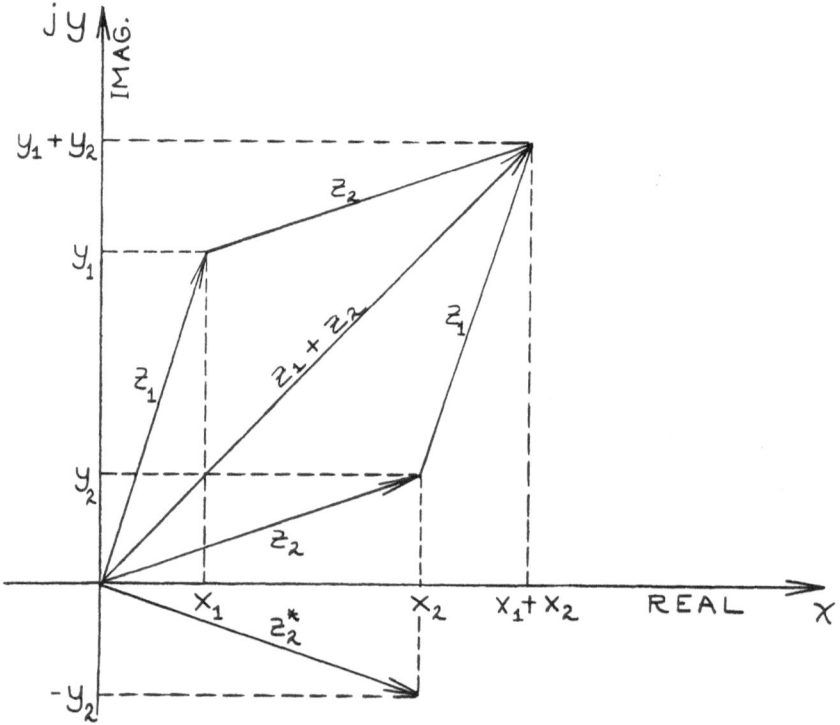

Figure A-1. Cartesian representation of complex numbers.

of vector addition for vectors in the x-y plane (see Appendix B), as shown in Figure A-1. The correspondence between complex numbers and two-dimensional vectors can sometimes be exploited for mathematical convenience.

The complex conjugate of a complex number z is obtained by changing the sign of the imaginary part. For example

$$z_1 = x_1 + jy_1, \qquad z_1{}^* = x_1 - jy_1;$$

$$z_2 = x_2 - jy_2, \qquad z_2{}^* = x_2 + jy_2. \qquad \text{(A-5)}$$

The absolute value or modulus of a complex number is the real number obtained from

$$|z| = \sqrt{z \cdot z^*} = \sqrt{x^2 + y^2}. \qquad \text{(A-6)}$$

Returning to the geometrical representation of Figure A-1, the complex conjugate is the reflection of the point (x, jy) about the x axis, and the modulus is the length of the line segment that connects the point (x, jy) to the origin.

It is often convenient to manipulate complex numbers in the polar representation, which is simply the two-dimensional graphical representation in polar coordinates. A complex number is now expressed

$$z = |z| \, (\cos\phi + j \sin\phi) = r \, (\cos\phi + j \sin\phi), \qquad \text{(A-7)}$$

where $r = |z| = \sqrt{x^2 + y^2}$ and $\tan \phi = y/x$. The geometrical interpretation of complex number multiplication and division is apparent when the polar representation is used:

$$z_1 \cdot z_2 = r_1 \, r_2 \, (\cos\phi_1 + j \sin\phi_1)(\cos\phi_2 + j \sin\phi_2);$$

$$= r_1 \, r_2[\cos(\phi_1 + \phi_2) + j \sin(\phi_1 + \phi_2)];$$

$$\frac{z_1}{z_2} = \frac{r_1}{r_2} \, [\cos(\phi_1 - \phi_2) + j \sin(\phi_1 - \phi_2)]; \qquad \text{(A-8)}$$

in which $r_1 = |z_1|$ and $r_2 = |z_2|$.

These results can be verified by using Eqs. (A-4) and the appropriate trigonometric identities. Eqs. (A-8) show that multiplication by a complex number of unit modulus and argument, ϕ, corresponds to a counterclockwise rotation by ϕ in the complex plane; division corresponds to clockwise rotation by the same angle. This property reemerges more explicitly when complex numbers are expressed in exponential form.

The following are Taylor expansions of several functions:

$$\sin x = x - \frac{x^3}{3!} + \frac{x^5}{5!} - \frac{x^7}{7!} + \ldots ; \qquad \text{(A-9a)}$$

$$\cos x = 1 - \frac{x^2}{2!} + \frac{x^4}{4!} - \frac{x^6}{6!} + \ldots ; \qquad \text{(A-9b)}$$

$$e^x = 1 + x + \frac{x^2}{2!} + \frac{x^3}{3!} + \ldots . \qquad \text{(A-9c)}$$

If $x = ju$ is formally substituted into Eq. (A-9c), then because $j^3 = -j, j^4 = 1$, $j^5 = j$, etc.,

$$e^{ju} = \left(1 - \frac{u^2}{2!} + \frac{u^4}{4!} - \frac{u^6}{6!} + \ldots \right) + j \left(u - \frac{u^3}{3!} + \frac{u^5}{5!} - \ldots \right),$$

or

$$e^{ju} \equiv \cos u + j \sin u \qquad \text{(A-10)}$$

which is the Euler-DeMoivre identity. The exponential representation of a complex number is obtained by combining Eq. (A-7) and the identity of Eq. (A-10):

$$z = x + jy = re^{j\phi},$$

$$r = |z|, \qquad \tan \phi = y/x; \qquad \text{(A-11)}$$

therefore,

$$z_1 \cdot z_2 = (r_1 \, e^{j\phi_1}) \cdot (r_2 \, e^{j\phi_2}) = r_1 \, r_2 \, e^{j(\phi_1 + \phi_2)},$$

and

$$\frac{z_1}{z_2} = \frac{r_1 \, e^{j\phi_1}}{r_2 \, e^{j\phi_2}} = \left(\frac{r_1}{r_2} \right) e^{j(\phi_1 - \phi_2)}, \qquad \text{(A-12a)}$$

which is consistent with Eq. (A-8).

From Eq. (A-10) the following interesting relations can be obtained:

$$e^{j\pi} = -1 ; \qquad e^{\pm j\pi/2} = \pm j ; \qquad e^{2jm\pi} = 1 \ (m = \text{integer}); \qquad \text{(A-13)}$$

and the general powers of z

$$z^p = (r\,e^{j\phi})^p = r^p\,e^{jp\phi}, \tag{A-14}$$

in which p is any real number. Likewise, the n nth roots of a complex number are

$$\sqrt[n]{z} = \sqrt[n]{r\,e^{j\phi}} = \sqrt[n]{r}\,e^{j[(\phi\,+\,2\pi m)/n]}$$

$$(m = 0,1\dots n-1). \tag{A-15}$$

For example, there are five fifth roots of -1: $e^{j\pi/5}$; $e^{3j\pi/5}$; $e^{j\pi}$; $e^{7j\pi/5}$; and $e^{9j\pi/5}$ that occur at equally spaced angular intervals around the unit circle in the complex plane, as shown in Figure A-2.

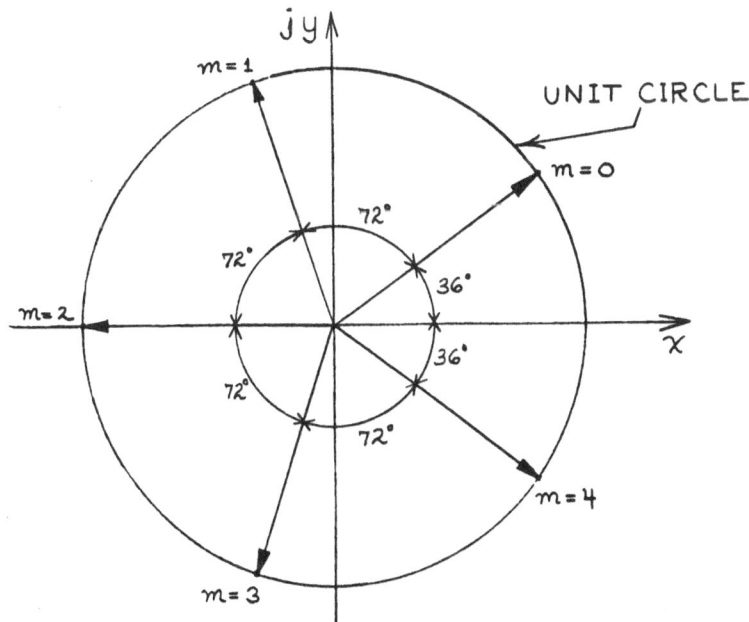

Figure A-2. The five roots of -1.

A.2 APPLICATION TO SINUSOIDAL PHENOMENA

Complex algebra provides an elegent description of sinusoidal phenomena. The functions $u_1(t)$ and $u_2(t)$, which are sinusoidal in time with frequency, f, and have real amplitudes, a, are

$$u_1(t) = a \sin(\omega t - \phi);$$

$$u_2(t) = a \cos(\omega t - \phi); \qquad \omega = 2\pi f. \qquad \textbf{(A-16)}$$

From Figure A-3, u_1 and u_2 are simply the x and y projections or components of the complex function

$$u(t) = a\, e^{j(\omega t - \phi)} = a\, e^{-j\phi}\, e^{j\omega t}. \qquad \textbf{(A-17)}$$

The complex amplitude is defined $A = a\, e^{-j\phi}$, so that

$$u(t) = A\, e^{j\omega t}, \qquad \textbf{(A-18)}$$

and

$$u_1(t) = Im\; u(t), \qquad u_2(t) = Re\; u(t), \qquad \textbf{(A-19)}$$

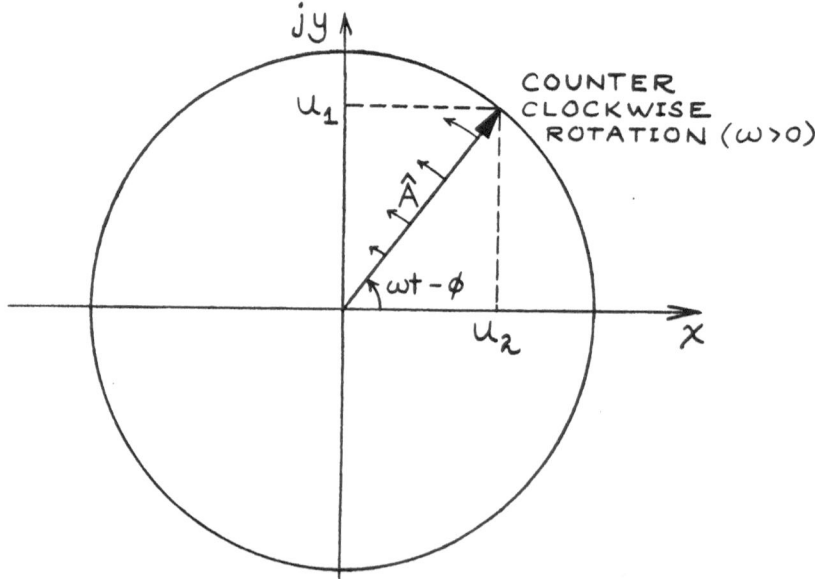

Figure A-3. Phasor representation of sinusoidal functions.

where *Im* is the imaginary part, and *Re* is the real part. The complex quantity $u(t)$, as indicated in Figure A-3, rotates counterclockwise in the complex plane with angular velocity, ω, and is called a phasor (not to be confused with phaser, which is a phase shifter). The exponential complex algebra notation simplifies mathematical manipulations in many cases, for example, the derivatives of u_1 and u_2 from Eqs. (A-16):

$$\frac{du_1}{dt} = \omega A \cos(\omega t - \phi) = \omega u_2;$$

$$\frac{du_2}{dt} = -\omega A \sin(\omega t - \phi) = -\omega u_1;$$

$$\frac{d^2 u_1}{dt^2} = -\omega^2 u_1; \qquad \frac{d^2 u_2}{dt^2} = -\omega^2 u_2 \text{ etc.} \qquad \textbf{(A-20)}$$

Clearly, keeping track of the derivatives and their relations to the original functions can potentially lead to confusion. On the other hand, if Eq. (A-18) is used:

$$\frac{du}{dt} = j\omega A e^{j\omega t} = j\omega u;$$

$$\frac{d^2 u}{dt^2} = -\omega^2 A e^{j\omega t} = (j\omega)^2 u;$$

$$\frac{d^m u}{dt^m} = (j\omega)^m u, \qquad m = \text{integer}; \qquad \textbf{(A-21)}$$

and furthermore, if $u_2 = Re\ u$, then for any m:

$$\frac{d^m u_2}{dt^m} = Re\ \frac{d^m u}{dt^m} = Re\ [(j\omega)^m u]. \qquad \textbf{(A-22)}$$

In Eqs. (A-21) and (A-22), m is an arbitrary positive or negative integer. Here, a negative derivative represents indefinite integration:

$$\frac{d^{-1} u}{dt^{-1}} \equiv \int u\, dt = u/j\omega,$$

$$\frac{d^{-1} u_2}{dt^{-1}} = \int u_2\, dt = Re\ (u/j\omega). \qquad \textbf{(A-23)}$$

Derivation of the results of Eqs. (A-21) through (A-23) is left as an exercise.

Because the phase, ϕ, in Eqs. (A-16) is arbitrary, either u_1 or u_2 can be used to describe any sinusoidal function of frequency, f. Alternatively, the real part of u can be used to describe any quantity that varies sinusoidally. For additional convenience, the prefix Re can be dropped, but it must be remembered that where u appears, $Re\ u$ is intended.

A.3   APPLICATION TO ELECTRIC CIRCUIT THEORY

Electrical circuit theory provides an example of the utility of complex algebra. For the circuit of Figure A-4, it is assumed that a sinusoidal current at frequency, f, is flowing. This current may be expressed

$$i(t) = Ie^{j\omega t}; \qquad \omega = 2\pi f, \qquad \textbf{(A-24)}$$

in which I is a complex amplitude; this complex function represents the real current

$$i(t) = |I| \cos (\omega t - \phi) = Re(Ie^{j\omega t}).$$

From AC circuit theory, the voltages across the various elements in terms of the current can be expressed:

$$v_R = Ri,$$

$$v_L = L \frac{di}{dt},$$

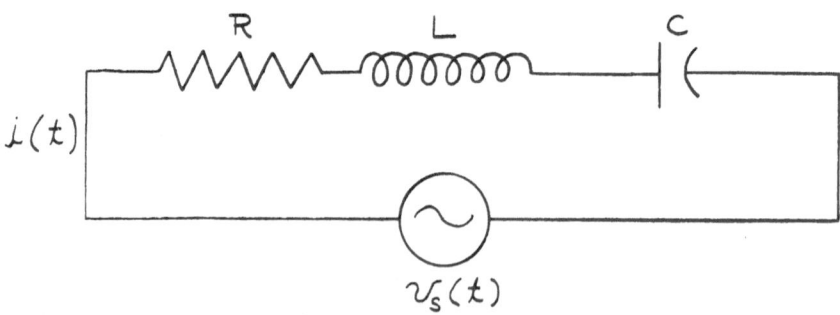

Figure A-4.   Simple AC circuit.

and

$$v_c = \frac{1}{c} \int i \, dt. \qquad \text{(A-25)}$$

Referring to Eqs. (A-22) through (A-24):

$$v_R = Ri = RIe^{j\omega t},$$

$$v_L = j\omega Li = j\omega LIe^{j\omega t},$$

and

$$v_c = \frac{1}{j\omega c} i = \frac{I}{j\omega c} e^{j\omega t}. \qquad \text{(A-26)}$$

If the source voltage is given by

$$v_s(t) = Ve^{j\omega t}, \qquad \text{(A-27)}$$

Kirchoff's voltage law requires that

$$v_s = Ve^{j\omega t} = v_R + v_L + v_c = \left(R + j\omega L + \frac{1}{j\omega c} \right) Ie^{j\omega t}$$

and the time-dependent factor, which is the same on both sides of this equation, can be dropped. The remaining relation between the complex voltage, V, and complex current, I, is expressed

$$V = ZI,$$

in which

$$Z = R + j\omega L + \frac{1}{j\omega c}$$

is the complex impedance. The complex relationship between voltage and current is illustrated in the phasor diagram of Figure A-5. Although the relative positions of the phasors are fixed, the whole group rotates with angular velocity, ω, due to the time factor $e^{j\omega t}$. The complex notation automatically keeps track of the relative phase relationships of the voltage contributions, thereby simplifying solution of the circuit problem.

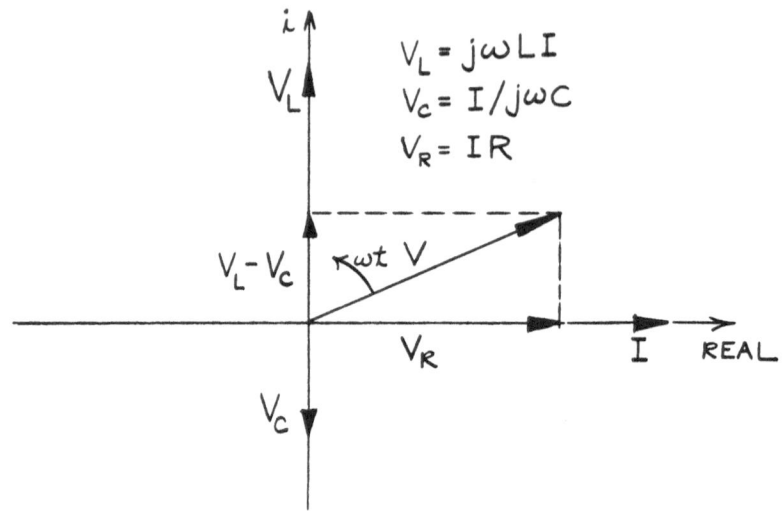

Figure A-5. Phasor diagram for voltages.

A.4 APPLICATION TO WAVE PHENOMENA

A sinusoidal plane wave with real amplitude, a, frequency, f, and phase velocity, c, which may be expressed

$$u(x,t) = a \cos [(\omega t - kx) - \phi]$$

$$\omega = 2\pi f, \qquad k = \omega/c = 2\pi/\lambda, \tag{A-27}$$

becomes, according to the rules established,

$$u(x,t) = Ae^{j\omega t} e^{-jkx},$$

in which $A = ae^{-j\phi}$, and the prefix Re has been omitted. This separation into time and space factors permits dropping the time factor altogether, as long as the calculations are restricted to sinusoidal waves of a single frequency. The notation

$$u(x) = Ae^{-jkx} \tag{A-28}$$

is therefore equivalent to that of Eq. (A-27), but far less cumbersome.

Economical notation is particularly desirable in the study of antennas and antenna arrays. For example, a uniformly illuminated linear phased array is shown in Figure A-6a. Uniform illumination means that all elements have the same amplitude, A, in this case. The total field amplitude at a point in the far field characterized by the off-broadside angle θ is given by

$$A_t\,(\xi) = \sum_{m=-n/2}^{n/2} Ae^{j\xi[m-(|m|/2m)]}, \qquad \text{(A-29)}$$

where

$$\xi = \frac{2\,\pi\,d}{\lambda}\,(\sin\,\theta\,-\,\sin\,\theta_s). \qquad \text{(A-30)}$$

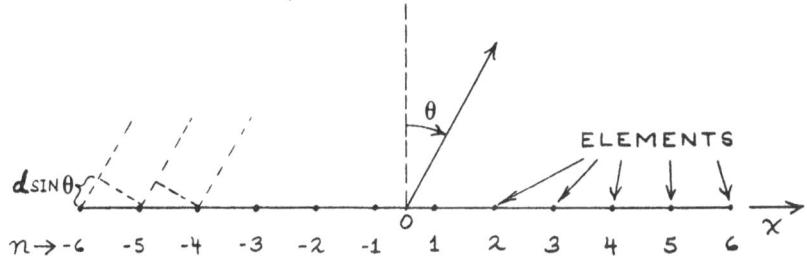

Figure A-6a.   Linear array schematic.

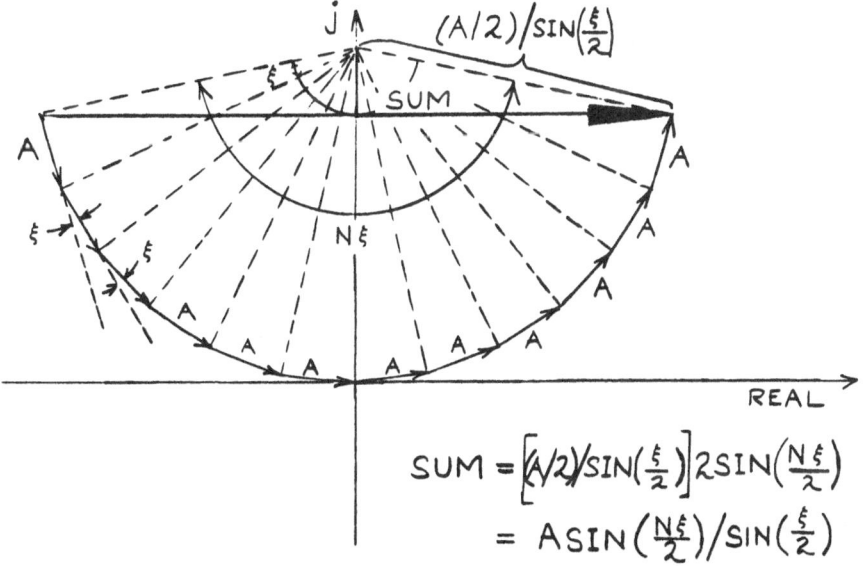

$$\text{SUM} = \left[(A/2)\!/\!\text{SIN}\!\left(\tfrac{\xi}{2}\right)\right]2\,\text{SIN}\!\left(\tfrac{N\xi}{2}\right)$$

$$= A\,\text{SIN}\!\left(\tfrac{N\xi}{2}\right)\!/\text{SIN}\!\left(\tfrac{\xi}{2}\right)$$

Figure A-6b.   Phasor diagram for total field amplitude.

Here d is the interelement spacing, λ is the wavelength, and θ_s is the beam steering angle. It is not necessary to understand the theory behind these results (described in Chap. 8) to appreciate the notational economy possible by use of the exponential complex representation. The phasor diagram of Figure A-6b shows how the amplitudes add for a particular value of ξ. From geometrical interpretation of the phasor diagram, the following is obtained for the sum:

$$A_t\,(\xi)\,=\,A\,\sin\left(\frac{n\xi}{2}\right)/\sin\left(\frac{\xi}{2}\right), \qquad \text{(A-31)}$$

which is the unnormalized array factor, a basic result. Phasor diagrams, although not often used for actual antenna pattern calculations, do provide very useful pictures that aid understanding of phase relationships.

For electromagnetic wave calculations, the field quantities are vectors, not scalars such as the function u discussed above. However, for antenna arrays, the scalar wave method (in which the field is treated as a scalar quantity and the field direction is inferred by other means such as symmetry), is quite useful. In general, however, the field vectors have components, each of which is a function of position and time. For sinusoidal time-varying fields, each component has a complex amplitude that can be represented by a phasor. The time factor $\exp(j\omega t)$ is usually not made explicit. Adding contributions from several sources at some point in space requires vector algebra, discussed in Appendix B.

Vector Algebra and Some Applications

B.1 DEFINITIONS AND ALGEBRAIC OPERATIONS

Vectors are used to represent quantities that have both magnitude and direction, such as displacement, force, and momentum. The electromagnetic field is described in terms of the vectors **E** and **H**, which are often accompanied by currents specified by **J**.

The vectors used most often are three-dimensional—they have three components. For this review, only three-dimensional vectors are described. These vectors are represented by arrows in three-dimensional space. The arrow analogy is useful for purposes of visualization; however, it can be misleading. For example, if an arrow is specified by a terminal point and head point in a coordinate system, rotation of the coordinate system will reorient the arrow. A physical vector, however, has a direction that must remain unchanged regardless of reorientation of the coordinate system. This concept is most easily visualized in a two-dimensional coordinate system, as illustrated in Figure B-1. The vector, **A** is specified in the unprimed coordinate system by its projections or components along the coordinate axes. If **A** has a magnitude (length) $|A| = a$, its x and y components are $a \times \cos\theta$ and $a \times \sin\theta$ respectively, by simple trigonometry. The same vector **A**, however, has new components $a \times \cos\theta'$ and $a \times \sin\theta'$ when considered in the primed coordinate system, which is rotated by $\theta - \theta'$ counterclockwise relative to the unprimed system. The essential point is this: In order for an arrow to maintain its orientation (and thereby represent a vector), its components must transform in a precise way if the coordinate system is changed. The basic idea is simple— a vector has a fixed direction because its components transform to maintain that direction.

Understanding that vectors have fixed directions, a coordinate system can be chosen by which to express the components of vectors. A system that simplifies the mathematics is usually selected. The most common coordinate system has three mutually perpendicular (orthogonal) axes, x, y, and z, as shown in Figure B-2. Vectors of unity magnitude or unit vectors, $\mathbf{x}_0$, $\mathbf{y}_0$, and $\mathbf{z}_0$, are directed along the x, y, and z axes, respectively. The coordinate system of Figure B-2 is right-handed, a property later defined in connection with the vector cross product. The length of the $\mathbf{A}$ projection on the x axis along projection lines perpendicular to $\mathbf{x}_0$ is a scalar quantity that is called the x component of $\mathbf{A}$ (compare Fig. B-1). Similarly, the y and z components correspond to the projected lengths of $\mathbf{A}$ on the y and z axes, respectively.

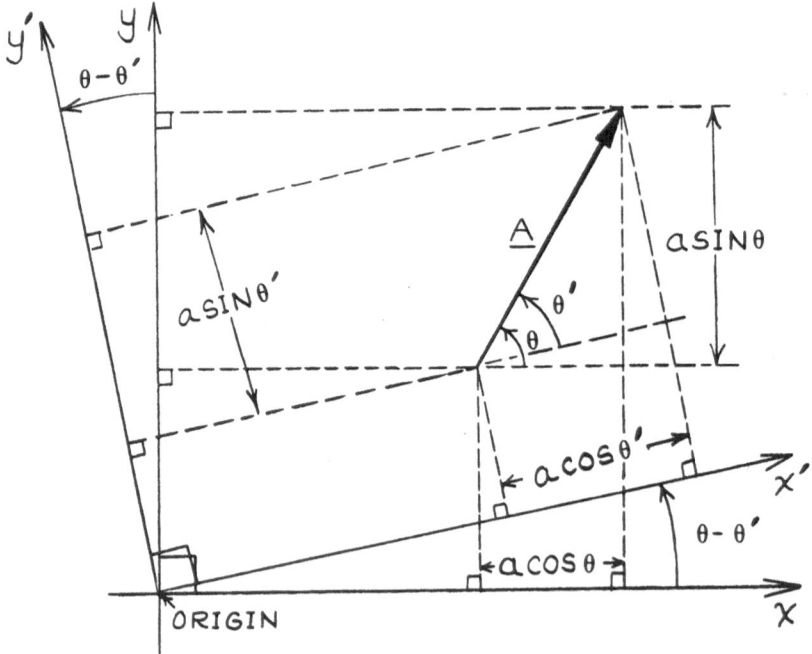

Figure B-1.   A vector's components in two coordinate systems.

Now, vector **A** can be expressed as a linear combination of the three unit or coordinate basis vectors:

$$\boxed{\mathbf{A} = a_1\,\mathbf{x}_0 + a_2\,\mathbf{y}_0 + a_3\,\mathbf{z}_0} \qquad \textbf{(B-1)}$$

Similarly, a second vector **B** may be expressed

$$\mathbf{B} = b_1\,\mathbf{x}_0 + b_2\,\mathbf{y}_0 + b_3\,\mathbf{z}_0.$$

To add or subtract vectors, the respective components are added or subtracted. Therefore,

$$\boxed{\mathbf{A} \pm \mathbf{B} = (a_1 \pm b_1)\,\mathbf{x}_0 + (a_2 \pm b_2)\,\mathbf{y}_0 + (a_3 \pm b_3)\,\mathbf{z}_0} \quad \textbf{(B-2)}$$

The sum vector, as shown in Figure B-2, is identical to the vector obtained by translating the tail of **B** to the head of **A**, in which case the sum vector

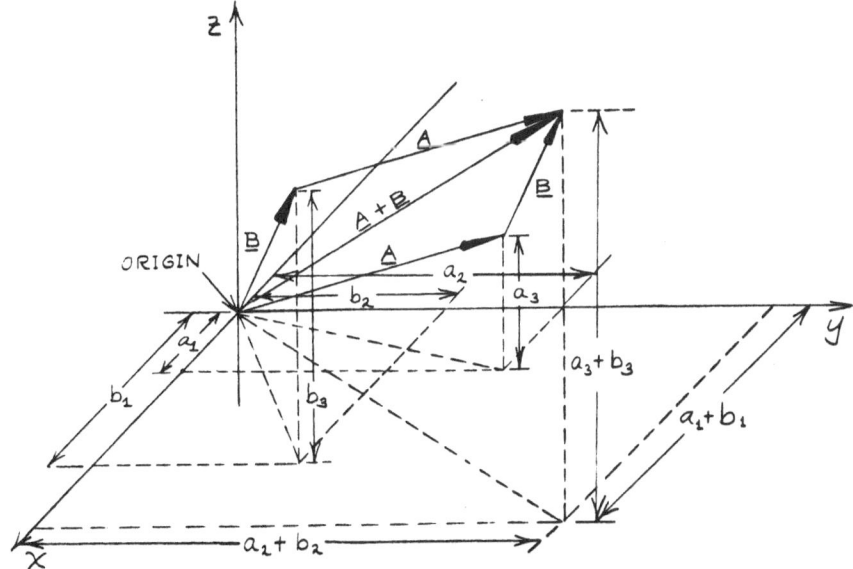

Figure B-2. Vector addition in a right-handed coordinate system.

extends from the tail of **A** to the head of **B**. From Eq. (B-2), the algebraic rules of commutivity, associativity, and distributivity apply to vector addition:

$$\mathbf{A} + \mathbf{B} = \mathbf{B} + \mathbf{A};$$

$$\mathbf{A} + (\mathbf{B} + \mathbf{C}) = (\mathbf{A} + \mathbf{B}) + \mathbf{C};$$

$$(m + n)\,\mathbf{A} = m\,\mathbf{A} + n\,\mathbf{A};$$

$$m\,(\mathbf{A} + \mathbf{B}) = m\,\mathbf{A} + m\,\mathbf{B}. \tag{B-3}$$

The length of a vector is a scalar quantity that, by the Pythagorean theorem, may be expressed in terms of the components:

$$|\mathbf{A}| = a = \sqrt{a_1^2 + a_2^2 + a_3^2}. \tag{B-4}$$

An important vector operation, the *dot* or *scalar* product, can be defined in geometrical terms: the scalar product of vectors **A** and **B** is equal to the length of vector **A** times the length of vector **B** times the cosine of the angle α between them:

$$\boxed{\mathbf{A} \cdot \mathbf{B} \equiv a\,b\,\cos\alpha} \tag{B-5}$$

By using the law of cosines as shown in Figure B-3, the dot product is expressed in terms of the vectors' components. From Figure B-3:

$$\mathbf{C} = \mathbf{A} - \mathbf{B} = (a_1 - b_1)\,\mathbf{x}_0 + (a_2 - b_2)\,\mathbf{y}_0 + (a_3 - b_3)\,\mathbf{z}_0 \tag{B-6}$$

and, from the law of cosines,

$$c^2 = a^2 + b^2 - 2\,a\,b\,\cos\alpha.$$

Because the last term on the right is essentially the dot product of **A** and **B**, by use of Eqs. (B-4) and (B-6):

$$\boxed{\mathbf{A} \cdot \mathbf{B} = a_1\,b_1 + a_2\,b_2 + a_3\,b_3} \tag{B-7}$$

The length of a vector can be expressed:

$$|\mathbf{A}| = \sqrt{\mathbf{A} \cdot \mathbf{A}}.$$

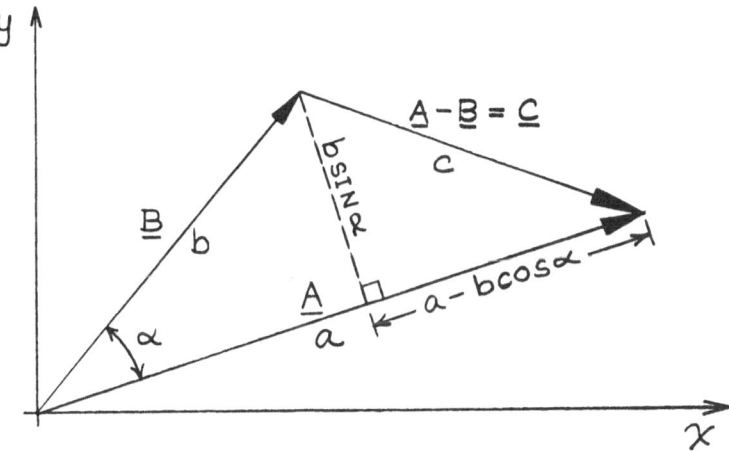

Figure B-3. Construction for derivation of dot product of A and B.

If $\mathbf{A} \cdot \mathbf{B} = 0$, the vectors $\mathbf{A}$ and $\mathbf{B}$ must be perpendicular (orthogonal) because, from definition (B-5), $\alpha = \pi/2$.

If a coordinate system is orthogonal, or has mutually perpendicular axes, then

$$\mathbf{x}_0 \cdot \mathbf{x}_0 = \mathbf{y}_0 \cdot \mathbf{y}_0 = \mathbf{z}_0 \cdot \mathbf{z}_0 = 1,$$

$$\mathbf{x}_0 \cdot \mathbf{y}_0 = \mathbf{y}_0 \cdot \mathbf{z}_0 = \mathbf{x}_0 \cdot \mathbf{z}_0 = 0, \qquad \text{(B-8)}$$

and the components of a vector $\mathbf{A}$ may be expressed

$$a_1 = \mathbf{A} \cdot \mathbf{x}_0, \qquad a_2 = \mathbf{A} \cdot \mathbf{y}_0, \qquad a_3 = \mathbf{A} \cdot \mathbf{z}_0. \qquad \text{(B-9)}$$

B.2 APPLICATIONS TO GEOMETRY AND ELECTROMAGNETIC WAVES

The equation of a plane is easy to derive by use of the dot product. From geometry, a plane is uniquely determined if any point in the plane and the plane's orientation are given. Thus, a plane can be specified in a coordinate system by providing the position vector, $\mathbf{r}$, of any point in the plane, and the unit vector normal to the plane, $\mathbf{n}$. If the vector $\mathbf{r}'$ is also a position vector of a point in the plane, then $\mathbf{r} - \mathbf{r}'$ must be a vector that lies in the plane. Therefore, because $\mathbf{n}$ is normal to the plane,

$$(\mathbf{r} - \mathbf{r}') \cdot \mathbf{n} = 0 \qquad \text{(B-10)}$$

is the vector equation of the plane. If the reference vector, $\mathbf{r}'$, is the position vector of that point in the plane closest to the origin of the coordinate system, then $\mathbf{r}'$ must be perpendicular to the plane, that is, $\mathbf{r}' = d\mathbf{n}$, in which d is the minimum distance between the plane and the origin. The vector diagram of Figure B-4 illustrates the geometrical relationships for this case. Eq. (B-10) is now replaced by

$$(\mathbf{r} - d\mathbf{n}) \cdot \mathbf{n} = 0$$

or

$$\mathbf{r} \cdot \mathbf{n} = d\mathbf{n} \cdot \mathbf{n} = d. \qquad (\text{B-11})$$

If

$$\mathbf{r} = x\,\mathbf{x}_0 + y\,\mathbf{y}_0 + z\,\mathbf{z}_0$$

and

$$\mathbf{n} = n_1\,\mathbf{x}_0 + n_2\,\mathbf{y}_0 + n_3\,\mathbf{z}_0,$$

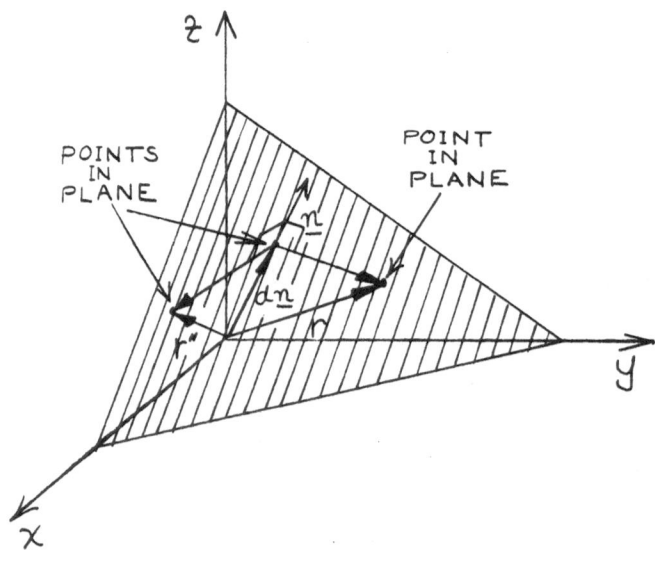

Figure B-4.   Vector construction for a plane.

Eq. (B-11) may be expressed in terms of the components:

$$x\, n_1 + y\, n_2 + z\, n_3 = d. \tag{B-12}$$

With this result, the general linear equation in three variables

$$a_1\, x + a_2\, y + a_3\, z = h \tag{B-13}$$

is the equation of a plane in Cartesian coordinates. Eqs. (B-12) and (B-13) are related by a multiplicative constant

$$a_1 = c\, n_1, \qquad a_2 = c\, n_2, \qquad a_3 = c\, n_3,$$

in which c is the constant. Because $\mathbf{n}$ is a unit vector, that is,

$$(n_1^2 + n_2^2 + n_3^2)^{1/2} = 1,$$

c must be given by

$$c = \sqrt{a_1^2 + a_2^2 + a_3^2},$$

and the minimum separation between the plane and the origin is, therefore

$$d = h/c = h/\sqrt{a_1^2 + a_2^2 + a_3^2}. \tag{B-14}$$

When the vector $d\mathbf{n}$ is associated with the wave vector $\mathbf{k}$,

$$\mathbf{k} = k_1\, \mathbf{x}_0 + k_2\, \mathbf{y}_0 + k_3\, \mathbf{z}_0,$$

where

$$|\mathbf{k}| = \sqrt{k_1^2 + k_2^2 + k_3^2} = 2\pi/\lambda,$$

then by analogy with Eq. (B-11),

$$\mathbf{k} \cdot \mathbf{r} = \text{constant}$$

specifies a plane of constant phase that is perpendicular to $\mathbf{k}$. The above expression for the phase is encountered in the plane wave function

$$u(\mathbf{k,r}) = (\text{constant})e^{-j\mathbf{k}\cdot\mathbf{r}}.$$

As we have seen, the dot product of two vectors is not a vector but a scalar. On the other hand, the *cross product* of two vectors is (for all practical purposes) a vector. The cross product, which is expressed

$$C = A \times B \qquad \text{(B-15)}$$

has the following properties:

- **C** is perpendicular to both **A** and **B**.

- The direction of **C** is given by the right-hand rule, (the thumb of the right hand indicates the direction of **C** when its fingers curl in the direction of rotation from **A** to **B** through the angle α ($\alpha < 180°$) included between **A** and **B**).

- The magnitude of the cross product vector is given by

$$|C| = c = a\, b\, \sin\alpha \qquad \text{(B-16)}$$

Figure B-5 helps to clarify the above definition and also provides a geometrical interpretation of the cross product. With reference to Eq. (B-16), the magnitude of **C** is numerically equal to the area of a parallelogram whose

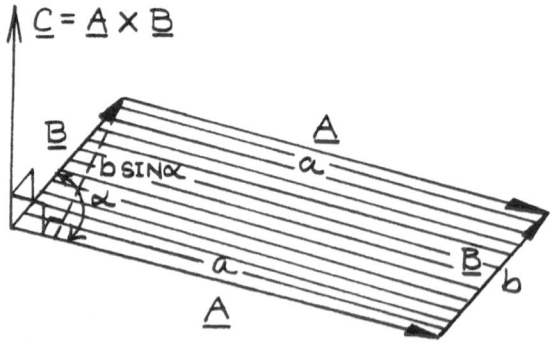

Figure B-5.   Cross product construction.

sides are formed by **A** and **B**, as shown in Figure B-5. Also, if **A** and **B** are parallel ($\alpha = 0$) or antiparallel ($\alpha = 180°$), their cross product equals zero, which is appropriate, because the direction of **C** is undefined under these circumstances.

Also, a right-handed coordinate system has basis vectors that obey

$$\mathbf{x}_0 \times \mathbf{y}_0 = \mathbf{z}_0, \quad \text{or} \quad \mathbf{z}_0 \times \mathbf{x}_0 = \mathbf{y}_0, \quad \text{or} \quad \mathbf{y}_0 \times \mathbf{z}_0 = \mathbf{x}_0.$$

If any one of the above equations is satisfied, the other two follow, and the system is right-handed. For a left-handed coordinate system,

$$\mathbf{y}_0 \times \mathbf{x}_0 = \mathbf{z}_0, \quad \text{or} \quad \mathbf{x}_0 \times \mathbf{z}_0 = \mathbf{y}_0, \quad \text{or} \quad \mathbf{z}_0 \times \mathbf{y}_0 = \mathbf{x}_0,$$

so that, for example, if $\mathbf{x}_0$ and $\mathbf{y}_0$ are given, $\mathbf{z}_0$ points in the opposite direction for the left-handed system. Note that commutivity does not hold for the cross product:

$$\mathbf{A} \times \mathbf{B} \neq \mathbf{B} \times \mathbf{A} = -(\mathbf{A} \times \mathbf{B}). \tag{B-17}$$

Thus, a change in the order of operation reverses the sign of the result.

The complete set of basis vector cross products (for a right-handed system) is, with reference to the above defining relations;

$$\mathbf{x}_0 \times \mathbf{x}_0 = 0, \quad \mathbf{x}_0 \times \mathbf{y}_0 = \mathbf{z}_0, \quad \mathbf{x}_0 \times \mathbf{z}_0 = -\mathbf{y}_0,$$

$$\mathbf{y}_0 \times \mathbf{x}_0 = -\mathbf{z}_0, \quad \mathbf{y}_0 \times \mathbf{y}_0 = 0, \quad \mathbf{y}_0 \times \mathbf{z}_0 = \mathbf{x}_0,$$

$$\mathbf{z}_0 \times \mathbf{x}_0 = \mathbf{y}_0, \quad \mathbf{z}_0 \times \mathbf{y}_0 = -\mathbf{x}_0, \quad \mathbf{z}_0 \times \mathbf{z}_0 = 0. \tag{B-18}$$

By using these basic results, it can be shown that

$$\mathbf{A} \times \mathbf{B} = (a_1 \mathbf{x}_0 + a_2 \mathbf{y}_0 + a_3 \mathbf{z}_0) \times (b_1 \mathbf{x}_0 + b_2 \mathbf{y}_0 + b_3 \mathbf{z}_0)$$

$$= c_1 \mathbf{x}_0 + c_2 \mathbf{y}_0 + c_3 \mathbf{z}_0,$$

where

$$c_1 = \begin{vmatrix} a_2 & a_3 \\ b_2 & b_3 \end{vmatrix} \qquad c_2 = - \begin{vmatrix} a_1 & a_3 \\ b_1 & b_3 \end{vmatrix} \qquad c_3 = \begin{vmatrix} a_1 & a_2 \\ b_1 & b_2 \end{vmatrix}.$$

$$\tag{B-19}$$

Thus, the cross product may be formally expressed by

$$C = A \times B = \begin{vmatrix} x_0 & y_0 & z_0 \\ a_1 & a_2 & a_3 \\ b_1 & b_2 & b_3 \end{vmatrix} \qquad \text{(B-20)}$$

if the determinant is expanded in terms of its first row.

The cross product is employed, for example, in specifying the boundary conditions of the electromagnetic field at the surface of a conductor, as discussed in Chapter 3:

$$n \times E = 0, \qquad n \times H = J_s, \qquad \text{(B-21)}$$

where n is a unit vector normal to the surface. The field, E, can be decomposed into two orthogonal vectors, $E = E_n + E_t$, in which E_n is normal to the surface, and E_t is tangential to the surface. Thus,

$$n \times E = n \times E_t + n \times E_n = n \times E_t,$$

because n and E_n are parallel. Therefore, the first boundary condition of Eq. (B-21) requires that the tangential component of E vanish. Similarly, the second boundary condition indicates that the component of H tangential to the surface is equal in magnitude to the surface current, J_s, and that the direction of J_s is perpendicular to H_t. The boundary conditions of Eq. (B-21) involve only the tangential components and do not involve the normal components of E and H.

Another application of the cross product is the Poynting vector

$$P = E \times H,$$

which indicates that the flux of electromagnetic energy is perpendicular to both the E and H fields. If E has a component that is parallel to H, that component does not contribute to the energy flux.

Because cross products and more complicated vector expressions do not appear often in the text, more detail is not provided. More involved expressions can always be verified or evaluated by working with the individual components, although this is sometimes tedious. For example, by use of expressions (B-7) and (B-19), the following identity can be verified:

$$D = A \times (B \times C) = B (A \cdot C) - C (A \cdot B). \qquad \text{(B-22)}$$

Because the dot products $A \cdot C$ and $A \cdot B$ are scalars, the identity indicates that D has components along B and C (unless A is perpendicular to one or both of these vectors).

APPENDIX **C**

The Fourier Integral and Its Application to Antennas

Chapter 4 noted that the far field due to a source that is distributed along a line is proportional to the Fourier integral (or transform) of the source distribution. Chapters 7 and 10 noted that the two-dimensional Fourier integral arises in the calculation of the far field patterns of plane aperture antennas. Because the Fourier transform is such a useful mathematical tool, it is reviewed here.

C.1 THE FOURIER SERIES

Fourier integrals may be regarded as generalizations of the Fourier series, developed by the French mathematical physicist Jean Baptiste Joseph Fourier. Fourier's analysis applies to periodic functions, which satisfy

$$F(t) = F(t + T) = F(t + 2T) = \ldots \quad \textbf{(C-1)}$$

for any value of the variable t, in which T is a constant called the period or repetition interval. An example of a periodic function is shown in Figure C-1. Fourier established that periodic functions of arbitrary shape can be decomposed into an infinite series of sinusoidal functions with periods that are related to the "fundamental" period, T, by $T/1$, $T/2$, $T/3$, etc.. Because the period is related to the frequency, f, and to the angular frequency, ω, by

303

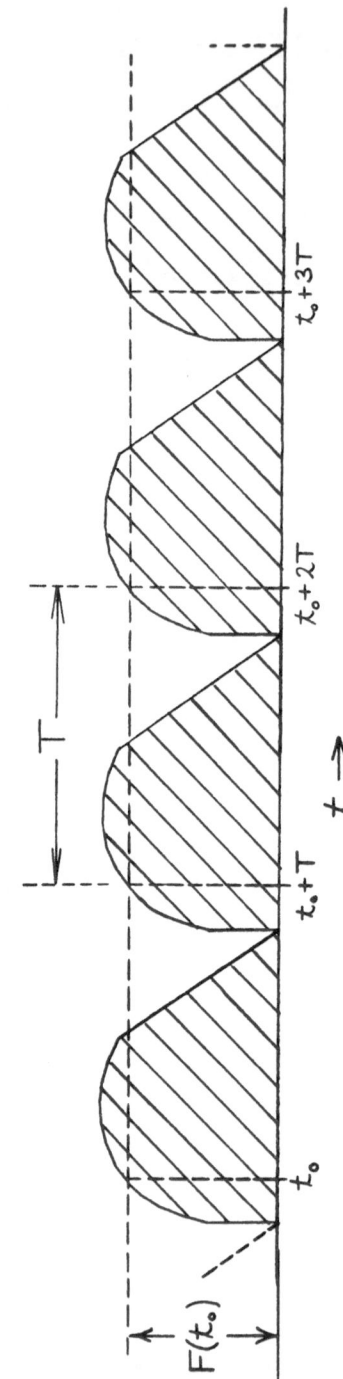

Figure C-1. Function with Period "T."

304

$$\omega = 2\pi f = \frac{2\pi}{T}, \qquad \text{(C-2)}$$

the Fourier series representation for the periodic function $F(t)$ may be expressed

$$F(t) = \frac{a_0}{2} + \sum_{n=1}^{\infty} [a_n \cos(n\omega t) + b_n \sin(n\omega t)]. \qquad \text{(C-3)}$$

The coefficients are given by

$$a_n = \frac{2}{T} \int_P F(t) \cos(n\omega t)\, dt, \qquad \text{(C-4a)}$$

$$b_n = \frac{2}{T} \int_P F(t) \sin(n\omega t)\, dt, \qquad \text{(C-4b)}$$

with the integration extending over one period. If it is assumed that Eq. (C-3) is valid, Eqs. (C-4) can be easily derived by use of the following identities

$$\int_P \sin(n\omega t)\, dt = 0, \qquad \int_P \cos(n\omega t)\, dt = \begin{cases} 0 \text{ for } n \neq 0 \\ T \text{ for } n = 0 \end{cases}$$

and the orthogonality relations

$$\int_P \sin(n\omega t) \cos(m\omega t)\, dt = 0; \qquad \text{for all n, m}$$

$$\int_P \sin(n\omega t) \sin(m\omega t)\, dt = \int_P \cos(n\omega t) \cos(m\omega t)\, dt \begin{cases} 0 \text{ for } n \neq m \\ T/2 \text{ for } n = m \end{cases}$$

$$\text{(C-5)}$$

Integration of Eq. (C-3) over one period of $F(t)$ yields

$$\int_P F(t)\, dt = \frac{a_0}{2} \int_P dt = \frac{Ta_0}{2}$$

or

$$a_0 = \frac{2}{T} \int_P F(t)\, dt,$$

which is just Eq. (C-4a) for $n = 0$. Similarly, if Eq. (C-3) is multiplied by $\cos(m\omega t)$ with $m \geq 1$ and integrated over one period, (C-4a) is obtained with $n \geq 1$. That repetition of this procedure with the factor $\sin(m\omega t)$ leads to Eq. (C-4b) can be verified.

The Fourier series expansion can be expressed in complex form by use of the identities

$$\cos(n\omega t) \equiv \frac{1}{2}(e^{jn\omega t} + e^{-jn\omega t}) \equiv \cos(-n\omega t),$$

$$\sin(n\omega t) \equiv \frac{1}{2j}(e^{jn\omega t} - e^{-jn\omega t}) \equiv -\sin(-n\omega t).$$

When inserted in Eq. (C-3) these give   .

$$F(t) = \frac{a_0}{2} + \frac{1}{2}\sum_{n=1}^{\infty}(a_n - jb_n)e^{jn\omega t} + \frac{1}{2}\sum_{n=1}^{\infty}(a_n + jb_n)e^{-jn\omega t}.$$

Because, by Eqs. (C-4), $a_n = a_{-n}$ and $b_n = -b_{-n}$ for $n \geq 1$ and $b_0 = 0$, the above result may be expressed

$$\boxed{F(t) = \sum_{n=-\infty}^{\infty} c_n e^{jn\omega t}} \qquad \textbf{(C-6a)}$$

with

$$\boxed{c_n = \frac{1}{2}(a_n - jb_n) = \frac{1}{T}\int_P F(t)e^{-jn\omega t}\,dt \qquad \text{for all } n} \qquad \textbf{(C-6b)}$$

This result is completely equivalent to expansion by Eqs. (C-3) and (C-4). Substitution of Eq. (C-6b) into Eq. (C-6a) yields an additional form

$$\boxed{F(t) = \frac{1}{\sqrt{2\pi}}\sum_{n=-\infty}^{\infty}\frac{2\pi}{T}e^{jn\omega t}\,\Phi(n\omega)} \qquad \textbf{(C-7a)}$$

where

$$\Phi(n\omega) = \frac{T}{\sqrt{2\pi}} c_n = \frac{1}{\sqrt{2\pi}} \int_P F(t) e^{-jn\omega t} \, dt \qquad \text{(C-7b)}$$

Shown thus far is how any one of the three equivalent Fourier series representations can be used to expand an arbitrary periodic function, $F(t)$, in terms of sinusoidal components. The frequencies of these component terms are all, except for the constant ($n = 0$) term, harmonics (i.e., integral multiples of the fundamental frequency, $1/T$). In general, periodic functions with sharp, angular discontinuities such as "square waves" or "sawteeth" have an infinite number of components, while for smooth functions the series may terminate. Antisymmetric functions, that is, functions for which

$$F(t) = -F(-t)$$

have series that are limited to sine terms (i.e., only b_n coefficients may be nonzero), while symmetric functions for which

$$F(t) = F(-t)$$

have series limited to cosine and constant terms (i.e., only a_n coefficients may be nonzero).

C.2 THE FOURIER INTEGRAL

The discrete function, $\Phi(n\omega)$, which is given in Eq. (C-7b), is referred to as the *spectral distribution*. This function specifies the amplitudes of the various harmonics in the frequency spectrum of $F(t)$. As the period, T becomes very large, the frequency separation of the components—the space on the frequency axis between adjacent harmonics—becomes very small, and $\Phi(n\omega)$ increasingly resembles a continuous *spectral density* function. In the limit, as $T \to \infty$, the function $F(t)$ is no longer periodic, and Eqs. (C-7) must be modified. The fundamental angular frequency is related to the period by Eq. (C-2). As T increases without limit, we make the transitions

$$\omega = \frac{2\pi}{T} \underset{T \to \infty}{\to} d\omega$$

and

$$n\omega = \frac{2\pi n}{T} \to nd\omega \to \omega,$$

because n can be arbitrarily large. Incorporation of these results into Eqs. (C-7) gives Fourier's integral theorem

$$F(t) = \frac{1}{\sqrt{2\pi}} \int_{-\infty}^{\infty} \Phi(\omega)e^{j\omega t}\, d\omega$$

$$\Phi(\omega) = \frac{1}{\sqrt{2\pi}} \int_{-\infty}^{\infty} F(t)e^{-j\omega t}\, dt \qquad \textbf{(C-8)}$$

The derivation of Eqs. (C-8) from (C-7) is not a rigorous mathematical proof, but is plausible. The integral transforms of Eq. (C-8) relate a time-varying quantity to its spectral density and vice versa. Of course, the Fourier transform's applicability is not limited to functions of time, $F(t)$, and frequency, $\Phi(\omega)$. In antenna theory, as seen in Chapters 4, 7, and 10, Fourier transforms relate source variables to far field angle variables. A discrete Fourier transform (i.e., a finite Fourier series) was used in Chapter 8 for array far field calculations.

The generalization of Eqs. (C-8) to two variable pairs, as required by aperture antenna theory, is quite straightforward:

$$F(x,y) = \frac{1}{2\pi} \int_{-\infty}^{\infty} \int_{-\infty}^{\infty} G(\xi,\eta)\, e^{-j(\xi x + \eta y)}\, d\xi d\eta$$

$$G(\xi,\eta) = \frac{1}{2\pi} \int_{-\infty}^{\infty} \int_{-\infty}^{\infty} F(x,y)e^{j(\xi x + \eta y)}\, dx\, dy \qquad \textbf{(C-9)}$$

For example, $F(x,y)$ might correspond to the aperture field distribution in the x-y plane, and

$$\xi = k \sin\theta \cos\phi, \qquad \eta = k \sin\theta \sin\phi$$

provide the relationships for the far field angle coordinates, θ and φ. The question of the existence of the Fourier integral does not arise in antenna theory, because apertures are always of finite extent.

In Chapter 10 it was shown that under certain symmetry conditions, the double Fourier integral can be simplified. The example considered was the axially symmetric source distribution over a circular aperture of radius, a. For that case the second double integral of Eqs. (C-9) becomes

$$G(\xi) = \int_0^a F(\rho) \, J_0 \, (\xi\rho) \, \rho \, d\rho \qquad \textbf{(C-10)}$$

where $\xi = k \sin \theta$ and

$$J_0 \, (\xi\rho) = \frac{1}{2\pi} \int_0^{2\pi} e^{j\xi\rho\cos u} \, du$$

is the Bessel function of zero order. This Fourier-Bessel integral is a useful special case because it reduces the double integral to a single integral over a function whose properties are well known.

C.3 APPLICATION TO SEVERAL ILLUMINATIONS

Some simple one-dimensional aperture distributions are illustrated in Figure C-2. Although these distributions are continuous, they may be periodically sampled to provide discrete distributions for a linear array of sources. As discussed in Chapter 8, the array factor for a collection of discrete sources is quite similar to the Fourier transform of the corresponding continuous distribution.

The uniform distribution of Figure C-2a has been discussed in connection with wire antennas in Chapter 4. The distribution is given by

$$F(x) = \begin{cases} 1 \, ; & -L/2 < x < L/2 \\ 0 \, ; & \text{elsewhere} \end{cases}$$

$$\textbf{(C-11a)}$$

The far field produced by such a source distribution is proportional to its Fourier integral

$$\Phi(\theta) = \frac{\displaystyle\int_{-L/2}^{L/2} e^{j2\pi \, x \, \cos\theta/2} \, dx}{\displaystyle\int_{-L/2}^{L/2} dx} = \sin u/u = \Phi(u) \qquad \textbf{(C-11b)}$$

where, for convenience, we denote

$$u = \pi L \cos\theta/\lambda. \qquad \textbf{(C-11c)}$$

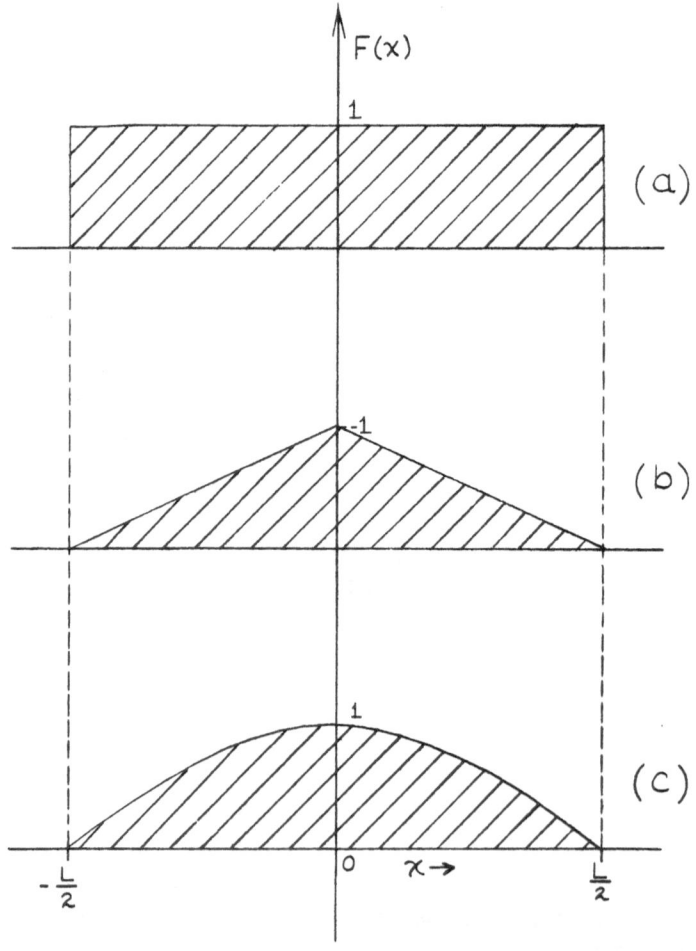

Figure C-2.   One-dimensional aperture distributions.

The field angle, θ, is measured from the positive x axis. The integral in the denominator of Eq. (C-11b) normalizes the pattern, so that $\Phi(0) = 1 = 0$ db. A central region of the power pattern, obtained from

$$S(u) = 10 \log[\Phi(u) \, \Phi^*(u)] \text{ (dB)}, \qquad \textbf{(C-11d)}$$

is shown in Figure C-3a. The relative power of the first side lobe is about −13.2 dB, and the side lobe envelope function is simply $1/u^2$. For a uniform distribution the Fourier transform of Eq. (C-11b) always yields a real number: therefore, the pattern is said to be real. Equivalently, a real (amplitude) pattern has a phase that is either 0° or 180° everywhere. Examination of Eq. (C-11b) reveals that the phase within the main lobe is 0°, and that the phase in successive side lobes is alternately 180° and 0°.

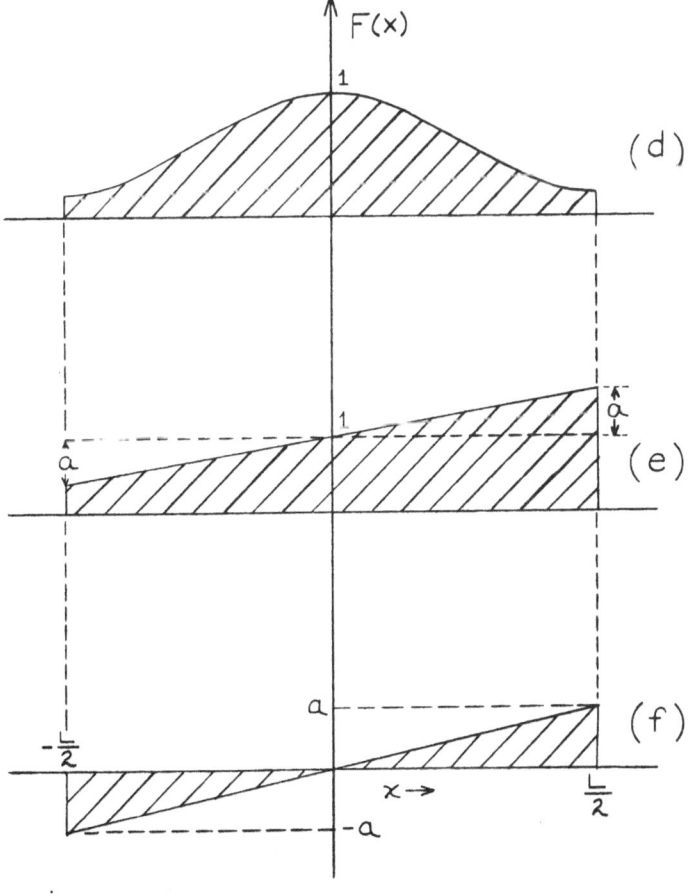

Figure C-2.   *(Cont.)*

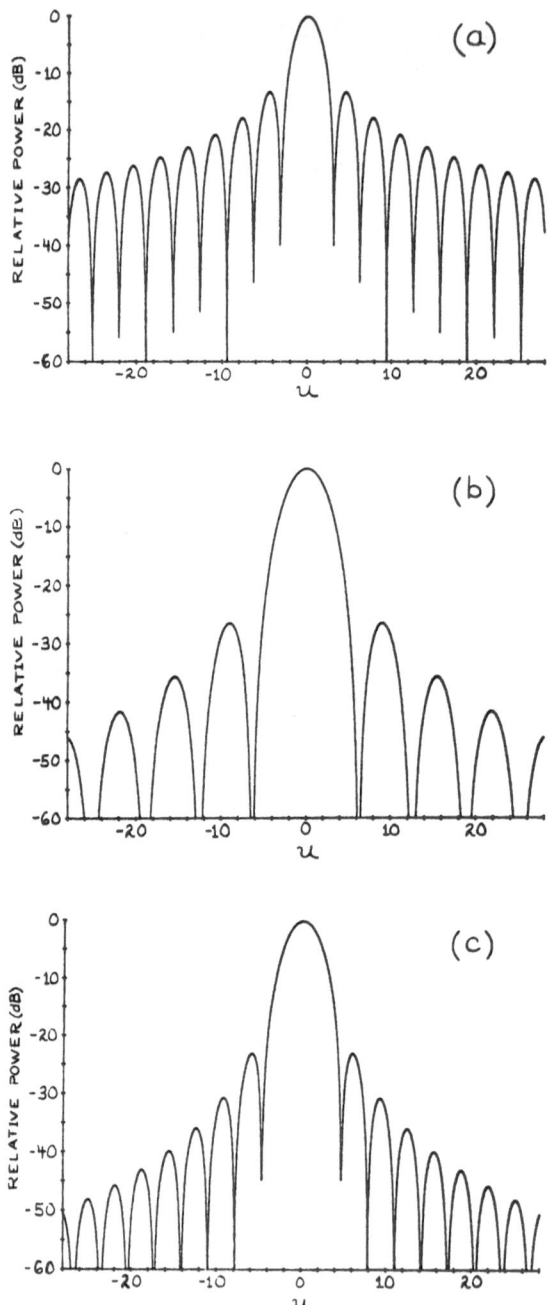

Figure C-3.   Power patterns for distributions of Figure C-2.

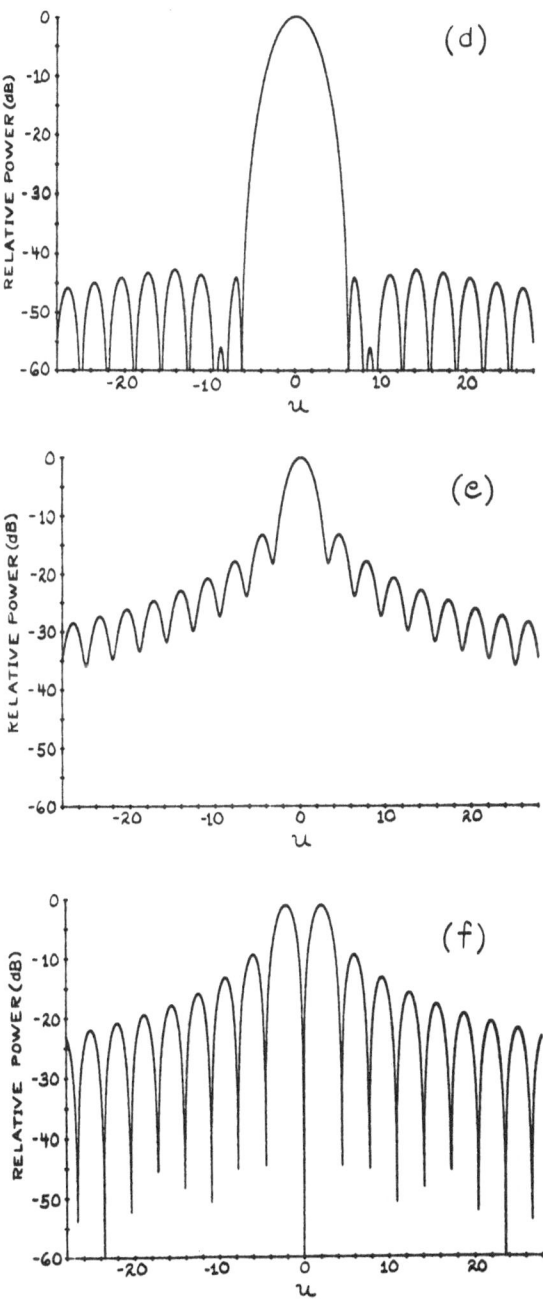

Figure C-3. (*Cont.*)

A triangular distribution defined by

$$F(x) = \begin{cases} 1 - 2x/L; & 0 \leqslant x < L/2 \\ 1 + 2x/L; & -L/2 < x < 0 \\ 0 \;\; ; & \text{elsewhere} \end{cases} \qquad \textbf{(C-12a)}$$

is illustrated in Figure C-2b. The normalized amplitude pattern is given by

$$\Phi(u) = 2 \left[\frac{1 - \cos u}{u^2} \right], \qquad \textbf{(C-12b)}$$

and a central section of the corresponding power pattern is shown in Figure C-3b. The relative power of the first side lobe is about -26.5 dB, and the main beam is about twice as wide as that of Figure C-3a. As noted in Chapters 7 and 10, the uniform distribution produces the highest gain (has the greatest aperture efficiency). The relation between gain and beam width is discussed at the end of Chapter 8. A broader beam, in general, corresponds to a lower gain. For the triangular distribution the power pattern side lobe structure is modified such that, relative to the uniform case, there are only half as many side lobes in a given section of u-space. Also, the side lobe envelope falls off as the reciprocal of u^4.

The cosine distribution of Figure C-2c has the form

$$F(x) = \begin{cases} \cos(\pi x/L); & -L/2 < x < L/2 \\ 0 \;\; ; & \text{elsewhere} \end{cases} \qquad \textbf{(C-13a)}$$

which has the Fourier transform

$$\Phi(u) = \frac{\cos u}{1 - (2/\pi)^2 u^2}. \qquad \textbf{(C-13b)}$$

The relative power of the first side lobe is approximately -23.0 dB, and the side lobe roll-off resembles that of the previous example. Although the main beam is significantly wider than that produced by the uniform distribution, the side lobe density is about the same.

A distribution of theoretical and practical value is the *Hamming*, or cosine-squared-on-a-pedestal distribution given by ($0 < a < 1$):

$$F(x) = \begin{cases} a + (1 - a) \cos^2 (\pi x/L) \;; & -L/2 < x < L/2 \\ 0 \;; & \text{elsewhere.} \end{cases} \qquad \textbf{(C-14a)}$$

An equivalent expression for the distribution in the nonzero region is

$$F(x) = 1/2 \left[(1 + a) + (1 - a) \cos(2\pi x/L) \right].$$

The Fourier transform, or amplitude pattern, is

$$\Phi(u) = \sin u/u + \left(\frac{1 - a}{1 + a} \right) \left[\frac{u \sin u}{\pi^2 - u^2} \right]. \qquad \text{(C-14b)}$$

The corresponding power pattern for $a = 0.08$ appears in Figure C-3d. The width of the main beam is nearly identical to that of Figure C-3b; however, unlike the previous examples, the side lobe envelope is relatively flat over the illustrated range. The first side lobe lies about 44 dB below the peak of the main beam, and the peak side lobe lies at -42.8 dB. The side lobe level is a function of the parameter a. Obviously, for $a = 1$ the uniform distribution pattern of Eq. (C-11b) is recovered. The popularity of the Hamming distribution accrues from its simplicity, flexibility, and excellent side lobe suppression.

Eqs. (C-11b), (C-12b), (C-13b), and (C-14b) seem indeterminate at those points for which the denominators are equal to zero. However, this difficulty is easily handled by application of L'Hospital's rule or series expansion for the numerator.

The distributions considered thus far are symmetrical about the origin, that is to say

$$F(x) = F(-x). \qquad \text{(C-15)}$$

Under these circumstances the Fourier transform may be expressed

$$\Phi(k_x) = \int_{-L/2}^{L/2} F(x) e^{jk_x x} \, dx = 2 \int_0^{L/2} F(x) \cos(k_x x) \, dx,$$

in which $k_x = 2\pi \cos\theta/\lambda$, so that the amplitude, $\Phi(k_x)$, is always real. The next distribution considered does not satisfy Eq. (C-15) and therefore has a complex amplitude pattern.

The linear distribution of Figure C-2e may be expressed

$$F(x) = \begin{cases} 1 + 2ax/L ; & -L/2 < x < L/2 \\ \\ 0 ; & \text{elsewhere.} \end{cases} \qquad \text{(C-16a)}$$

The amplitude pattern, which is complex for this asymmetrical distribution, is given by

$$\Phi(u) = \sin u/u + ja \,[\sin u/u^2 - \cos u/u]. \qquad \textbf{(C-16b)}$$

Note that for $a = 0$ the pattern of Eq. (C-11b) is recovered, as expected. The power pattern, obtained from (C-11d), is, of course, real and is shown in Figure C-3e for $a = 0.4$. Note that the pattern remains symmetrical, despite the asymmetry of the distribution function. Apart from null fill-in, this pattern is virtually identical to that of Figure C-3a, which represents the pattern of the uniform distribution. The existence of the imaginary component in Eq. (C-16b) not only fills in the uniform pattern nulls, but also causes the pattern phase to vary with u (i.e., the angular position in the far field). For larger values of a the side lobe peaks remain fixed, but the nulls fill in more completely.

The last distribution example is a difference distribution closely related to the linear distribution. As shown in Figure C-2f, the distribution has odd symmetry about the origin,

$$F(x) = -F(-x) \qquad \textbf{(C-17)}$$

for all x. All difference distributions satisfy Eq. (C-17). For the present case

$$F(x) = \begin{cases} 2ax/L \,; & -L/2 < x < L/2 \\ 0 \,; & \text{elsewhere} \end{cases}$$

$$\textbf{(C-18a)}$$

and

$$\Phi(u) = 2j[\sin u/u^2 - \cos u/u], \qquad \textbf{(C-18b)}$$

which is essentially the imaginary component of the linear distribution amplitude pattern of Eq. (C-16b). The nulls of this pattern occur at the peaks of the previous pattern—a result that accounts for the match of the side lobe peaks of parts a and e of Figure C-3. The difference pattern is also characterized by the central null, which is expected on the basis of symmetry. The difference pattern is purely imaginary, which means that the pattern phase is either $90°$ or $-90°$ with respect to the associated sum pattern. Calculation of the sum pattern for this case is left as an exercise.

Design of a Circular Parabolic Reflector Antenna

This appendix, in contrast to Appendixes A, B, and C—the mathematical appendixes—provides a fairly detailed example of how to design a simple parabolic dish reflector antenna in order to meet given specifications. Because these antennas are very common and their designs are not as simple as those of dipole or linear array antennas, this appendix may be helpful to engineers who, although not antenna specialists, may occasionally have to select a parabolic dish antenna for a particular purpose.

The basic theory of reflector antennas can be found in Chapter 10. Section 10.3 deals with the common case of circular apertures. It is shown there that the far field of such an aperture, which is given by a Fourier integral of the illumination, may be reduced to a single integral shown in Eq. (10-21) in the case of circular symmetry. Yet, to perform calculations the illumination must be known.

In this example, it is assumed that the illumination is Gaussian, which means that it has the form

$$E_x(\rho) = E_o \exp \left[-B(\rho/a) \right]^2 \qquad \textbf{(D-1)}$$

Where a is the aperture radius, and B is a constant. The ratio of the field at the aperture edge to the field at the center is

$$E_x(a)/E_x(o) = e^{-B}.$$

The edge taper in decibels is given by

$$T = |20 \log E_x(a)/E_x(o)| = 8.68 \, B \qquad \text{(D-2)}$$

and is an important design parameter.

The assumption underlying Eq. (D-1) is a somewhat crude approximation, but it enables computation of $F(\theta)$ from Eq. (10-21) with good accuracy for various values of B. The result of this computation is summarized in Figure D-1. Shown on the horizonal axis is the edge taper, T. The vertical scale on the right shows the normalized quantity $\theta D/\lambda$, where the spherical angle θ is shown in Figure 10-4; $D = 2a$ is the aperture diameter and λ the wavelength. From the labeled curves (3-dB point, first null, first side lobe, etc.), the far field pattern of a circular aperture can be constructed with a given Gaussian illumination. For example, when $T = 12$ (i.e. when $B = 1.38$) and for $D/\lambda = 20$, the 3-dB point occurs at $\theta D/\lambda = 34$ as read on the vertical scale at the right. Thus the half-power beam width is $2 \times 34/20 = 3.4°$. The first null occurs at $\theta D/\lambda = 90$, which is $90/20 = 4.5°$ removed from the peak of the main beam. The peak of the first side lobe occurs at about $5.5°$ from the main beam, and its power is about -26.3 dB below the main beam. (The decibel scale for the first side lobe is superimposed on the curve that shows the first side lobe location.) Similarly, the second null is at about $6.8°$. The second side lobe is at $8.2°$ and is about -31 dBmb (i.e., decibel relative to the main beam).

Figure D-2 has efficiency curves in the upper part. The aperture efficiency is highest (1.0) at uniform illumination ($T = 0$ dB) and diminishes with growing T. On the other hand, the power spillover from the feed horn increases with decreasing T and goes to zero as T becomes very large. The total efficiency is the product of these two factors and has a peak of about 82% at $T = 11$ dB.

An a priori design of a circular reflector antenna is rather simple. Given wavelength λ, desired gain, beam width, and side lobe level, the reflector diameter and the illumination edge taper can be chosen.

Following is an example of how to choose an antenna that would give a gain of at least 35 dB and a beam width of no more than $3.0°$ at $f = 6$ GHz,

$$f = 6 \text{ GHz} \rightarrow \lambda = 5 \text{ cm}$$

$$G_{\text{max}} = 4\pi \frac{A_{ef}}{\lambda^2} = \eta \left(\frac{\pi D}{\lambda} \right)^2.$$

Because there is no requirement on side lobe level, the highest efficiency is chosen (i.e., $T = 11$ db, $\eta = 82\%$), and

PATTERN OF TRUNCATED CIRCULAR GAUSSIAN APERTURE

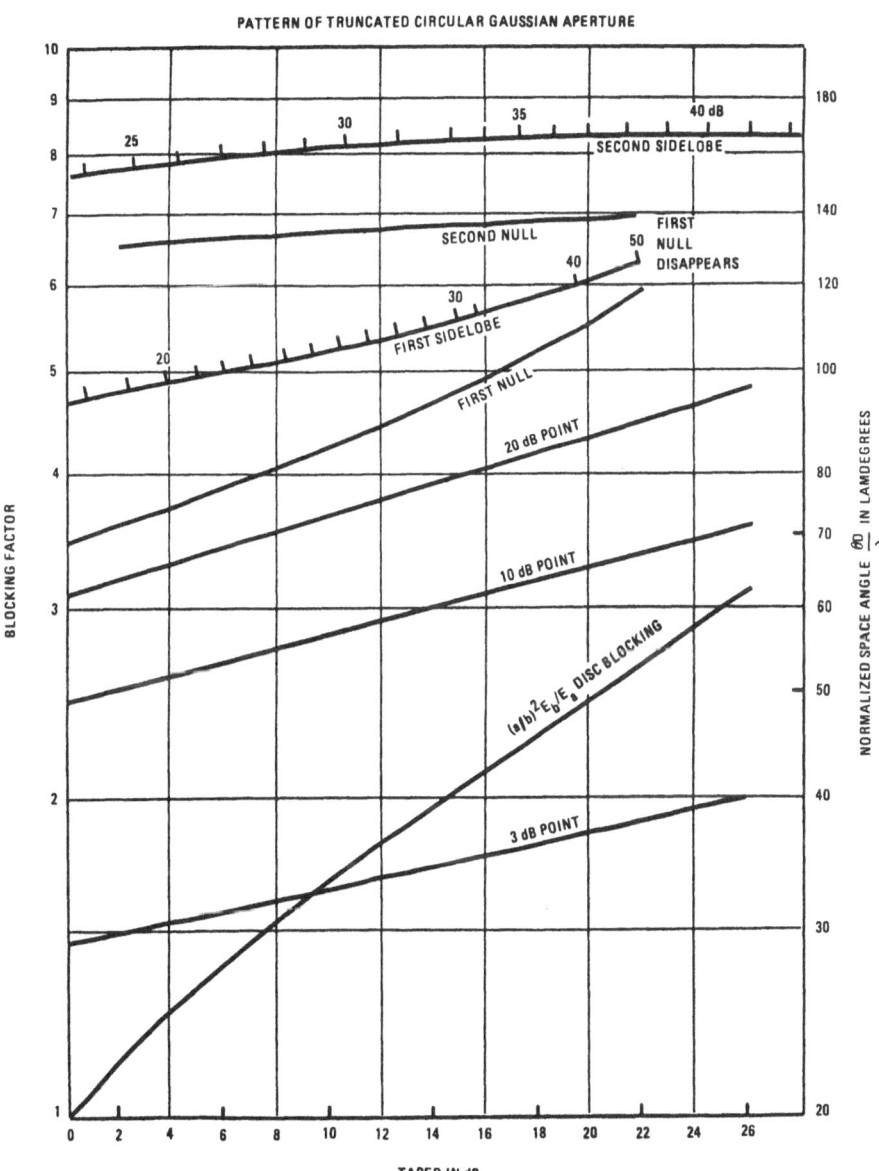

Figure D-1. Nomogram for constructing circular aperture far field patterns.

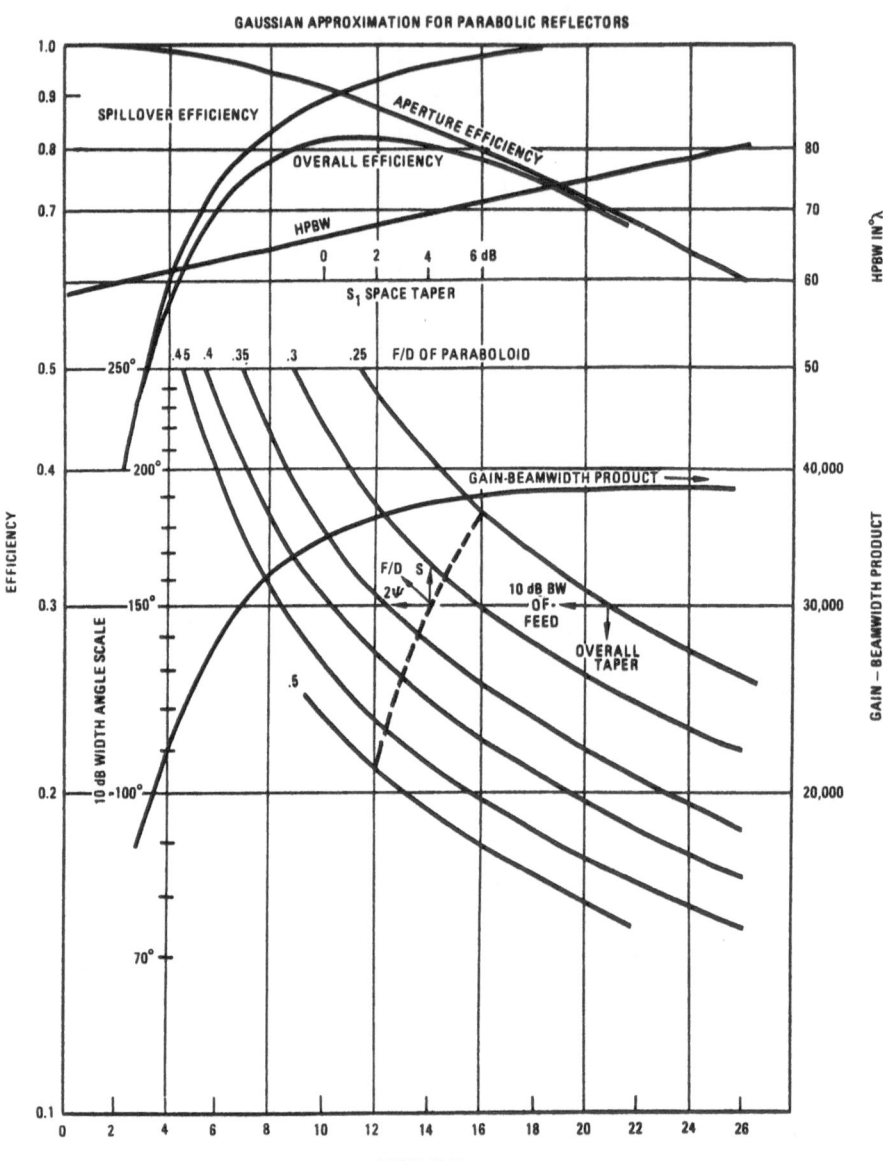

Figure D-2. Nomogram for circular aperture efficiency and feed design.

$$G_{max} = 0.82 \left(\frac{\pi D}{5}\right)^2 = 35 \text{ dB} = 3162$$

$$D = 5/\pi \sqrt{3162/0.82} = 98.8 \text{ cm} \cong 1.0 \text{ m.}$$

To check the beam width, Figure D-1 is consulted for the 3-dB point at $T = 11$,

$$\theta D/\lambda = 34 \rightarrow BW = 2\theta = 2 \times 34 \times 5/100 = 3.4°.$$

The beam of a 1.0 reflector is too wide. To get a 3.0° beam at $\lambda = 5$ cm and $T = 11$ dB,

$$\theta D/\lambda = 1.5 \times D/5 = 34 \rightarrow D = 113 \text{ cm.}$$

The corresponding gain is

$$G = 0.82 (113\pi/5)^2 = 4134 = 36.2 \text{ dB.}$$

A smaller reflector can be used by choosing $T = 8$ dB. The efficency (from Fig. D-2) is now 79%. This would yield

$$\theta D/\lambda = 32 \quad \text{and} \quad D = 1.07 \text{ m}$$
$$G = 0.79 (107 \, \pi/5)^2 = 3570 = 35.5 \text{ dB.}$$

It is not practical, however, to choose a lower T, because a large feed horn, large blockage, and very high side lobes would result.

The reflector shape (i.e., F/D) is arbitrary and can be chosen between 0.25 and 0.5. A small F/D (0.25 for example) reduces somewhat the blockage effects. This is shown in detail later.

The design problem becomes more difficult when estimating the actual side lobe levels (including the feed blockage effect), or when attempting a low side lobe design. For example, if the requirement that the side lobe level should be -30 dBmb or better is added to the previous problem, Figure D-1 shows immediately that a higher edge taper is needed. Checking the levels of first and second side lobes at various values of T shows that specifications cannot be met with an edge taper of less than 16 dB. $T = 16$ will give nominal first and second side lobes of -31.8 dB and 34 dB, respectively. The blockage effect of the feed will further degrade this. Starting with $T = 16$ gives $\theta D/\lambda = 35$ and $\eta = 0.77$, which yields $D = 1.166$ m. Choosing $D = 1.2$ m gives

$$G = 0.77 (120\pi/5)^2 = 4377 = 36.4 \text{ dB.}$$

This shows that it is easy to meet the 35-dB gain requirement. The difficult point will be the −30-dB side lobe requirement.

The feed horn is designed with the help of the lower part of Figure D-2 and with Figure D-3. From the family of curves labeled F/D of Paraboloid and the angle scale on the left of Figure D-2, the angle 2ψ subtended by the reflector as seen from the focus is found (see Fig. 10-7). The first choice would be $F/D = 0.5$, because this is a shallow reflector and the least expensive choice. The point of intersection of the dotted curve with the $F/D = 0.5$ curve is projected to the left, giving an angle of about 106° degrees, or $\psi = 53°$.

Figure D-3 shows the universal radiation patterns for horns that are based on Eq. (7-10). From them the horn dimensions A and B are found as follows. The 16-dB edge taper is a combined result of the feed horn radiation pattern and the space taper (i.e., the effect of the distance from the focus to the reflector, which is smallest at the center and largest at the edge). The point of intersection of the dotted curve with the $F/D = 0.5$ curve in Figure D-2 is

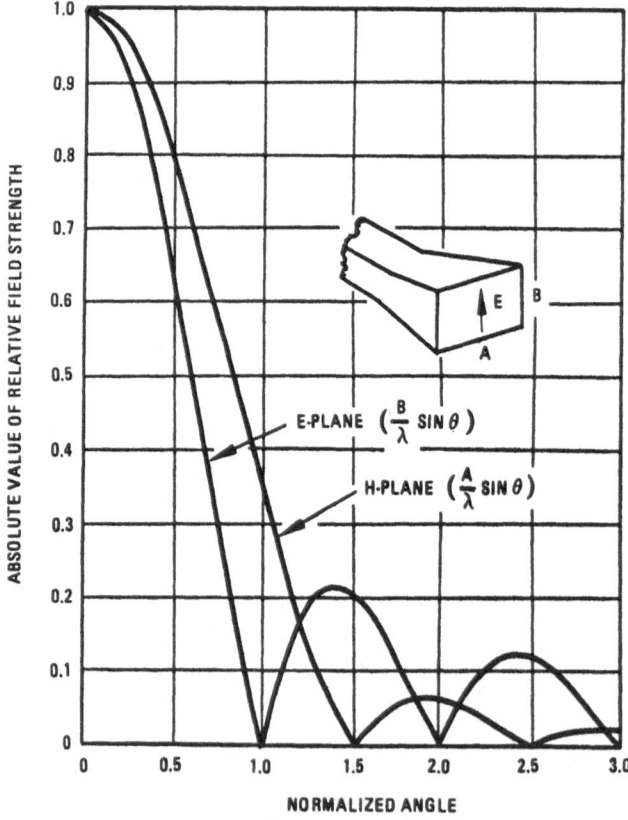

Figure D-3.   E and H plane radiation patterns for sectoral or pyramidal horns.

projected upward to the linear scale labeled "space taper." The space taper is found to be 2 dB, thus the feed has to provide a 14-dB ratio between a center ray (*OA* in Fig. 10-7) and an edge ray (*OB* in that figure). If

$$20 \log \left| \frac{E_{center}}{E_{edge}} \right| = 14,$$

then $E_{edge} = 0.2 \, E_{center}$.

The vertical scale in Figure D-3 gives field strength ratios. For a value of 0.2,

$$\frac{B}{\lambda} \sin \psi \cong 0.8 \qquad \text{in the } E \text{ plane,}$$

$$\frac{A}{\lambda} \sin \psi \cong 1.15 \qquad \text{in the } H \text{ plane.}$$

Thus, for $\psi = 53°$ and $\lambda = 5$ cm, the values $B = 5$ cm, $A = 7.2$ cm are found, and the horn may be assumed equivalent to a blocking disk of diameter

$$d = \sqrt{A^2 + B^2} = 8.8 \text{ cm.}$$

The feed horn blockage was estimated to be equivalent to a disk of diameter 8.8 cm; the effect from the curve labeled "disk blocking" is found on the lower part of Figure D-1. The following are identified:

E_a = peak aperture field,
E_b = peak blockage field,
a = reflector diameter, and
b = blockage diameter.

At $T = 16$ dB, the left-hand scale shows a blocking factor of 2.2. Thus,

$$E_b = 2.2 \, E_a \, (b/a)^2 = 2.2 \times \left(\frac{8.8}{120} \right)^2 E_a = 0.0118 \, E_a.$$

To this field must be added the field of the first side lobe, whose unperturbed intensity was -31.8 dB*mb* or

$$20 \log |E_{SL}/E_{mb}| = -31.8 \rightarrow |E_{SL}| = 0.0257 \, |E_{mb}|.$$

The perturbed first side lobe is now

$$|E'_{SL}| = (0.0257 + 0.0118) \, |E_{mb}| = 0.0375 \, |E_{mb}|,$$

or

$$20 \log |E'_{SL}/E_{mb}| = -28.5 \text{ dB.}$$

The first side lobe has increased due to the blockage effect from -31.8 to -28.5 dB, and the antenna design has to be changed to meet the -30-dB requirement. A physical rationale for this phenomenon can be drawn from Figure D-4. The effects of aperture blockage are shown in two ways: on the right there is a blocked aperture illumination—a combination of an unblocked reflector aperture minus the feed blockage, which is equivalent to a large aperture in antiphase to another (small) aperture representing blockage by a round centered feed. The curves on the left of Figure D-4 show the far field effects: the reflector aperture produces a high, narrow main beam and a side lobe pattern. The feed blockage produces a much lower and broader radiation pattern, in antiphase to the reflector pattern. When the effects are summed, the blockage is seen to reduce the main beam and the even side lobes (second, fourth, etc.). At the same time it enhances the odd side lobes (first, third, etc.), because it is in phase with them. The loss in main beam gain is usually not severe because it is proportional to the ratio of aperture areas (reflector

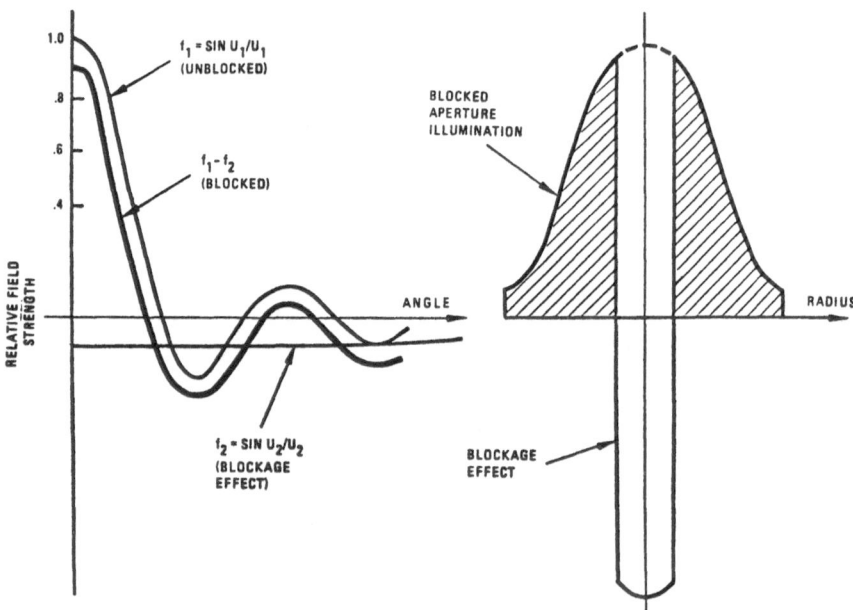

Figure D-4.   The effects of a centered blockage on a circular aperature's radiation pattern.

to feed). The effect on side lobes, however, can be severe, especially in a low side lobe design.

A deeper reflector, using $f/D = 0.25$, requires a less directive horn for the feed. Repeating the procedure yields a smaller horn with less blockage and a perturbed first side lobe of -29.5 dB. This may not be quite good enough, and a larger edge taper, perhaps $T = 17$ dB, might be necessary and the procedure repeated.

Clearly, the process shown is rather inaccurate and laborious. The given design curves are sufficient for designing a simple antenna, such as for a home satellite receiver. More sophisticated design procedures use computer programs that are based on the theory of Chapter 10 and this appendix or on some more accurate mathematical models. What the design problem exemplifies is that antenna specifications (frequency, gain, beam width and side lobe levels) determine system dimensions (diameter, focal length, and feed dimensions), provided some reasonable assumptions are made about the illumination.

As mentioned before, the Gaussian illumination is a somewhat crude approximation, but will suffice for simple antenna design. More accurate methods have been developed for the design of high-performance reflector antennas. Usually an iterative procedure is necessary to arrive at a design that meets all the requirements. The given curves in Figures D-1 through D-3 may be useful to quickly check a set of antenna specifications in order to determine whether or not they make sense and to obtain some preliminary ideas about an antenna's dimensions.

Not covered in this appendix are several phenomena that affect reflector antenna performance, such as depolarization, edge diffraction, the effect of supporting strutts, and radomes. Also, the design of multiple reflector systems (e.g., Cassegrain or Gregorian) and offset reflector systems were not described. These require a level of sophistication that is beyond the scope of this introductory book.

Annotated Bibliography

The following citations are for books on general antenna theory and design. With the exception of Blake's book, they are considerably more advanced than this text.

C. A. Balanis, *Antenna Theory: Analysis and Design*, New York: Harper & Row, 1982.
A textbook for graduate students of electrical engineering. Requires a good background in electromagnetic theory.

L. V. Blake, *Antennas*, Dedham, MA: Artech House, 1984.
A new edition of a book that first appeared in 1966. The level is elementary, with minimal mathematical prerequisites.

R. E. Collin and F. J. Zucker (eds.), *Antenna Theory*, 2 vols., New York: McGraw-Hill, 1969.
Each chapter is written by a recognized authority. This comprehensive advanced text covers most of the important areas in antenna engineering.

R. S. Elliott, *Antenna Theory and Design*, Englewood Cliffs, N.J.: Prentice-Hall, 1981.
Similar to Balanis in scope and level.

R. C. Johnson and H. Jasik (eds.), *Antenna Engineering Handbook*, 2nd Ed., New York: McGraw-Hill, 1984 (1961).
A comprehensive, somewhat more applications-oriented handbook for antenna engineers.

A. W. Rudge, K. Milne, A. D. Olver, and P. Knight (eds.), *The Handbook of Antenna Design*, 2 vols., London: P. Peregrinus, 1983.
A comprehensive, theoretical handbook for antenna engineers.

S. Silver, *Microwave Antenna Theory and Design*, London: P. Peregrinus, 1984.
A new edition of a classic text that first appeared as part 12 of the 27-volume Radiation Laboratory Series published by McGraw Hill (New York) in 1947.

L. Stutzman and G. A. Thiele, *Antenna Theory and Design*, New York: Wiley, 1981.
Similar to the two textbooks by Balanis and Elliott in scope and level.

The following citations are for books that were published during the past 20 years. They either cover a specific area of antenna theory and design or are not as widely used as those in the first list.

N. Amitay, V. Galindo, and Chen Pang Wu, *Theory and Analysis of Phased Array Antennas*, New York: Wiley-Interscience, 1972.
I. J. Bahl and P. Bhatia, *Microstrip Antennas*, Dedham, MA: Artech House, 1980.
R. G. Brown (and others). *Lines, Waves, and Antennas: The Transmission of Electric Energy*, New York: Ronald Press, 1973.
M. L. Burrows, *ELF Communications Antennas*, Stevenage, England: P. Peregrinus, 1978.
L. Cantor, *How to Select and Install Antennas*, Rochelle Park, N.J.: Hayden Book Co., 1978.
P. J. B. Clarricoats and A. D. Olver, *Corrugated Horns for Microwave Antennas*, London: P. Peregrinus, 1984.
R. E. Collin, *Antennas and Radiowave Propagation*, New York: McGraw-Hill, 1985.
S. Cornbleet, *Microwave Optics: The Optics of Microwave Antenna Design*, New York: Academic Press, 1976.
J. E. Cunningham, *The Complete Broadcast Antenna Handbook: Design, Installation, Operation and Maintenance*, Blue Ridge Summit, PA, 1977.
S. W. Gibson, *Radio Antennas*, Reston, Va.: Reston Pub. Co., 1983.
R. C. Hansen, *Microwave Scanning Antennas*, 3 vols., New York: Academic Press, 1964–1966.
R. C. Hansen (compiler), *Significant Phased Array Papers*, Dedham, MA: Artech House, 1973.
J. S. Hollis, T. J. Lyon, and L. Clayton, Jr. (eds.), *Microwave Antenna Measurements*, Atlanta: Scientific-Atlanta, 1970.
IEEE Standard Test Procedures for Antennas, New York: Institute of Electrical and Electronics Engineers (distributed in cooperation with Wiley-Interscience), 1979.
J. R. James, P. S. Hall, and C. Wood, *Microstrip Antenna: Theory and Design*, New York: Peregrinus, 1981.
E. V. Jull, *Aperture Antennas and Diffraction Theory*, New York: P. Peregrinus, 1981.
J. King, *The Practical Aerial Handbook*, London: Newnes-Butterworths, 1970.
R. W. P. King and C. W. Harrison, Jr., *Antennas and Waves: A Modern Approach*, Cambridge, MA: M.I.T. Press, 1969.
J. A. Kuecken, *Exploring Antennas and Transmission Lines by Personal Computer*, New York: Van Nostrand Reinhold, 1986.

P. E. Law, *Shipboard Antennas*, Dedham, MA: Artech, 1983.

Kai Fong Lee, *Principles of Antenna Theory*, New York: Wiley, 1984.

M. T. Ma, *Theory and Application of Antenna Arrays*, New York, Wiley, 1974.

T. S. M. Maclean, *Principles of Antennas: Wire and Aperture*, New York: Cambridge University Press, 1986.

T. Milligan, *Modern Antenna Design*, New York: McGraw-Hill, 1985.

W. I. Orr, *Simple, Low-Cost Wire Antennas for Radio Amateurs*, Wilton, CT: Radio Publications, 1972.

W. I. Orr and S. D. Cowan, *The Radio Amateur Antenna Handbook*, Wilton, CT: Radio Publications, 1978.

B. D. Popovic, M. D. Dragovic, and A. R. Djordjevic, *Analysis and Synthesis of Wire-Antennas*, New York: Research Studies Press, 1982.

D. M. Pozar, *Antenna Design Using Personal Computers*, Dedham, MA: Artech House, 1985.

V. H. Rumsey, *Frequency Independent Antennas*, New York: Academic Press, 1966.

W. V. T. Rusch and P. D. Potter, *Analysis of Reflector Antennas*, New York: Academic Press, 1970.

B. D. Steinberg, *Principles of Aperture and Array System Design: Including Random and Adaptive Arrays*, New York: Wiley, 1976.

C. H. Walter, *Traveling Wave Antennas*, New York: Dover Publications, 1970, 1965.

J. D. Walton, Jr. (ed.), *Radome Engineering Handbook: Design and Principles*, New York: M. Dekker, 1970.

W. L. Weeks, *Antenna Engineering*, New York: McGraw-Hill, 1968

B. S. Westcott, *Shaped Reflector Antenna Design*, New York: Wiley, 1983.

E. A. Wolff, *Antenna Analysis*, New York: Wiley, 1966.

P. J. Wood, *Reflector Antenna Analysis and Design*, New York: P. Peregrinus, 1980.

Many books on electromagnetic theory and radar contain chapters on antennas. The following list provides a sample. These books use electromagnetic theory and are limited to a few selected antenna topics.

E. C. Jordan and K. G. Balmain, *Electromagnetic Waves and Radiating Systems*, 2nd ed., Englewood Cliffs, NJ: Prentice Hall, 1968.
A textbook for advanced undergraduate or beginning graduate electrical engineering students.

J. D. Kraus, *Electromagnetics*, 3rd ed., New York: McGraw Hill, 1984.
A very good textbook for advanced undergraduate electrical engineering students.

S. Ramo, J. R. Whinnery, and T. VanDuzer, *Fields and Waves in Communication Electronics*, 2nd ed., New York: John Wiley, 1984.
This undergraduate text has been popular for over 40 years. The first edition was published in 1944.

M. I. Skolnik (ed.), *Radar Handbook*, New York: McGraw Hill, 1970.
A reference book for radar engineers (fairly advanced level).

Index